ACMTAA-2012

Advanced Composite Materials and Technologies for Aerospace Applications

Richard Day   Sergey Reznik

ACMTAA-2012

Proceedings of the Second International Conference on

Advanced Composite Materials and Technologies for Aerospace Applications

June 11-13, 2012, Wrexham, North Wales, United Kingdom

Edited by

Prof Richard Day
Glyndŵr University

Prof Sergey Reznik
Bauman Moscow State Technical University

2012 • Wrexham • Glyndŵr University

Proceedings of the Second International Conference on
Advanced Composite Materials and Technologies for Aerospace Applications
June 11-13, 2012, Wrexham, North Wales, United Kingdom

Edited by

Prof Richard Day
Glyndŵr University

Prof Sergey Reznik
Bauman Moscow State
Technical University

Published in 2012 by
Glyndŵr University
Plas Coch
Mold Road
Wrexham
LL11 2AW
United Kingdom
www.glyndwr.ac.uk

ISBN 978-0-946881-76-5

Printed and bound in the United Kingdom

The Second International Conference on
Advanced Composite Materials and Technologies for Aerospace Applications

June 11-13, 2012, Wrexham, North Wales, United Kingdom

The conference is organised by

Bauman Moscow State Technical University, Russia
Glyndŵr University, UK

The conference is supported by

Institution of Engineering and Technology (IET) – North Wales Network

The conference co-chairmen

Prof Richard Day
Glyndŵr University, UK

Prof Sergey Reznik
Bauman Moscow State Technical University, Russia

International Programme Committee

Dr Michael J. Elwell
Huntsman Advanced Materials
Switzerland

Prof Igor Emri
University of Ljubljana
Slovenia

Prof Bronwyn Fox
Deakin University
Australia

Prof Gennady A. Frolov
Institute for Problems of Materials Science
Ukraine

Prof Xiaodong He
Harbin Institute of Technology
China

Prof Alma Hodzic
University of Sheffield
UK

Prof Xiao (Matthew) Hu
Nanyang Technological University
Singapore

Dr Fawad Inam
Glyndŵr University
UK

Prof Galina Malysheva
Bauman Moscow State Technical University
Russia

Prof Steve Ogin
University of Surrey
UK

Prof Vladimir Tarasov
Bauman Moscow State Technical University
Russia

Prof James Thomason
University of Strathclyde
UK

Prof Valery P. Timoshenko
NPO Molniya
Russia

Local Organising Committee

Mr Olivier Durieux (Chair)
Dr Yuriy Vagapov
Mr Cedric Belloc
Mrs Olga Edwards

4

Content

Composite Wing for Reusable Space Vehicles

Sergey Reznik, Tatiana Ageyeva, Inna Shafikova

Rocket and Spacecraft Composite Structures Department, Faculty of Special Machinery, Bauman Moscow State Technical University, 5 2nd Baumanskaya Street, Moscow, 105005, Russia

Abstract: The winged structural layout of a reusable space vehicle (RSV) proposed ensures comfortable flight conditions along maneuverability in the atmosphere. These advantages can be most effectively be implemented by means of composite materials structures, since composite materials possess high specific stiffness and strength. In order to solve a number of engineering problems, one of them being the design of RSV wing, it is feasible to use hybrid composites. The authors proposed a methodology for the optimal design of a RSV composite wing.

Key Words: Composites, Reusable space vehicle, Wing.

1. Reusable space vehicles. Engineering excellence assessment

Space tourism has come to reality after space tourists' flights to the ISS on board Soyuz-TM space vehicles and 2004 suborbital missions by SpaceShipOne. For the space tourism industry one of the urgent issues is to design state -of-the-art, safe, power-efficient and comfortable space vehicles (Reznik and Ageyeva, 2010).

A number of factors have driven the recent development of the space tourism industry: the high level of aircraft manufacturing technologies, air transportation becoming an everyday phenomenon, significant experience of manned space missions, a growing number of extreme sports fans and affluent people who can afford a space flight. The Russian space tourism industry might be stimulated by the diversification of the military industry, the reduction of the armed forces and the resulting available workforce.

Space tourism vehicles will need to be highly reliable and power efficient for the purposes of multiple long-term uses, to provide best comfort, to be environmentally friendly and simple in operation and maintenance. These requirements turn the construction of the tourist class reusable space vehicle (TC RSV) and of the appropriate infrastructure into a complex scientific and engineering challenge.

When designing TC RSV to meet the above requirements, at the initial stages it is important to acquire data on the designes of close analogues. At present there exist more than 30 RSV designs distinguished according to their purpose (experimental/military/transport/tourist), trajectory (suborbital/orbital flight), launch type (ground/air), landing type (parachute/aerobraking/rocket engine aided/aircraft-type), launch vehicle type (launch plane/air balloon/booster), layout (single-stage/multistage; tandem-staging/cluster; winged/wingless), control system (manned/automatic/hybrid), sustainer type (jet propulsion/liquid fuel/solid fuel/hybrid/composite rocket engine).

The engineering design parameters of a RSV include mass ratio $\mu_r = M_r/M_0$, where M_r denotes remaining mass; M_0 – launch mass; payload mass ratio $\mu_{pl} = M_{pl}/M_0$, where M_{pl} denotes payload mass; propulsion system mass ratio $\gamma_{ps} = M_{ps}/P_{ps}$, with M_{ps} denoting the mass of the propulsion system (loaded); P_{ps} denoting thrust. Interpretation of the information from the literature provided data on re-

maining mass ratio, payload mass ratio and propulsion system mass ratio for a number of vehicles, similar in purpose and design to space tourism vehicles. It should be noted that for the orbital RSV remaining mass was assumed to be the orbital mass, while for the suborbital the same as the mass of the landed RSV. To calculate γ_{ps} the engine was assumed to be loaded with the rocket propellant components.

As follows from the comparative analysis, state-of-the-art reusable space vehicles are characterized by high values of the parameters μ_r and μ_{pl}. This tendency can be accounted for, firstly, by new approaches to TCRSV design and, secondly, by the use of new materials. In most projects preference is given to the winged configuration with horizontal take-off and landing (Saenger, X-34, Astroliner, Ascender, MAKS, SpaceShipOne) or with vertical take-off and horizontal aircraft-type landing (Space Shuttle, Buran-Energia, Hermes-Ariane, Venture Star, Oduvanchik (Fig. 1). This configuration becomes particularly suitable for TCRSV purposes not only in terms of maneuverability, but also due to comfortable g environments.

Each of the above configurations has its advantages and disadvantages. For instance, in the case of horizontal take-off or landing the single-stage RSV mass would hardly exceed 400-500 ton, consequently, the payload mass would not exceed 15-30 ton. RSVs of this type would benefit from hybrid propulsion systems which are still to be tested operating in turbojet or direct-flow modes in various flight phases.

Polymer composite materials (PCM) are planned to be extensively used in RSV to enhance mass performance and increase durability of the structure. According to Andreeva (2001) decreasing structural mass of the Space Shuttle orbital RSV by 1 kg results in $10,000-15,000 savings, for a supersonic passenger liner the economy is $200-500. It should also be noted that PCM parts will produce no more than 30% waste by manufacturing, while conventional aerospace materials waste is 4-12 times the mass of the item itself. Aerospace industry mostly employs high-modulus carbon fibres (CF), which in addition to high specific strength properties possess a very low, practically zero linear thermal expansion coefficient critical for stable-size structures. Despite those advantages, high-modulus carbon fibre composites still remain relatively expensive owing to the high cost of the fibre (Meleshko and Polovnikov, 2007). This problem can be partially solved by use of hybrid PCM which use glass fibre along with the carbon fibres as a filler. It is assumed that these hybrid PCM will be used for manufacturing wings and fuselages, as well as load-bearing elements (spars, stiffening ribs, etc.).

2. TCRSV Oduvanchik

Specialists at Bauman MSTU are currently working on the project of multipassenger winged TCRSV Oduvanchik (Reznik and Stepanischev, 2009; Ageyeva et. al. 2010). The purpose of the spacecraft is both orbital and suborbital missions.

Oduvanchik wing can be made of multi-layer hybrid PCM with excellent mechanical properties. The challenge of manufacturing a hybrid composite structure is to find the balanced combination of different types of materials (for instance, glass fibre plastics and carbon fibre plastics, to select the lamination angle, to estimate the proportion of each material in PCM. The wing design having been determined, thermal and strength calculations are carried out and the stress-strain behaviour of the object is determined. The results are used to draw conclusions with regard to the performance under predetermined loading conditions. Thus, to find the design that will have optimal mass and cost and meet the strength criteria it is essential to perform a number of computations while modifying its structure. The number of calculations is large even for a limited number of layers, so it is feasible to employ numerical methods and computerize the procedure of the multilayered construction optimization.

The process of the TC RSV Oduvanchik design required several assumptions to be made. The wing is assumed to be isolated from the frame, the finite span as a thin multi-layered trapezoidal plate. The skins form the outer surface whose quality for a certain degree determines aerodynamic performance. Multi-layered skin possesses high lateral stiffness, and as a consequence, high critical stresses compared to single-layer skin. Owing to this fact multi-layered skin does not require frequent stringers and allows for a considerable reduction of the amount of ribs and hence a more lightweight construction. Such a skin has superior thermal insulating properties, which makes it suitable for wings subjected to intensive aerodynamic heating. Moreover, the absence of riveting enhances the surface quality.

First and foremost, the skin must ensure the wing structure has the required strength and stiffness. According to the degree of involvement all wings are subdivided into single-spar wings and monoblock wings. The TC RSV Oduvanchik wing is a single-spartype with a relatively thin skin which only receives the torque moment and a small part of the intersecting force. The bending moment is received by the carbon fibre spar, while polyurethane (PU) foam-filled sandwich panels are used as skins. In sandwich panels the surface layers are spaced apart to increase the moment of inertia and, consequently, to increase beam stiffness relative to the neutral axis of the structure.

Each constituent part of the "sandwich" has its own specific function. The surface layers receive tensile and compressive loads; the filler must be sufficiently rigid and

Figure 1. Tourist Class Reusable Space Vehicle Oduvanchik.

shift resistant to prevent surface layers from sliding against each other. Thus, the shear stiffness of the core enables the upper and the lower layers to act as a single entity. For the core and the surface layers to act integrally, it is essential that the binder provides the transfer of the shear forces between them.

3. Selecting quality criteria and constraints for the composite wing design

Weight geometry and cost performance are of utmost importance for the load-bearing unit of the vehicle. This paper employs the following design parameters: the number of glass fibre and carbon fibre layers (n_{gf}, n_{cf}) and the proportion of the layers with $0°$, $\pm45°$, $90°$ angle (v_0, $v_{\pm45}$, v_{90}). The mass of the hybrid skin was determined by means of the following formulae:

$$\bar{M} = \sum_{n=1}^{N} \int_{h_{n-1}}^{h_n} \left(\iint_S \bar{\rho}^{(n)} dS \right)$$

$$\bar{\rho}^{(n)} = v_m^{(n)} \bar{\rho}_m^{(n)} + \sum_{k=1}^{K^{(n)}} v_f^{(n)} \bar{\rho}_f^{(n)}$$

$$v_m^{(n)} = 1 - \sum_{k=1}^{K^{(n)}} v_f^{(n)}$$

with $\bar{M}$ for the mass of the multilayered hybrid skin: $\bar{\rho}^{(n)}$ for the volume density of the nth layer; $\bar{\rho}_m^{(n)}$, $\bar{\rho}_f^{(n)}$ for the volume densities of the matrix materials and the fibres of the Kth family in the nth layer; $K^{(n)}$ for the number of fibre families in the nth layer; $v_m^{(n)}$, $v_f^{(n)}$ for the proportion of the matrix and fibre in the nth layer.

The cost of the skin was calculated according to the formulae:

$$\bar{P} = \sum_{n=1}^{N} \int_{h_{n-1}}^{h_n} \left(\iint_S \bar{p}^{(n)} dS \right)$$

$$\bar{P}^{(n)} = v_m^{(n)} \bar{\rho}_m^{(n)} \bar{P}_m^{(n)} + \sum_{k=1}^{K^{(n)}} v_f^{(n)} \bar{\rho}_f^{(n)} \bar{P}_f^{(n)}$$

with P for the cost of the skin; $\bar{P}^{(n)}$ for the unit cost of the nth layer; $\bar{P}_m^{(n)}$, $\bar{P}_f^{(n)}$ for the unit costs of the matrix materials and the fibres of the Kth family in the nth layer.

Based on the wing's operating conditions a number of compliance constraints were introduced.

8

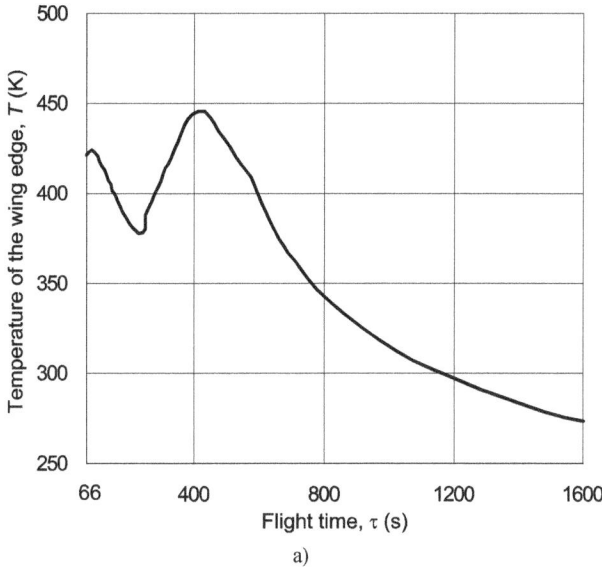

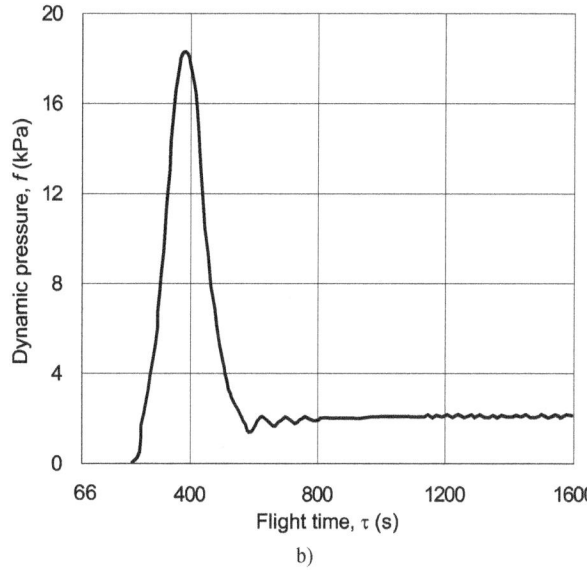

Figure 2. Trajectory parameters vs. flight time τ: a) is temperature of the wing edge T; b) is dynamic pressure f.

$$\bar{J} = W\left(x^*\right)$$

$$\bar{J} = W_0\left(x^*\right)$$

$$J = \frac{\bar{J}}{\bar{J}_0}$$

$$\left(x^*\right) \in S$$

with $\left(x^*\right)$ the maximum deflection point coordinate; $\bar{J}$, $\bar{J}_0$ for the deflection vectors of the skin in question and the reference skin respectively.

The multi-criterion objective of optimal wing design was to indentify the set of design variables:

$$n = \left\{n_{cf}; n_{gf}; v_0; v_{\pm45}; v_{90}\right\}$$

with $n_{cf(gf)}$ for the amount of carbon fibre (glass fibre) layers, v_0, $v_{\pm45}$, v_{90} for the proportion of layers with $0°$, $\pm45°$, $90°$ angles:

$$\bar{F} \to \min_n$$

with $\bar{F} = \left\{F_1; F_2; F_3; ...; F_s\right\}$ for the vector-valued construction quality criterion, including S scalar-valued functional $F_i = F_i\left(\bar{M}; \bar{C}; \bar{J}\right)$. Scalar-valued functionals are defined as functions of predetermined functional characteristics of mass, cost and compliance – $\bar{M}, \bar{C}, \bar{J}$ respectively.

In order to solve the optimization problem the following algorithm was suggested and implemented:

1) A Matlab program was created which by searching all possible layers combinations in a multi-layered package determines its stiffness matrix and subsequently records the results in a text file;

2) A finite-element wing model was created in Femap and the loads acting upon it were assigned;

3) Data from Matlab text file was transferred to Femap by means of a visual Basic program, the finite element calculation was performed. After each iteration the results (maximum deflection values) were recorded in a new text file.

4) Data was transferred from the 3rd stage Matlab file, fitness function values were calculated:

$$G(n) = \left(1 - \frac{\bar{M}(n)}{\bar{M}_{ref}}\right)^2 + \left(1 - \frac{\bar{P}(n)}{\bar{P}_{ref}}\right)^2 + \left(1 - \frac{\bar{J}(n)}{\bar{J}_{ref}}\right)^2$$

with $\bar{M}_{ref}, \bar{P}_{ref}, \bar{J}_{ref}$ as, respectively, mass, cost and deflection of the reference covering (reference covering is assumed to be a wing with a fully glass fibre skin);

5) The fitness function maximum was determined by means of the Matlab built-in module of 'gatool'. Each calculated fitness function value corresponds to the skin with a specific structure and reinforcement pattern.

The Femap system was employed to create the finite-element wing model according to the preset geometric dimensions, moreover, with both the skin and aileron formed by the shell elements. After that the model was broken down into LAMINATE-type finite elements with Layup structure – a set of layers of materials with varied properties and varied anisotropy axis directions. The number of finite elements was 6829 and the number of nodes was 6990. The skin was assumed to comprise six bottom layers of carbon fibre and glass fibre, a 20 mm thick PU foam layer and six top layers of carbon fibre and glass fibre. Then the boundary conditions (the wing was assumed to be fixed to the fuselage) and the loads acting at the stage of entry into the atmosphere were established.

4. Loads acting on the structure

Parameters of the TC RSV Oduvanchik are defined by means of ballistic calculations, which are carried out by means of computer programs that simulate the motion of the RSV centre of mass in the course of the flight (start, booster-assisted acceleration, suborbital flight, lower atmosphere entry, descent to the landing ground, final manoeuvring, approach and landing). The RSV stays at 110 kilometers' altitude from 3 to 5 minutes, which is the time when the tourists are exposed to zero gravity conditions. In order to achieve the maximum cost efficiency of the project it is necessary to ensure that the desired path flight is carried out at the lowest possible speed at the end of the boost phase. It helps minimize weight and cost of the rocket booster, and at the same time it determines large values of the trajectory angle at the beginning of the passive part of the flight and, therefore, the dynamic pressure peaks f (Fig. 2a), and the TC RSV surface temperature T (Fig. 2b). Strength analysis of the TC RSV wing was performed for the case of simultaneous exposure to the maximum pressure and maximum temperature for three minutes. The angle of attack was taken to be 35°, with the upstream stream velocity being 1200 meters per second. The maximum temperature and pressure acting on the wing of the TC RSV were localized on its windward side constituting respectively 16 850 Pa and 527 K. Thus, the operating temperature level determines the choice of polymeric matrix materials for the wing skin. The epoxy resin softening point is in the range of 533 K, which makes it suitable for manufacturing components, operating at temperatures which do not exceed this value.

5. The optimization results

Maximum deflection values were determined for a wing with various combinations of materials and various lamination angles. It was assumed that the proportion of each layer with a definite angle (0°, ±45° and 90°) should not be less than 10%. Fig. 3 presents some of the results obtained.

$$\frac{\Delta M}{\Delta P} = \frac{|M_i - M_{ref}|}{|P_i - P_{ref}|}$$

with P_i for the cost of the ith design, P_{ref} for the cost of the reference design (reference wing is assumed to be a wing with a fully glass fibre skin), M_i for the ith design mass, M_{ref} for reference design mass.

The bending stiffness of the multilayer skin is determined not only by the lamination angles, but by the position of each layer in a multilayer package as well, i.e. on Z-coordinate (see Fig. 4) and is defined by the following formula:

$$EI_{skin} = \frac{1}{3}\sum b(x) \cdot b_{xx}^i(\alpha)(z_i^3 - z_{i-1}^3)$$

with $b(x)$ for the wing chord,

$$b_{xx}(\alpha) = g_{xx}(\alpha) + \frac{g_{xy}^2(\alpha)}{g_{xx}(\alpha)}$$

with $g_{xx}(\alpha)$, $g_{xy}(\alpha)$, $g_{yy}(\alpha)$ for a monolayer stiffness matrix coefficient:

$$G_{xy}(\alpha) = \begin{bmatrix} g_{xx} & g_{xy} & g_{xs} \\ g_{yx} & g_{yy} & g_{ys} \\ g_{sx} & g_{sy} & g_{ss} \end{bmatrix}$$

Thus, the bending stiffness of the skin can be enhanced by placing layers with higher elasticity properties, in this case carbon fibre, on the outer surface of the wing, while glass fibre layers should be positioned on the inner surface.

The results obtained permit to select the optimal structure and the optimal reinforcement pattern for a hybrid composite wing (Tab. 1). The thickness of the carbon fibre or glass fibre monolayer constitutes 0.0003 m, PU foam thickness – 0.02 m. Glass fibre layers must be positioned internally, with the carbon fibre externally.

6. Experimental verification

Experimental verification of the wing is a significant challenge due to the large dimensions of the structure (the wingspan exceeding 8 m). As a result, the need arises to

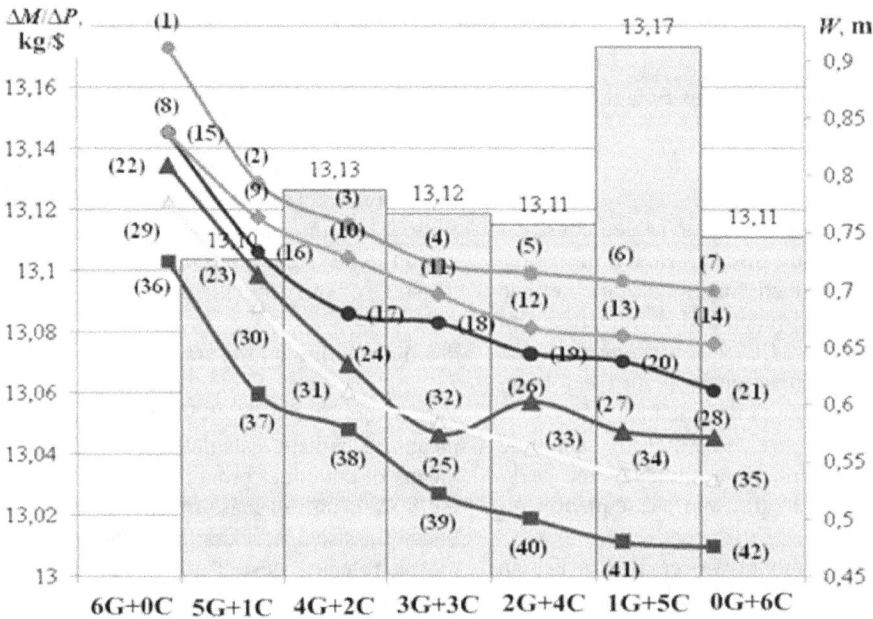

Figure 3. Wing deflections vs. different glass fibre reinforced plastic (G) and carbon fibre reinforced plastic (C) layers proportion and different 0/±45/90 layers proportion.

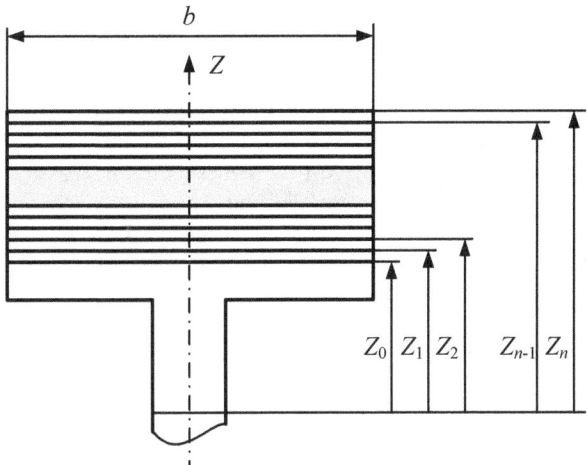

Figure 4. Multilayer wing skin with spar.

validate thermal and strength tests of scaled, cheaper materials models. For this purpose similarity theory methods are traditionally used (Shapovalov, 1990).

The wing construction is formed by a spar and skins which comprise alternating layers of glass fibres, carbon fibres and filler (Fig. 5). In the first approximation the power load is assumed to be uniformly distributed over the wing surface (Fig. 6), with the properties of the composite materials, such as thermal conductivity coefficients, linear thermal expansion coefficients and density, independent of temperature. Thermal conductivity coefficients in the x-y plane are assumed to equal each other.

Conforming to the dimensional theory methods (Shapovalov, 1990) the main heating and deformation parameters of a full-scale design of the wing (prototype) and its scaled model under thermal and power loads can be stated as:

$$\delta,\, T,\, l,\, h,\, x,\, z,\, T_0,\, \tau,\, q,\, f,\, \sigma,\, E,\, \alpha_x,\, \alpha_z,\, \lambda_x,\, \lambda_z,\, C_V,\, A,\, \varepsilon,\, v,$$

Table 1
Deflection, mass and cost values for the optimal wing structures

Lamination pattern	Maximum deflection W, m	Wing mass m_w, kg	Wing cost P_w, USD
$\left[0_3^g / \pm 45_2^c / 90^c\right]$	0.522	1008	18940
$\left[0^g / 0_2^c / \pm 45^g / \pm 45^c / 90^c\right]$	0.5	966	20410
$\left[0^g / 0_2^c / \pm 45_2^c / 90^c\right]$	0.488	926	21870

with δ for load induced deflection of the construction, m; l for length, m; h for thickness, m; x, z for coordinate, m; T for temperature, K; T_0 for initial temperature, K; τ for load time, sec; q for the thermal flux density, W/m^2; f for power load, Pa; σ for the ultimate strength of the material, Pa; E for elasticity modulus of the material, Pa; α_x, α_z for, respectively, longitudinal and transverse thermal expansion, 1/K; λ_x, λ_z for, respectively, longitudinal and transverse thermal conductivity coefficients, W/(m·K); C_V for volumetric heat capacity, J/(m^3·K); A for absorptivity; ε for emissivity; v for Poison ratio.

In order to determine the dimensionless group, which establish relationships between the parameters of the prototype and model, the following steps are to be carried out:

1) Creating a dimensions matrix, determining its rank and the total number of relevant dimensionless groups.

2) Creating a system of algebraic equations for the unknown exponents of the dimensionless complexes. Compiling a decision matrix. Determining the structure of the dimensionless complexes.

3) Recording the functional relationship between the defined and defining dimensionless groups.

4) Determining the scale ratio of the model's and prototype's dimensions.

Within the framework of the classical similarity theory the above procedures will result in a fundamental system of dimensionless groups:

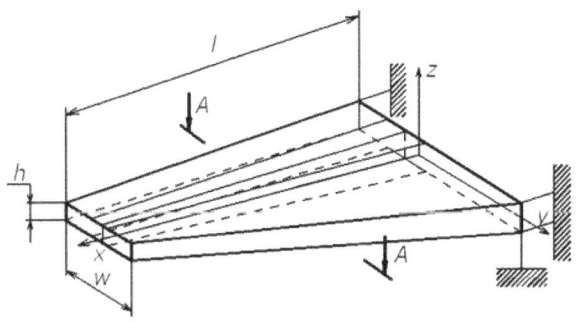

Figure 5. Geometrical model of the wing

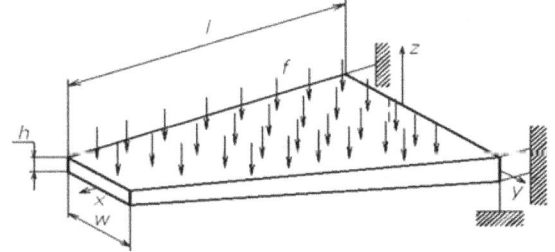
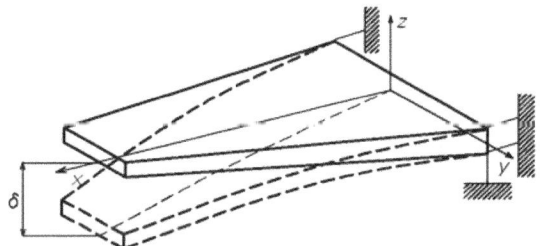

Figure 6. Load conditions.

$$S_1 = \frac{T}{T_0} = \theta = idem; \ S_2 = \frac{\delta}{h} = \varphi = idem; \ S_3 = \frac{x}{l};$$

$$S_4 = \frac{z}{h}; \ S_5 = \frac{qh}{\lambda z T_0} = Ki = idem;$$

$$S_6 = \frac{\tau \lambda z}{C_V h^2} = Fo = idem; \ S_7 = \frac{\alpha_z}{\alpha_x} = idem;$$

$$S_8 = \frac{\lambda_x}{\lambda_z} = idem; \ S_9 = \alpha_x T = idem; \ S_{10} = \alpha_z T = idem;$$

$$S_{11} = \frac{\sigma}{E} = idem; \ S_{12} = \frac{f}{E} = idem; \ S_{13} = \frac{h}{l} = idem;$$

$$S_{14} = A = idem; \ S_{15} = \varepsilon = idem; \ S_{16} = v = idem \tag{1}$$

where S_1 stands for the dimensionless temperature, S_2 for dimensionless deflection, S_3, S_4 for dimensionless coordinates, S_5 for Kirpichev criterion, S_6 for Fourier criterion, S_7 for the relative criterion of linear thermal expansion, S_8 for relative heat conductivity coefficient, S_9, S_{10} for heat-induced distortion criteria, S_{11} for dimensionless critical stress (relative characteristic of the resistance of the material), S_{12} for the relative power load, S_{13} for the relative thickness.

The solution to the problem can be presented in the dimensionless form:

$$\theta = \frac{T}{T_0} = F_1 \left(\frac{qh}{\lambda_z T_0}; \frac{\tau \lambda_z}{C_V h^2}; \frac{\alpha_x}{\alpha_z}; \frac{\lambda_x}{\lambda_z}; \right.$$
$$\left. \frac{x}{l}; \frac{z}{h}; \frac{h}{l}; \alpha_x T; \alpha_z T; \frac{\sigma}{E}; \frac{f}{E}; A; \varepsilon; v \right)$$

$$\varphi = \frac{\delta}{z} = F_2 \left(\frac{qh}{\lambda_z T_0}; \frac{\tau \lambda_z}{C_V h^2}; \frac{\alpha_z}{\alpha_x}; \frac{\lambda_x}{\lambda_z}; \right.$$
$$\left. \frac{x}{l}; \frac{z}{h}; \frac{h}{l}; \alpha_x T; \alpha_z T; \frac{\sigma}{E}; \frac{f}{E}; A; \varepsilon; v \right) \tag{2}$$

where F_1 and F_2 are functions to be determined as the model is being tested.

According to the ratios obtained, for the thermal and stress-strain states of the model (M) and the prototype (P) to be similar the following conditions need to be fulfilled:

$$\left(\frac{qh}{\lambda_z T_0} \right)_M = \left(\frac{qh}{\lambda_z T_0} \right)_P; \ \left(\frac{\tau \lambda_z}{C_V h^2} \right)_M = \left(\frac{\tau \lambda_z}{C_V h^2} \right)_P;$$

$$\left(\frac{\alpha_x}{\alpha_z} \right)_M = \left(\frac{\alpha_x}{\alpha_z} \right)_P; \ \left(\frac{x}{l} \right)_M = \left(\frac{x}{l} \right)_P; \ \left(\frac{z}{h} \right)_M = \left(\frac{z}{h} \right)_P;$$

$$\left(\frac{\lambda_x}{\lambda_z} \right)_M = \left(\frac{\lambda_x}{\lambda_z} \right)_P; \ (\alpha_x T)_M = (\alpha_x T)_P;$$

$$(\alpha_z T)_M = (\alpha_z T)_P; \ \left(\frac{\sigma}{E} \right)_M = \left(\frac{\sigma}{E} \right)_P;$$

$$\left(\frac{f}{E} \right)_M = \left(\frac{f}{E} \right)_P; \ \left(\frac{h}{l} \right)_M = \left(\frac{h}{l} \right)_P;$$

$$A_M = A_P; \ \varepsilon_M = \varepsilon_P; \ v_M = v_P$$

By employing the invariance of the defining similarity criteria the following ratios can be obtained:

$$q_M = \frac{q_P \lambda_{zM} T_{0M}}{\lambda_{zP} T_{0P}} \frac{h_P}{h_M} \tag{3}$$

$$\tau_M = \tau_P \frac{\lambda_{zP}}{(C_V)_P} \frac{(C_V)_M}{\lambda_{zM}} \frac{h_M^2}{h_P^2} \tag{4}$$

As follows from (3) and (4) a five-fold reduction in the length (l) the model in comparison to the prototype shall result accordingly in a five-fold reduction in thickness (h). Then, for the same- materials model and at the identical initial temperature the heat flux to the model should be increased five-fold, while the test timecan be reduced 25 fold. Such a significant increase in heat flux will cause a significant change in temperature of the material of the model, so the assumptions introduced while formulating the problem will be violated.

It is obvious, however, that simulation test challenges are determined by the necessity to proportionally change the length and thickness of the wing, which, among other things, can prove to be problematic for thin skins. The solution can be found in the affine similarity theory, which involves the introduction of several linear scales, and thus allows for maintaining the thickness h constant while reducing the length l of the specimen. In this case, $h_M = h_P$, and the relationships (1)-(2) retain their form, and expression (3) and (4) assume the following form:

$$q_M = \frac{q_P \lambda_{zM} T_{0M}}{\lambda_{zP} T_{0P}} \tag{5}$$

$$\tau_M = \tau_P \frac{\lambda_{zP}}{(C_V)_P} \frac{(C_V)_M}{\lambda_{zM}} \tag{6}$$

Judging by expression (5), the heat flux to the model can be decreased by lowering the initial temperature and reducing the thermal conductivity of the material. For example, if the entire model is manufactured from glass fibre instead of carbon fibre (as in case with prototype), the $\lambda_{zM}/\lambda_{zP}$ ratio will be close to 0.5 (Meleshko and Polovnikov, 2007), which means that the heat flux to the model can be reduced two-fold even if $T_{0M} = T_{0P}$. Moreover, the load application time, as follows from (6), should increase by a factor of 2 $(C_V)_M/(C_V)_P$ times. The specific heat capacities of carbon fibre and glass fibre at room temperature are fairly similar, but their densities are different. If the glass fibre density is assumed to be 2000 kg/m^3, with carbon fibre density of 1600 kg/m^3, the volumetric heat capacity ratio will constitute 1.25. Consequently, if, unlike the prototype, the model is made entirely of glass fibre, the load time will increase approximately 2.5 times owing to the differences in the thermal physical properties.

Qualitative results are consistent with the results of finite element modelling.

7. Conclusion

The developed algorithm makes it possible to determine the optimal design in terms of strength, mass and cost of the TC RSV wing skin structure. The results show the feasibility of use glass fibre and carbon fibre reinforced plastic simultaneously in the same wing structure. The minimization of wing mass provides the reduction of the necessary rocket engine power, reduction of its mass, dimensions and overall cast of the TC RSV.

Experimental verification of the results obtained is based on similarity theory and assumes using of scaled models made of cheaper materials and effected by decreased heat flux for longer time. Some results of the current work are obtained in the TC RSV Oduvanchik project.

8. Acknowledgements

The authors would like to thank Dr. E. N. Dudar (JSC "NPO "Molniya") for support in ballistic and aerodynamic researches and Dr. K.V. Mikhailovskiy (Bauman MSTU) for help with FEM analysis.

References

Ageyeva, T.G., Dudar, E.N. and Reznik, S.V. (2010). Complex method of design of wing structure for reusable space vehicle, *Aerospace Technology*, No. 10, pp. 3-8. (in Russian)

Andreeva, A.V. (2001). *Fundamentals of Physics and Chemistry of Composite Materials*. Moscow: IPRGR. (in Russian)

Meleshko, A.I. and Polovnikov, S.P. (2007). *Carbon, Carbon Fibres, Carbon Composites*. Moscow: SAINS-PRESS. (in Russian)

Reznik, S.V. and Ageyeva, T.G. (2010). Comparison analysis of technological perfection of reusable space vehicles, *Bauman MSTU Reporter*, special issue *Actual Problems of Russian Cosmonautics* dedicated to Bauman MSTU 180th anniversary, pp. 19-34. (inRussian)

Reznik, S.V. and Stepanischev, N.A. (2009). Design an engineering solutions for light weight tourist class reusable space vehicles, *ActualProblems of Russian Cosmonautics: Proceedings of the 33rd Academic Readings on Cosmonautics*, pp. 71-73. (in Russian)

Shapovalov, L.A. (1990). *Simulation in Problems of Mechanics of Structural Elements*. Moscow: Mechinostroeniye. (in Russian)

Vibration and Buckling of Cross-Ply Composite Beams using Refined Shear Deformation Theory

Thuc Vo, Fawad Inam

School of Mechanical, Aeronautical and Electrical Engineering, Glyndŵr University, Plas Coch, Mold Road, Wrexham, LL11 2AW, UK

Abstract: Vibration and buckling analysis of cross-ply composite beams using refined shear deformation theory is presented. The theory accounts for the parabolical variation of shear strains through the depth of beam. Three governing equations of motion are derived from the Hamilton's principle. The resulting coupling is referred to as triply coupled vibration and buckling. A two-noded C^1 beam element with five degree-of-freedom per node is developed to solve the problem. Numerical results are obtained for composite beams to investigate modulus ratio on the natural frequencies, critical buckling loads and load-frequency interaction curves.

Key Words: Composite beams, Refined shear deformation theory, Triply coupled vibration and buckling.

1. Introduction

Structural components made with composite materials are increasingly being used in various engineering applications due to their attractive properties in strength, stiffness, and lightness. Understanding their dynamic and buckling behaviour is of increasing importance. The classical beam theory (CBT) known as Euler-Bernoulli beam theory is the simplest one and is applicable to slender beams only. For moderately deep beams, it overestimates buckling loads and natural frequencies due to ignoring the transverse shear effects. The first-order beam theory (FOBT) known as Timoshenko beam theory is proposed to overcome the limitations of the CBT by accounting for the transverse shear effects. Since the FOBT violates the zero shear stress conditions on the top and bottom surfaces of the beam, a shear correction factor is required to account for the discrepancy between the actual stress state and the assumed constant stress state. To remove the discrepancies in the CBT and FOBT, the higher-order beam theory (HOBT) is developed to avoid the use of shear correction factor and have a better prediction of response of laminated beams. The HOBTs can be developed based on the assumption of higher-order variations of in-plane displacement or both in-plane and transverse displacements through the depth of the beam. Many numerical techniques have been used to solve the dynamic and/or buckling analysis of composite beams using HOBTs. Some researchers studied the free vibration characteristics of composite beams by using finite element (Chandrashekhara and Bangera, 1992; Marur and Kant, 1996; Shi and Lam, 1999; Murthy et. al., 2005; Subramanian, 2006). Khdeir and Reddy (1996, 1997) developed analytical solutions for free vibration and buckling of cross-ply composite beams with arbitrary boundary conditions in conjunction with the state space approach. Analytical solutions were also derived by Kant et al. (1997, 2001) and Zhen and Wanji (2008) to study vibration and buckling of composite beams. By using the method of power series expansion of displacement components, Matsunaga (2001) analysed the natural frequencies and buckling stresses of composite beams. Aydogdu (2005, 2006a, 2006b) carried out the vibration and buckling analysis of cross-ply and angle-ply with different sets of boundary conditions by using Ritz method. Jun et al. (2009, 2011)

introduced the dynamic stiffness matrix method to solve exactly the free vibration and buckling problems of axially loaded composite beams with arbitrary lay-ups. Although the HOBTs offer a slight improvement in accuracy compared to the FOBT, they are computationally more demanding due to higher-order terms included in the theory. Hence, there is a scope to develop accurate refined shear deformation theory which is simple to use.

In this paper, which is extended from previous research (Vo and Thai, 2012), vibration and buckling analysis of composite beams using refined shear deformation theory is presented. The displacement field of the present theory is chosen based on the following assumptions: (1) the axial and transverse displacements consist of bending and shear components in which the bending components do not contribute toward shear forces and, likewise, the shear components do not contribute toward bending moments; (2) the bending component of axial displacement is similar to that given by the CBT; and (3) the shear component of axial displacement gives rise to the higher-order variation of shear strain and hence to shear stress through the depth of the beam in such a way that shear stress vanishes on the top and bottom surfaces. The most interesting feature of this theory is that it satisfies the zero traction boundary conditions on the top and bottom surfaces of the beam without using shear correction factors. The three governing equations of motion are derived from the Hamilton's principle. The resulting coupling is referred to as triply coupled vibration and buckling. A two-noded C^1 beam element with five degree-of-freedom per node which accounts for shear deformation effects and all coupling coming from the material anisotropy is developed to solve the problem. Numerical results are obtained for

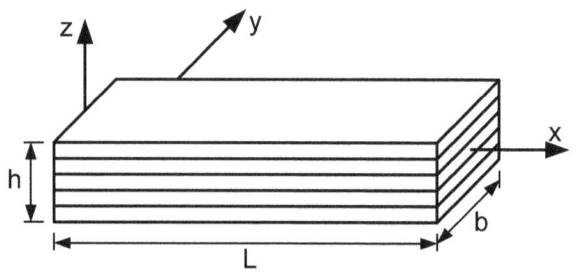

Figure 1. Geometry of a laminated composite beam.

composite beams to investigate effects of fiber orientation and modulus ratio on the natural frequencies, critical buckling loads and load-frequency interaction curves as well as corresponding mode shapes.

2. Kinematics

A laminated composite beam made of many plies of orthotropic materials in different orientations with respect to the x-axis, as shown in Fig. 1, is considered.

Based on the assumptions made in the preceding section, the displacement field of the present theory can be obtained as:

$$U(x,z,t) = u(x,t) - z\frac{\partial w_b(x,t)}{\partial x} +$$
$$+ z\left[\frac{1}{4} - \frac{5}{3}\left(\frac{z}{h}\right)^2\right]\frac{\partial w_s(x,t)}{\partial x} \qquad (1)$$

$$V(x,z,t) = 0$$
$$W(x,z,t) = w_b(x,t) + w_s(x,t)$$

where u is the axial displacement along the mid-plane of the beam, w_b and w_s are the bending and shear components of transverse displacement along the mid-plane of the beam, respectively. The non-zero strains are given by:

$$\varepsilon_x = \frac{\partial u}{\partial x} = \varepsilon_x^\circ + z\kappa_x^b + f\kappa_x^s$$
$$\gamma_{xz} = \frac{\partial w}{\partial x} + \frac{\partial u}{\partial z} = (1 - f')\gamma_{xz}^\circ = g\gamma_{xz}^\circ \qquad (2)$$

where

$$f = z\left[-\frac{1}{4} + \frac{5}{3}\left(\frac{z}{h}\right)^2\right]$$
$$g = 1 - f' = \frac{5}{4}\left[1 - 4\left(\frac{z}{h}\right)^2\right] \qquad (3)$$

and $\varepsilon_x^\circ, \gamma_{xz}^\circ, \kappa_x^b, \kappa_x^s$ and κ_{xy} are the axial strain, shear strains and curvatures in the beam, respectively defined as:

$$\varepsilon_x^\circ = u'$$
$$\gamma_{xz}^\circ = w_s'$$
$$\kappa_x^b = -w_b'' \qquad (4)$$
$$\kappa_x^s = -w_s''$$

where differentiation with respect to the x-axis is denoted by primes (').

3. Variational Formulation

In order to derive the equations of motion, Hamilton's principle is used:

$$\delta\int_{t_1}^{t_2}(K - U - V)dt = 0 \qquad (5)$$

where U is the variation of the strain energy, V is the variation of the potential energy, and K is the variation of the kinetic energy.

The variation of the strain energy can be stated as:

$$\delta U = \int_v (\sigma_x\delta\varepsilon_x + \sigma_{xz}\delta\gamma_{xz}) dv =$$
$$= \int_0^l (N_x\delta\varepsilon_z^\circ + M_x^b\delta\kappa_x^b + M_x^s\delta\kappa_x^s + Q_{xz}\delta\gamma_{xz}^\circ) dx \qquad (6)$$

where N_x, M_x^b, M_x^s and Q_{xz} are the axial force, bending moments and shear force, respectively, defined by integrating over the cross-sectional area A as:

$$N_x = \int_A \sigma_x dA$$
$$M_x^b = \int_A \sigma_x z dA$$
$$M_x^s = \int_A \sigma_x f dA \qquad (7)$$
$$Q_{xz} = \int_A \sigma_{xz} g dA$$

The variation of the potential energy of the axial force can be expressed as:

$$\delta V = -\int_0^l P_0\left[\delta w_{b'}(w_{b'} + w_{s'}) + \delta w_{s'}(w_{b'} + w_{s'})\right]dx \qquad (8)$$

The variation of the kinetic energy is obtained as:

$$\delta K = \int_v \rho_k(\dot{U}\delta\dot{U} + \dot{V}\delta\dot{V} + \dot{W}\delta\dot{W})dv =$$
$$= \int_0^l [\delta\dot{u}(m_0\dot{u} - m_1\dot{w}_b' - m_f\dot{w}_s') +$$
$$+ \delta\dot{w}_b'(-m_1\dot{u} + m_2\dot{w}_b' + m_{fz}\dot{w}_s') +$$
$$+ \delta\dot{w}_s m_0(\dot{w}_b + \dot{w}_s) + \qquad (9)$$
$$+ \delta\dot{w}_s'(-m_f\dot{u} + m_{fz}\dot{w}_b' + m_{f2}\dot{w}_s')]dx$$

where the differentiation with respect to the time t is denoted by dot-superscript convention and ρk is the density of a kth layer and $m_0, m_1, m_2, m_f, m_{fz}$ and m_{f2} are the inertia coefficients, defined by:

$$m_f = -\frac{m_1}{4} + \frac{5}{3h^2}m_3$$
$$m_{fz} = -\frac{m_2}{4} + \frac{5}{3h^2}m_4 \qquad (10)$$
$$m_{f2} = \frac{m_2}{16} - \frac{5}{6h^2}m_4 + \frac{25}{9h^4}m_6$$

where

$$(m_0, m_1, m_2, m_3, m_4, m_6) = \int_A \rho_k(1, z, z^2, z^3, z^4, z^6)dA \qquad (11)$$

By substituting Eqs. (6), (8) and (9) into Eq. (5), the following weak statement is obtained:

$$0 = \int_{t_2}^{t_2}\int_0^l [\delta\dot{u}(m_0\dot{u} - m_1\dot{w}_b' - m_f\dot{w}_s') +$$
$$+ \delta\dot{w}_b m_0(\dot{w}_b + \dot{w}_s) +$$
$$+ \delta\dot{w}_b'(-m_1\dot{u} + m_2\dot{w}_b' + m_{fz}\dot{w}_s') +$$
$$+ \delta\dot{w}_s m_0(\dot{w}_b + \dot{w}_s) + \qquad (12)$$
$$+ \delta\dot{w}_s'(-m_f\dot{u} + m_{fz}\dot{w}_b' + m_{f2}\dot{w}_s') +$$
$$+ P_0\left[\delta w_{b'}(w_{b'} + w_{s'}) + \delta w_{s'}(w_{b'} + w_{s'})\right] -$$
$$+ N_x\delta u' + M_x^b\delta w_b'' + M_x^s\delta w_s'' - Q_{xz}\delta w_s']dxdt$$

15

4. Constitutive equations

The stress-strain relations for the kth lamina are given by:

$$\sigma_x = \bar{Q}_{11}\gamma_x$$
$$\sigma_{xz} = \bar{Q}_{55}\gamma_{xz} \qquad (13)$$

where $\bar{Q}_{11}$ and $\bar{Q}_{55}$ are the elastic stiffnesses transformed to the x-direction. More detailed explanation can be found in (Jones, 1999).

The constitutive equations for bar forces and bar strains are obtained by using Eqs. (2), (7) and (13):

$$\begin{Bmatrix} N_x \\ M_x^b \\ M_x^s \\ Q_{xz} \end{Bmatrix} = \begin{bmatrix} R_{11} & R_{12} & R_{13} & 0 \\ & R_{22} & R_{23} & 0 \\ & & R_{33} & 0 \\ \text{sym.} & & & R_{44} \end{bmatrix} \begin{Bmatrix} \varepsilon_x^\circ \\ \kappa_x^b \\ \kappa_x^s \\ \gamma_{xz}^\circ \end{Bmatrix} \qquad (14)$$

where R_{ij} are the laminate stiffnesses of general composite beams and given by:

$$R_{11} = \int_y A_{11}\,dy$$

$$R_{12} = \int_y B_{11}\,dy$$

$$R_{13} = \int_y \left(-\frac{B_{11}}{4} + \frac{5}{3h^2}E_{11} \right) dy$$

$$R_{22} = \int_y D_{11}\,dy \qquad (15)$$

$$R_{23} = \int_y \left(-\frac{D_{11}}{4} + \frac{5}{3h^2}F_{11} \right) dy$$

$$R_{33} = \int_y \left(\frac{D_{11}}{16} - \frac{5}{6h^2}F_{11} + \frac{25}{9h^4}H_{11} \right) dy$$

$$R_{44} = \int_y \left(\frac{25}{16}A_{55} - \frac{25}{2h^2}D_{55} + \frac{25}{h^4}F_{55} \right) dy$$

where A_{ij}, B_{ij} and D_{ij} matrices are the extensional, coupling and bending stiffness and E_{ij}, F_{ij}, H_{ij} matrices are the higher-order stiffnesses, respectively, defined by:

$$\left(A_{ij}, B_{ij}, D_{ij}, E_{ij}, F_{ij}, H_{ij} \right) = \int_z \bar{Q}_{ij}\left(1, z, z^2, z^3, z^4, z^6 \right) dz \quad (16)$$

5. Governing equations of motion

The equilibrium equations of the present study can be obtained by integrating the derivatives of the varied quantities by parts and collecting the coefficients of δu, δu_b and δu_s:

$$N_x' = m_0\ddot{u} - m_1\ddot{w}_b' - m_f\ddot{w}_s';$$

$$M_x^{b''} - P_0\left(w_{b''} + w_{s''} \right) =$$
$$= m_0\left(\ddot{w}_b + \ddot{w}_s \right) + m_1\ddot{u}' - m_2\ddot{w}_b'' - m_{fz}\ddot{w}_s''; \qquad (17)$$

$$M_x^{s''} + Q_{xz}' - P_0\left(w_{b''} + w_{s''} \right) =$$
$$= m_0\left(\ddot{w}_b + \ddot{w}_s \right) + m_f\ddot{u}' - m_{fz}\ddot{w}_b'' - m_{f2}\ddot{w}_s''$$

The natural boundary conditions are of the form:

$$\delta u : N_x$$

$$\delta w_b : M_x^{b'} - P_0\left(wb' + ws' \right) - m_1\ddot{u} + m_2\ddot{w}_b' + m_{fz}\ddot{w}_s'$$

$$\delta w_b' : M_x^b \qquad (18)$$

$$\delta w_s : M_x^{s'} + Q_{xz} - P_0\left(wb' + ws' \right) - m_f\ddot{u} + m_{fz}\ddot{w}_b' + m_{f2}\ddot{w}_s'$$

$$\delta w_s' : M_x^s$$

By substituting Eqs. (4) and (14) into Eq. (17), the explicit form of the governing equations of motion can be expressed with respect to the laminate stiffnesses R_{ij}. Eq. (17) is the most general form for axial-flexural coupled vibration of axially loaded of composite beams, and the dependent variables, u, w_b and w_s are fully coupled.

6. Finite element formulation

The present theory for composite beams described in the previous section was implemented via a displacement based finite element method. The variational statement in Eq. (12) requires that the bending and shear components of transverse displacement w_b and w_s be twice differentiable and C^1-continuous, whereas the axial displacement u must be only once differentiable and C^0-continuous. The generalized displacements are expressed over each element as a combination of the linear interpolation function Ψ_j for u and Hermite-cubic interpolation function ψ_j for w_b and w_s associated with node j and the nodal values:

$$u = \sum_{j=1}^{2} u_j \Psi_j$$

$$w_b = \sum_{j=1}^{4} w_{bj}\psi_j \qquad (19)$$

$$w_s = \sum_{j=1}^{4} w_{sj}\psi_j$$

Substituting these expressions in Eq. (19) into the corresponding weak statement in Eq. (12), the finite element model of a typical element can be expressed as the standard eigenvalue problem:

$$\left([K] - P_0[G] - \omega^2[M] \right)\{\Delta\} = \{0\} \qquad (20)$$

where $[K]$, $[G]$ and $[M]$ are the element stiffness matrix, the element geometric stiffness matrix and the element mass matrix, respectively. In Eq.(20), $\{\Delta\}$ is the eigenvector of nodal displacements corresponding to an eigenvalue:

$$\{\Delta\} = \{u \; w_b \; w_s\}^{\mathrm{T}} \qquad (21)$$

7. Numerical examples

In this section, a number of numerical examples are presented and analysed for verification the accuracy of the present theory in predicting the natural frequencies, critical buckling loads and corresponding mode shapes of composite beams with arbitrary lay-ups. All laminate are of equal thickness and made of the same orthotropic material, whose properties are as follows:

Material I:

$$E_1 / E_2 = open,$$
$$G_{12} = G_{13} = 0.6E_2$$
$$G_{23} = 0.5E_2 \qquad (22)$$
$$v_{12} = 0.25$$

Material II:

$$E_1 / E_2 = open$$
$$G_{12} = G_{13} = 0.5E_2$$
$$G_{23} = 0.2E_2 \qquad (23)$$
$$v_{12} = 0.25$$

For convenience, the following non-dimensional terms are used in presenting the numerical results:

$$\overline{P}_{cr} = \frac{P_{cr}L^2}{E_2 bh^3}$$
$$\overline{\omega} = \frac{\omega L^2}{h}\sqrt{\rho E_2} \qquad (24)$$

As the first example, vibration and buckling analysis of a symmetric and an anti-symmetric cross-ply composite beam with simply-supported boundary condition is performed. Material I and II with $E_1/E_2 = 10$ and 40 are used. The fundamental natural frequencies and critical buckling loads for different span-to-height L/h ratios are compared with exact solutions (Khdeir and Reddy, 1994; Khdeir and Reddy, 1997) and the finite elements results (Murthy et al., 2005; Aydogdu, 2005; Aydogdu, 2006a) in Tables 1 and 2. An excellent agreement between the predictions of the present model and the results of the other models mentioned can be observed.

Material I with $E_1/E_2 = 40$ is chosen to show the effect of the axial force on the fundamental natural frequencies of beam with various L/h ratios (Fig. 2). It can be seen that the change of the natural frequency due to axial force is noticeable. The natural frequency diminishes when the axial force changes from tensile to compressive, as expected. It is obvious that the natural frequency decreases with the increase of axial force, and the decrease becomes more quickly when the axial force is close to critical buckling load. For an anti-symmetric cross-ply lay-up, with $L/h = 5$, 10 and 20, at about P = 3.903, 4.936 and

Table 1.
Effect of span-to-height ratio on the fundamental natural frequencies of a simply-supported cross-ply composite beam (Material I with $E_1/E_2 = 40$).

Lay-ups	Reference	L/h		
		5	10	20
$[0^0/90^0/0^0]$	Murthy et al. (2005)	9.207	13.614	-
	Khdeir and Reddy (1994)	9.208	13.614	-
	Aydogdu (2005)	9.207	-	16.337
	Present	9.206	13.607	16.327
$[0^0/90^0]$	Murthy et al. (2005)	6.045	6.908	-
	Khdeir and Reddy (1994)	6.128	6.945	-
	Aydogdu (2005)	6.144	-	7.218
	Present	6.058	6.909	7.204

Table 2.
Effect of span-to-height ratio on the critical buckling loads of a simply-supported cross-ply composite beam (Material I and II with $E_1/E_2 = 40$).

Lay-ups	Reference	L/h		
		5	10	20
Material I				
$[0^0/90^0/0^0]$	Khdeir and Reddy (1997)	8.613	18.832	-
	Aydogdu (2006a)	8.613	-	27.084
	Present	8.609	18.814	27.050
$[0^0/90^0]$	Aydogdu (2006a)	3.906	-	5.296
	Present	3.903	4.936	5.290
Material II				
$[0^0/90^0/0^0]$	Aydogdu (2006a)	5.896	-	24.685
	Present	5.895	14.857	24.655
$[0^0/90^0]$	Aydogdu (2006a)	3.376	-	5.225
	Present	3.373	4.697	5.219

5.290, respectively, the natural frequencies become zero which implies that at these loads, bucklings occur as a degenerate case of natural vibration at zero frequency. It also means that the buckling loads of composite beams under axial force can be also obtained indirectly through vibration problem by increasing the axial force until the corresponding natural frequency vanishes. Besides, Fig. 2 explains the duality between the buckling load and natural frequency. In order to show the effect of material anisot-

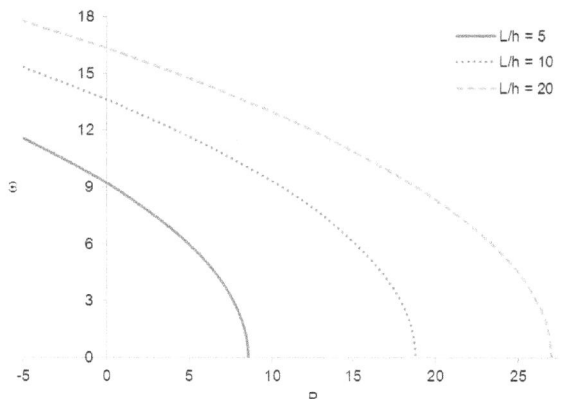

a) Symmetric cross-ply lay-up ($[0^0/90^0/0^0]$)

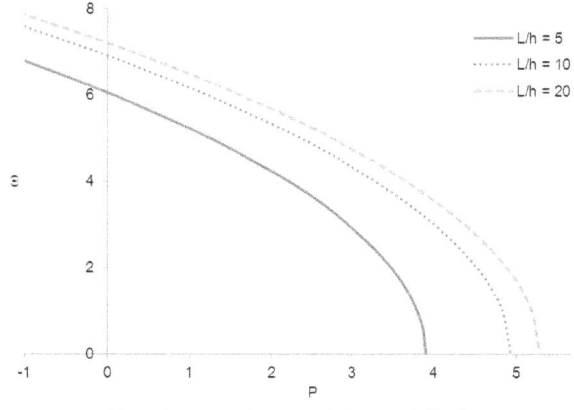

b) Anti-symmetric cross-ply lay-up ($[0^0/90^0]$)

Figure 2. Load-frequency curves of a simply-supported cross-ply composite beam ($L/h = 5$, 10 and 20).

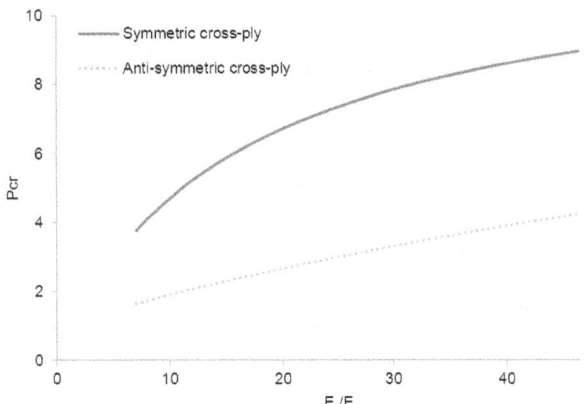

Figure 3. Effect of material anisotropy on the critical buckling loads of a simply-supported cross-ply composite beam ($L/h = 5$).

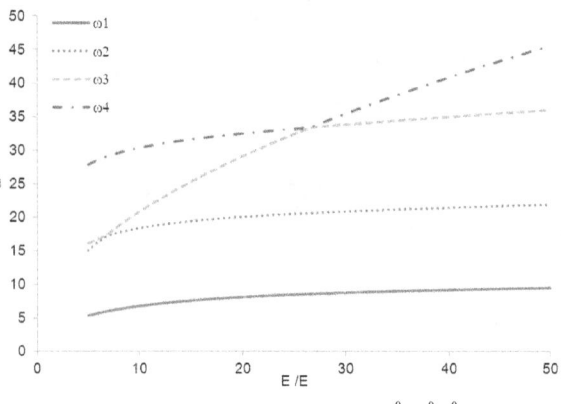

a) Symmetric cross-ply lay-up ($[0^0/90^0/0^0]$)

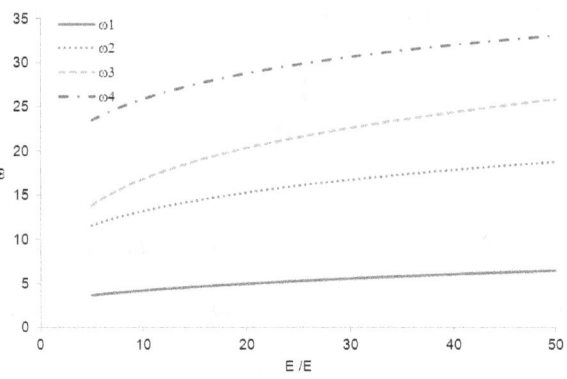

b) Anti-symmetric cross-ply lay-up ($[0^0/90^0]$)

Figure 4. Effect of material anisotropy on the first four natural frequencies of a simply-supported cross-ply composite beam ($L/h = 5$).

ropy (E_1/E_2) on the critical buckling loads and the first four natural frequencies of a symmetric and an anti-symmetric cross-ply lay-up, a simply-supported composite beam with $L/h = 5$ is performed. It is observed that the critical buckling loads and natural frequencies increase with increasing orthotropy (Figs. 3 and 4). For a symmetric cross-ply lay-up, as ratio of E_1/E_2 increases, the order of the second and third vibration mode as well as the third and fourth vibration mode changes each other at $E_1/E_2 = 7$ and 27, respectively (Fig. 4).

8. Conclusions

A two-noded C^1 beam element of five degree-of-freedom per node is developed to study the vibration and buckling behaviour of cross-ply composite beams using refined shear deformation theory. This model is capable of predicting accurately the natural frequencies, critical buckling loads and load-frequency interaction curves. It accounts for the parabolical variation of shear strains through the depth of the beam, and satisfies the zero traction boundary conditions on the top and bottom surfaces of the beam without using shear correction factor. The present model is found to be appropriate and efficient in analyzing vibration and buckling problem of cross-ply composite beams.

Reference

Aydogdu, M. (2005). Vibration analysis of cross-ply laminated beams with general boundary conditions by Ritz method. *International Journal of Mechanical Sciences*, Vol. 47, No 11, pp. 1740-1755.

Aydogdu, M. (2006a). Buckling analysis of cross-ply laminated beams with general boundary conditions by Ritz method. *Composites Science and Technology*, Vol. 66, No 10, pp. 1248-1255.

Aydogdu, M. (2006b). Free vibration analysis of angle-ply laminated beams with general boundary conditions. *Journal of Reinforced Plastics and Composites*, Vol. 25, No 15, pp. 1571-1583.

Chandrashekhara, K. and Bangera, K. (1992). Free vibration of composite beams using a refined shear flexible beam element. *Computers and Structures*, Vol. 43, No 4, pp. 719-727.

Chandrashekhara, K., Krishnamurthy, K. and Roy, S. (1990). Free vibration of composite beams including rotary inertia and shear deformation. *Composite Structures*, Vol. 14, No 4, pp. 269-279.

Chen, W.Q., Lv, C.F. and Bian, Z.G. (2004). Free vibration analysis of generally laminated beams via state-space-based differential quadrature. Composite Structures, Vol. 63, No 3-4, pp. 417-425.

Jones, R.M. (1999). *Mechanics of Composite Materials*, Philadelphia: Taylor and Francis.

Jun, L. and Hongxing, H. (2011). Free vibration analyses of axially loaded laminated composite beams based on higher-order shear deformation theory. *Meccanica*, Vol. 46, pp. 1299-1317.

Jun, L., Xiaobin, L. and Hongxing, H. (2009). Free vibration analysis of third-order shear deformable composite beams using dynamic stiffness method. *Archive of Applied Mechanics*, Vol. 79, pp. 1083-1098.

Kant, T., Marur, S.R. and Rao, G. (1997). Analytical solution to the dynamic analysis of laminated beams using higher order refined theory. *Composite Structures*, Vol. 40, No 1, pp. 1-9.

Kant, T. and Swaminathan, K. (2001). Analytical solutions for free vibration of laminated composite and sandwich plates based on a higher-order refined theory. *Composite Structures*, Vol. 53, pp. 73-85.

Khdeir, A.A. and Reddy, J.N. (1994). Free vibration of cross-ply laminated beams with arbitrary boundary conditions. *International Journal of Engineering Science*, Vol. 32, No. 12, pp. 1971-1980.

Khdeir, A.A. and Reddy, J.N. (1997). Buckling of cross-ply laminated beams with arbitrary boundary conditions. *Composite Structures*, Vol. 37, No 1, pp. 1-3.

Krishnaswamy, S., Chandrashekhara, K. and Wu, W.Z.B. (1992). Analytical solutions to vibration of generally layered composite beams. Journal of Sound and Vibration, Vol. 159, No 1, pp. 85-99.

Marur, S.R. and Kant, T. (1996). Free vibration analysis of fiber reinforced composite beams using higher order theories and finite element modelling. *Journal of Sound and Vibration*, Vol. 194, No 3, pp. 337-351.

Matsunaga, H. (2001). Vibration and buckling of multilayered composite beams according to higher order deformation theories. *Journal of Sound and Vibration*, Vol. 246, No 1, pp. 47-62.

Murthy, M.V.V.S., Mahapatra, D.R., Badarinarayana, K. and Gopalakrishnan, S. (2005). A refined higher order finite element for asymmetric composite beams. *Composite Structures*, Vol. 67, pp. 27-35.

Shi, G. and Lam, K.Y. (1999). Finite element vibration analysis of composite beams based on higher-order beam theory. *Journal of Sound and Vibration*, Vol. 219, No 4, pp. 707-721.

Subramanian, P. (2006). Dynamic analysis of laminated composite beams using higher order theories and finite elements. *Composite Structures*, Vol. 73, No 3, pp. 342-353.

Vo T.P. and Thai, H.T. (2012). Static behaviour of composite beams using various refined shear deformation theories. *Composite Structures*, in press.

Zhen, W. and Wanji, C. (2008). An assessment of several displacement-based theories for the vibration and stability analysis of laminated composite and sandwich beams. *Composite Structures*, Vol. 84, No 4, pp. 337-349.

Reliability Issues in Composite Thermal Protection Structures of Reusable Space Vehicles

Valery Timoshenko

Rocket and Spacecraft Composite Structures Department, Faculty of Special Machinery, Bauman Moscow State Technical University, 5 2nd Baumanskaya Street, Moscow, 105005, Russia

Abstract: Carbon based thermal protection composite structures are widely used in extremely heated surfaces of reusable space vehicles such as nose caps and wing leading edges. Any mechanical damage to such structures may lead to a disaster during its re-entry flight to Earth. This paper presents analysis of workability for some common types of thermal protection elements after their mechanical damage. Strategies for improving thermal protection structures for reusable space vehicles are considered in this paper. These structures would potentially have greater reliability and will be able to ensure safe descent from the orbit, even after mechanical damage.

Key Words: Thermal protection, Composite structures, Reusable Space Vehicles, Reliability.

1. Introduction

One of the most important research areas in the near Earth space exploration is the creation of Reusable Space Vehicles (RSVs). In long term, such RSVs will provide prompt and more economical means of space transportation. Various defense institutions show active interest in the development of RSVs. Once fully developed, such RSVs can be used for operations on Earth surface ecological monitoring, emergency rescue of astronauts and for the luxurious space tourism.

Despite the ending of space shuttle exploitation programme, research work on the development of new RSVs (including testing) is rapidly growing. For example, the unmanned space vehicle X-37B and suborbital system SpaceShipTwo (SS2), now operational, have incorporated many features from the main space shuttle program. These vehicles create the base for multipurpose new generation RSV (manned and unmanned).

Production of RSV requires significant financial investments, which would demand many countries to work together. Moreover, a number of technical problems for ensuring safety of manned flights have no acceptable solutions under cost-efficiency criteria up to now.

Disasters like Challenger and Columbia have raised numbers of concerns regarding greater reliability, not only for American RSVs, but also for similar aerospace systems developed in Russia and other countries. Some of the most important issues with these space ships are crew rescue during the critical stages of flight (launch stage and orbit entry) and safe return to earth in case of thermal protection system (TPS) damage.

Improving reusable TPS' reliability, inevitably having compromises and restrictions, decreases payload mass and reduces the efficiency of the flight program. For improving reliability and safety of space flights, following general guidelines should be considered:

- Revision in manned flight programmes should be made and special diagnostic equipment should be incorporated in order to provide detailed inspection of the TPS elements immediately after critical stages of space and atmosphere journeys, i.e. flight orbit insertion.

- Development of the repairing technologies and equipment for repair of small thermal protection damages during space flight.
- Carefully selecting appropriate orbital trajectories and crew number, with taking into account the possibility of astronauts transfer to the orbital station or to the other rescue vehicle, in case their safe return to Earth is impossible.
- Applying new technical solutions and materials, which would ensure reusable exploitation of TPS without significant damages. Such solutions and materials should also provide guaranteed single return of the space vehicle to Earth without damaged thermal protection.

The first two guidelines have received significant importance for space shuttle flights after the loss of Columbia. Possibility of rescue vehicles flights is, at present, far away from the practical implementation, while projects on such rescue means are under investigation since long time (Umansky, 2003).

Radical modernisation of space vehicles used for regular flights is practically impossible. Therefore, all technical problems, associated with improving of TPS reliability, should be solved at the design stage of newly created RSV.

2. Reliability of basic TPS types

Despite the enormous research interest for manned and unmanned space vehicles, the number of TPS types used is relatively small. If we exclude too sophisticated and unreliable types of TPS from our selection, we would only have four basic types of high temperature passive TPS: ablative, ceramic, metallic and carbon based.

Ablative types of thermal protection, based on polymer composites, are the most reliable TPSs. This is confirmed from more than 50 years of experience. They have been used for many manned and unmanned space descent vehicles like Vostok, Voskhod, Gemini, Apollo, Soyuz, Bor-4s, Bor-5 and others (Semenov, 1996, Lozino-Lozinsky and Timoshenko, 1998).

Tiled TPS, based on fibrous ceramic materials, were used in aerospace systems like Energiya-Buran. Despite

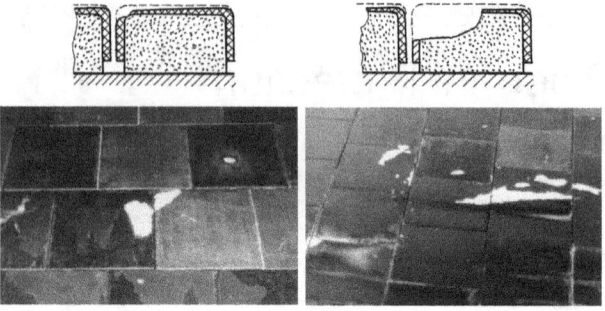

Figure 1. Typical damage in TPS of Buran.

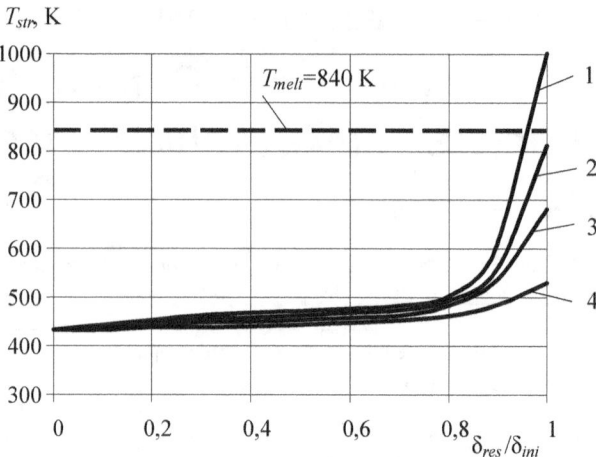

Figure 2. Dependence of maximal temperatures of the Buran metallic structure with extent of TPS tile damage: 1 corresponds to δ_{ini}=67 mm for T_{aw}=1520 K; 2 – to δ_{ini}=63 mm for T_{aw}=1320 K; 3 – to δ_{ini}=40 mm for T_{aw}=1170 K; 4 – to δ_{ini}=24 mm for T_{aw}=810 K.

of their low erosion resistance and high cost of repair works between flights, tiled TPSs are characterised for their high level of survivability after receiving significant mechanical damage. This means, that even after local damage appears on any tile, or any single TPS tile is detached from the space ship, the space vehicle will be able to perform safe return to Earth.

Fig. 1 shows basic damage types on Buran TPS tiles. Defects caused by impacts or erosive actions of hard objects (rain, hail, dust etc.) at different stages of RSV (launch or landing) are also very visible (Fig. 1).

Due to the fibre reinforcement in ceramic material, TPS tiles do not break into several parts (i.e. brittle failure) upon mechanical impact. Usually TPS fibrous material crumbles only in the limited area near the zone of direct impact. The nearest analogy of such failure may be the behaviour of a dense compressed snowdrift under similar mechanical impact.

Several parametric calculations were performed to estimate the damage of the TPS tile at operational temperature of RSV Buran's metallic thin wall structure. In these calculations, the thickness of TPS tile recession δ_{res} was increased in several steps from zero (undamaged tile) to 100% (full recession), while adjacent tiles were considered undamaged.

Maximal temperatures of the Buran's metallic structure $T_{str,max}$ depending on extent of TPS tile damage $\delta_{res}/\delta_{ini}$ are presented in Fig. 2. Here, δ_{ini} is the initial design value of the tile thickness, which depends on maximal level of the external heat flux q_{aw}, expressed in terms of adiabatic wall temperature T_{aw}, ($q_{aw} = \varepsilon\sigma T_{aw}^4$).

It can be observed that in the worst case scenario, the temperature of the metallic structure (aluminium alloy D-

16T) remains below its melting temperature T_{melt} = 840 K as long at the thickness of residual part of the tile is higher then 5% of its initial value. This worst case corresponds to the maximal allowable operational temperature of TPS tiles T_{aw} = 1520 K in the areas directly adjacent to the carbon-carbon hot structures of nose cap and wing leading edge.

Even in case of the complete loss of a single TPS tile, the damage to RSV metallic airframe structure is only possible in the limited areas where the most intensive external heat fluxes acts. This can be seen in the extensive calculations and experimental investigations presented by Timoshenko (1993) in Fig. 3. Two basic cases of the tile loss considered are, the loss of tile together with felt insulation pad and tile detaching without felt pad. In the latter case, the residual felt pad layer works as some kind of thermal insulation even after its thermal degradation and helps to reduce the value of $T_{str,max}$.

Flight test results of large-scale model Bor-4 and RSV Buran comprehensively confirmed the sufficiently high reliability of TPS based on ceramic tiles (Timoshenko, 1999).

After the flight of Buran, it was found that only 7 ceramic tiles have been lost. Fig. 4 shows that no damage to the airframe skin had occurred on the vertical fin. Even on the forward leading edge of the fin, where half of felt pad

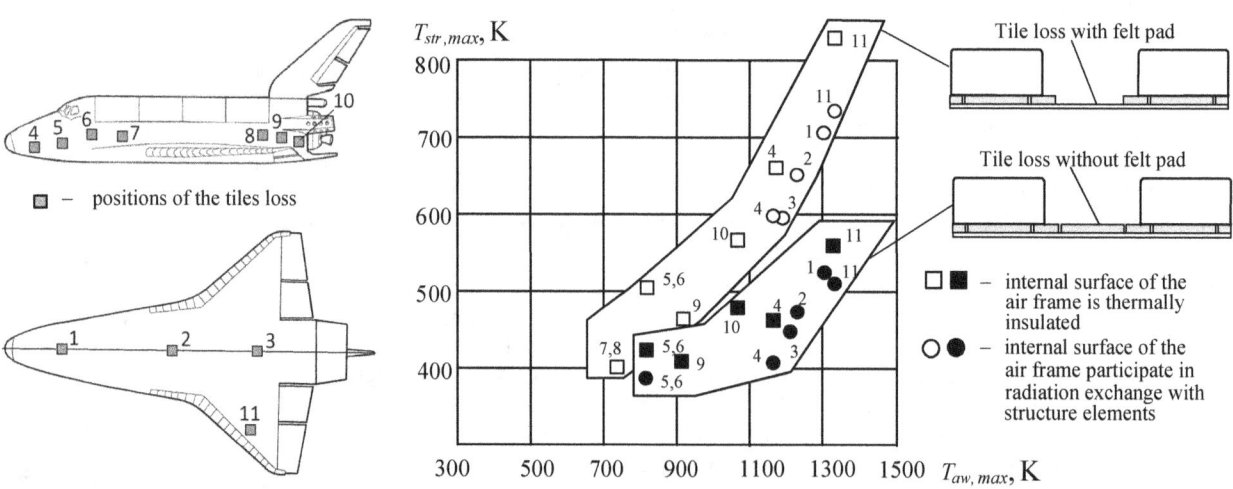

Figure 3. Maximal temperatures of the Buran airframe structure $T_{str,\,max}$ at the locations of possible tile loss.

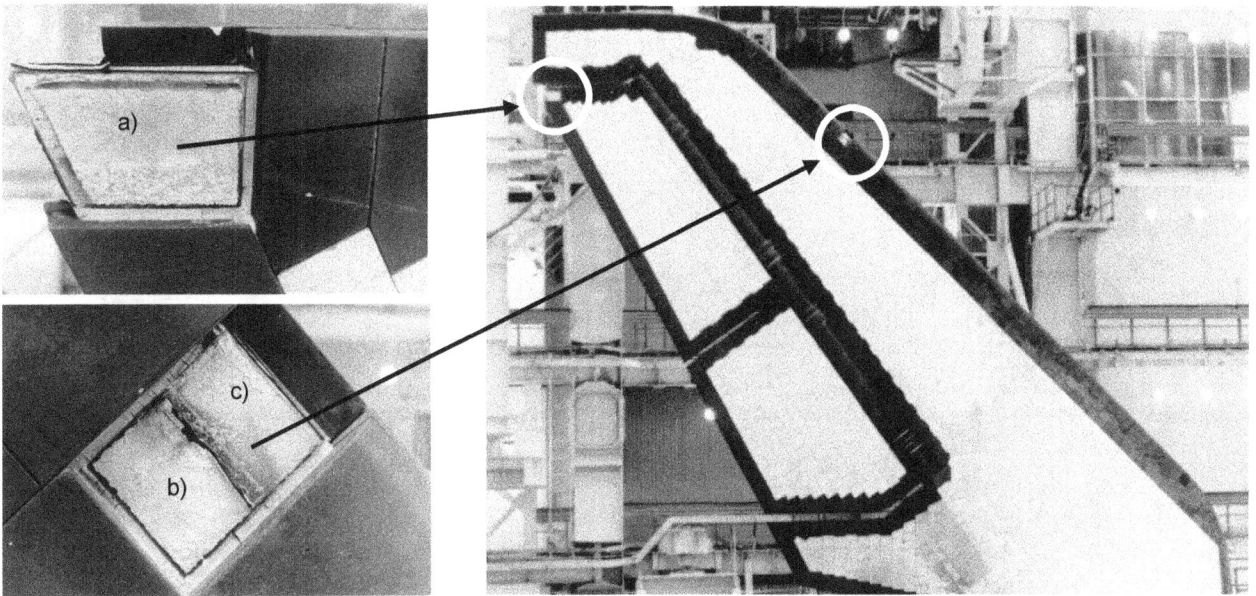

Figure 4. Damage free airframe showing the locations of tile loss on the Buran fin: a) and b) felt pad remained intact; and c) damaged felt pad.

was removed together with tile, no appreciable airframe damage was found.

Fig. 5 shows the area on the lower surface of the Buran's left wing where three TPS tiles were lost. At those locations, aerodynamic heat fluxes achieve maximal allowable level for TPS tiles application. Nevertheless, only partial melting of the Buran airframe was observed.

Development of metallic TPSs started well before ceramic tiles. Nevertheless, to date, the use of metallic TPS has been limited to areas of relatively low exploitation temperatures about 1300 K (Blosser, 1996).

The thickness of metallic TPS structures has to be very thin (0.05-0.1 mm) because of the significant higher densities of metallic TPSs (nickel-based alloys). In addition, a major problem is the structure warping at high temperature heating as well as the loss of integrity of welded or brazed joints at high-cycle thermal loads.

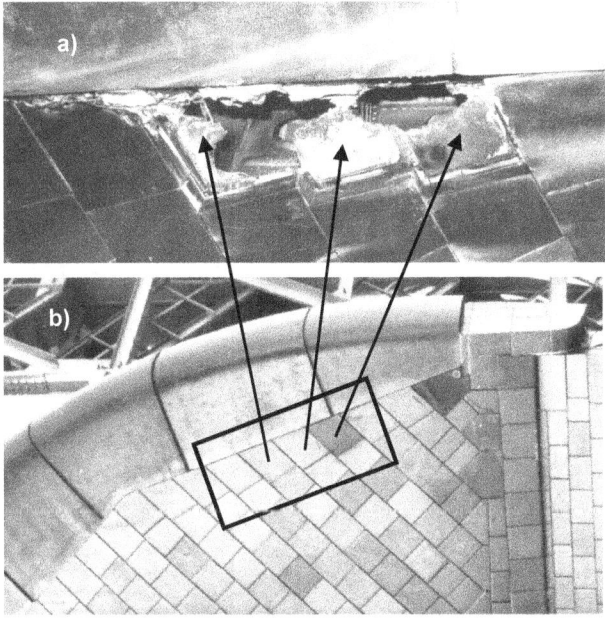

Figure 5. Local damage to Buran's airframe at the sites of 3 tiles loss on the lower surface of the left wing: a) airframe damage after flight; and b) the same damage free area before the flight.

Use of carbon-carbon (C-C) or carbon-silicon carbide (C-SiC) composite materials seems to be a good opportunity for increasing the operational temperatures limit up to the level of 2000-2200 K. However, at these conditions, the risk of RSV disaster due to TPS damage drastically increases.

It should be noted here that in case of Columbia, a cryogenic insulation fragment, torn off from the fuel tank, hit one of the Wing Leading Edges (WLEs) and significantly damaged the carbon-carbon section, which caused one of the major disasters in the history of space exploration. Results of experimental investigations by Gehman (2003) confirmed the possibility of such leading edge damage. As shown in Fig. 6, air plasma, after penetrating inside the leading edge through damaged carbon-carbon shell, burnt the thin-walled metallic envelops made of Inconel alloys and containing low-density fibrous thermal insulation Cerachrome for thermal protection of aluminum forward wing structure. Due to its high catalycity, Inconel alloy was destroyed under the action of air plasma at relatively low temperatures about 1500 K. Afterwards the airflow disperses loose fibres of thermal insulation and initiated further thermal destruction of the structure.

It should be noted here, that the design of Buran's WLE is having greater reliability. This is due to adjacent C-C sections, which are assembled with an overlap. Also, there are no low breaking strength carbon-carbon T-seals installed between them. Moreover, internal insulation of the forward wing structure under C-C sections is made of ceramic tiled thermal protection (Fig. 7), which can withstand heating up to 1850 K for a single return flight to Earth. Therefore, in case of a similar WLE damage, Buran would have more chances for survival.

3. Main principles of TPS reliability improvement

WLE sections and nose cap are made from carbon-based composite materials. They are relatively large objects with typical dimensions about 0.5-1.5 m. Therefore, in case of mechanical damage, their repair or replacement in the orbital flight conditions is nonrealistic. The only way

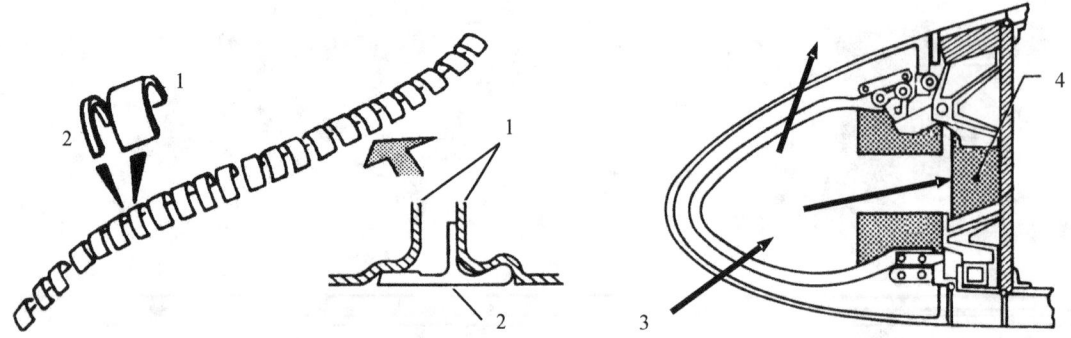

Figure 6. Schematics of Columbia's WLE and its destruction after penetration of plasma flow: 1) main C-C sections of WLE; 2) C-C T-seal; 3) main flow direction of plasma after WLE damage; and 4) Inconel 601 envelope and Cerachrom thermal insulation.

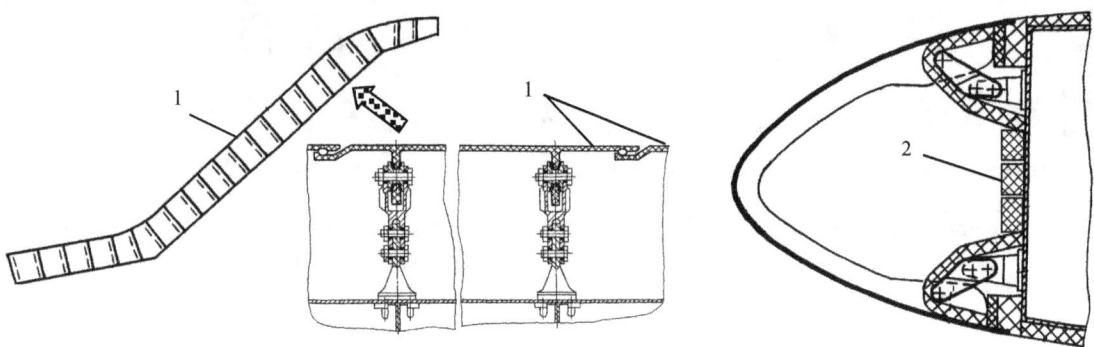

Figure 7. Scheme of the Buran wing leading edge: 1) main C-C sections of WLE; and 2) ceramic TPS tiles.

to solve the problem is to develop reusable TPS constructions, which after damage is still able to provide safe return to Earth.

Basic principle for making such constructions consists of doubling the thermal protection functions at the cost of increase in the vehicle's mass.

As a classic example, the structure of nose cap (Fig. 8) of the experimental space vehicle Bor-4 can be seen. This vehicle was used for testing of all basic elements of the Buran thermal protection (Lukashevitch and Afanasiev, 2009). In case of external shell damage to the carbon-carbon material, the internal layer made of fibreglass reinforced ablative composite material protects the vehicle and provides safe return to Earth.

Fig. 9 is one of the possible variants of WLE for RSV. It is of relatively small dimensions as compared to that of Buran, where it is difficult to use conventional joints for attachment of thermal resistant carbon-based structures. The internal front part of the wing is made of a composite ablative material PKT-FL, which is used for the thermal protection of Souyz descent module. The thermal protection tiles made of lightweight ceramic (TZMK-1700) hav-

ing Al_2O_3 fibres were adhesively bonded over the ablative material. This material is stable up to temperatures above 2000 K, but cannot operate as reusable if directly exposed to air plasma. Outer layer of the leading edge is made of C-SiC reusable composite material having maximum operational temperature above 2000 K.

The proposed assembly of the WLE ensures reusable exploitation of RSV if external C-SiC composite remains undamaged. In case of C-SiC shell damage, the first level of the dual thermal protection is provided by tiled thermal protection from TZMK-1700 material that have certain resistance against mechanical impacts as shown earlier (Fig. 1). If this layer is also damaged, the ablative material PKT-FL will protect the vehicle. Therefore, due to double protection, the safety of the main RSV structure would significantly improve. The proposed work with dual thermal protection system will probably require more detailed design investigations and significant experimental researches. Nevertheless, this approach seems to be the most reasonable one for producing new-generation manned RSVs having superior thermal protection systems.

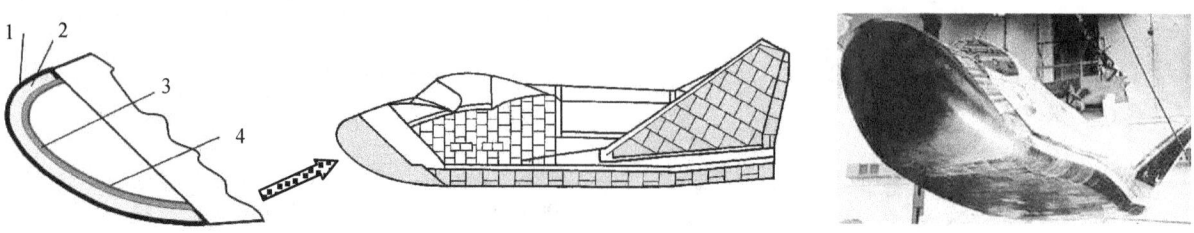

Figure 8. Schematics of dual thermal protection for nose cap of Bor-4 space vehicle: 1) external shell of Gravimol C-C material; 2) light weight soft fibrous insulator; 3) fibreglass type ablative material; and 4) carrying structure made from aluminium alloy.

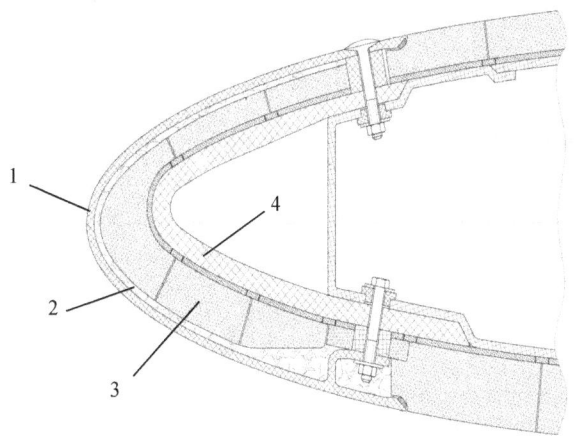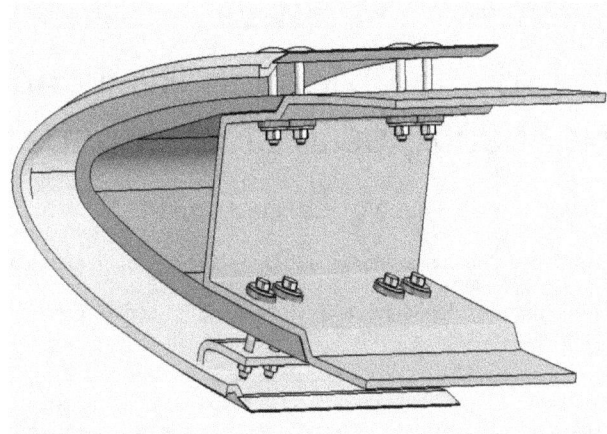

Figure 9. Schematics of dual thermal protection for RSV's WLE: 1) outrer shell of C-SiC composite material;
2) intermediate soft layer of Al_2O_3 fibres; 3) ceramic tiles made from TZMK-1700 material; and 4) composite ablative material PKT-FL.

References

Blosser, M.L. (1996). Development of metallic thermal protection systems for the reusable launch vehicle. *NASA technical memorandum 110296*. Langley Research Center, Hampton, Virginia, 23681-0001. [online]. Available at: http://techreports.larc.nasa.gov/ltrs/PDF/NASA-96-tm11-0296.pdf.

Gehman, H. (eds.) (2003). Columbia accident investigation board: Final report, Vols. 1-6. [online]. Available at: http://caib.nasa.gov

Lozino-Lozinsky, G.E. and Timoshenko, V.P. (1998). Lessons learned from the Bor flight campaign, in *Proceedings of the 3rd European Symposium on Aerothermo-dynamics for Space Vehicles*, November 24-26, Noordwijk, The Netherlands, pp. 675-683.

Lukashevitch, V.P. and Afanasiev, I.B. (2009). *Space Wings*. Moscow: Lenta Stranstviy.

Semenov, Y.P. (eds.) (1996). *Rocket-Space Corporation Energiya by S.P Korolev*, Moscow: Mensovpolygraph.

Timoshenko, V.P. (1993). Buran main thermal protection components, in *Proceedings of the 1st ESA/ESTEC Workshop on Thermal Protection Systems*, May 5-7, Noordwijk, The Netherlands, pp. 40-57.

Timoshenko, V.P., (1999). Evaluation of real TPS performances on the base of Bor/Buran experience, in *Proceedings of the International Symposium on Atmospheric Reentry Vehicles and Systems*, March 16-18, Arcachon, France, p. 5.

Umansky, S.P. (2003). *Launch Vehicles and Launch Sites*. Moscow: Restart.

Improving Oxidation Resistance of Carbon Nanotube Nanocomposites for Aerospace Applications

Fawad Inam[1], Thuc Vo[1], Saman Kumara[2]

[1] Advanced Composite Training and Development Centre and School of Mechanical and Aeronautical Engineering, Glyndŵr University, Plas Coch, Mold Road, Wrexham, LL11 2AW, UK
[2] Department of Mechanical and Aerospace Engineering, Tokyo Institute of Technology, 2-12-1 Ookayama, Meguro-ku, Tokyo, 152-8550, Japan

Abstract: Carbon nanotubes (CNTs) based materials possess strong potential to substitute various functional materials developed exclusively for aerospace applications. However, because of the low oxidation temperature of CNTs (400-500 °C), using CNT based ceramic nanocomposites in high temperature applications can be problematic. Making ceramic-CNT nanocomposites by atomic layer deposition (ALD) method and field assisted sintering technology (FAST) is a good route to improve oxidative stability of CNTs. In this study, thermo-gravimetric analysis (TGA) of alumina coated CNTs (prepared by ALD) and alumina-CNT nanocomposites (prepared by FAST) were carried out. 16% improvements were observed in the oxidation resistance for alumina-CNT nanocomposites prepared by ALD and SPS techniques. Different strategies to improve oxidation resistance are discussed.

Key Words: Carbon nanotubes, Ceramic nanocomposites, Oxidation resistance.

1. Introduction

Since the accidental re-discovery (Monthioux and Kuznetsov, 2006) of carbon nanotubes (CNTs) by Iijima (1991), it was expected that they would play an important role in the development of ceramic nanocomposites. Adding CNTs resulted significant improvements in the electrical conductivity (Thostenson et al., 2005; Dou et al., 2006), thermal conductivity (Jiang and Gao, 2008) and the fracture toughness (Zhan et al., 2003; Balazsi et al., 2005) of ceramic nanocomposites. Ceramics are very common in various high temperature aerospace applications (Barsoum et al., 1997). At high temperatures, in order to achieve maximum benefits provided by CNTs, it is crucial to retain un-attacked raw CNTs in the composites. In this way the toughening effects characteristic to micron-scale fiber composites could be explored as well. However, because of their low oxidation temperature resistance, using ceramic-CNT nanocomposites at high temperatures is an obstacle for their commercial success. Fabricating ceramic-CNT nanocomposite by atomic layer deposition (ALD) method and spark plasma sintering (SPS) provide shielding that improves oxidative stability of the encapsulated CNTs. These two strategies are investigated in this work.

ALD provides an ideal method for depositing ultrathin films on high aspect ratio surfaces as it is independent of line of sight and self-limiting (Hakim et al., 2007). Sequential surface chemical reactions deposit highly conformal films with precise control at the atomic scale (Hakim et al., 2007). The method has been shown to be a viable technique to deposit a coating on a single CNT without adversely affecting its inherent properties (Farmer and Gordon, 2006, Zhan et al., 2008).

The SPS technique is a pressure assisted fast sintering method based on high-temperature plasma momentarily generated in the gaps between powder materials by electrical discharge during on–off direct current pulsing, which causes localized high temperatures (Yamamoto et al., 2006). Other conventional methods like hot-pressing involves longer durations and high temperatures that dam-age CNTs, leading to a decrease or total loss of reinforcing effects without producing fully dense nanocomposites (Flahaut et al., 2000; Wood, 2003; Zhan et al., 2003; Balazsi et al., 2005). SPS is always carried out in vacuum on inert gas and CNTs have been proven to be stable under temperatures as high as 2430°C in these environments (Heer and Ugarte, 1993; Ma et al., 1998). Zhang et al. fabricated bulk CNT samples by SPS and confirmed the preservation of the phase structure and the diameter of the cylindrical tubules at high temperatures of up to 2000°C (Zhang et al., 2005). During SPS, CNTs can carry current densities up to 10^9–10^{10} A/cm^2 (compared to a typical value of 10^5–10^6 A/cm^2 for superconductors), and proved their stability for extended periods of time (Wei et al., 2001; Dou et al., 2006), which makes them an ideal reinforcement for ceramics matrices like boron carbide and boron nitride that are difficult to sinter in short durations.

Most of previous reports about oxidation resistance of CNTs are based on polymer based CNT composites (Kashiwagi et al., 2002; Yang et al., 2005; Costache, 2006; Bocchini et al., 2007; Kong and Zhang, 2008). Yuen et al. (2008) prepared TiO_2 coated CNTs - epoxy nanocomposite and reported significant improvement in the mechanical properties as compared to the uncoated CNTs - epoxy nanocomposite. For improving field emission characteristics, ceramic layers were coated on CNTs (Heo et al., 2002; Son et al., 2003; Chakrabarti et al., 2007; Pan et al., 2007), but the authors did not discuss stability against oxidation for the coated CNTs. There are many reports discussing ceramic coatings on CNTs for enhancing electronic and electrical properties (Wind et al.; 2002, Kawasaki et al.; 2008, Cao et al.; 2004, Fu et al., 2006; Bachtold et al., 2001; Javey et al., 2004). Cao et al. (2004) coated CdS and Fu et al. (2006), Bachtold et al. (2001) and Javey et al. (2004) coated alumina on CNTs for improving electronic properties only. But to the best of authors' knowledge, this is the first paper on the positive effect of ceramics on the oxidation resistance of CNTs. Wang et al. (2006) coated 10 nm of silicon layer and reported an improvement of 18.4 % in oxidation resistance. However, superior features of CNTs may not be fully exploited with such a thick coating. Li et al. (2008) reported that the increase in Ni catalyst during CVD growth improves the oxidation temperature of Multiwall CNTs. The catalyst composition had great effect on the CNT structure and stability, which is the focus for large-scale CNT synthesis (Li et al., 2008). CNTs were distinctly observed by Laha et al. (2004) after plasma spraying blended powder (Al-Si-CNTs) on a rotating metallic mandrel. In the same report, CNTs were sprayed at very high temperatures (9700°C – 14700°C), but for very short durations. This conference paper reports the high temperature shielding effect of alumina on CNTs in the coated and the nanocomposite systems, fabricated by ALD and SPS respectively. The oxidation resistance of CNTs in different systems was characterized thermogravimetric analysis (TGA).

2. Experimental procedure

2.1 Material

DMF was supplied by Sigma-Aldrich, UK. The CNTs used in this study were commercially available as "Multiwall carbon nanotubes (MWNTs), NC-7000" from Nanocyl Inc., Belgium. They were synthesized by the catalytic CVD method and have an entangled cotton-like form. The CNTs had an average outer diameter of 9.5 nm (10 graphitic shells), length of up to ~1.5 microns and density of 1.66 g/cm^3. The alumina matrix used in this study was commercially available "544833 aluminum oxide" nanopowder from Sigma-Aldrich, UK. As supplied by the supplier, the main features of this product are: gamma phase; particle size < 50 nm; surface area 35-43 m^2/g; melting point 2040°C; and density 3.97 g/cm^3. Another batch of multiwall CNTs was supplied by Nanodynamics Inc., USA (figure 1a). Theses CNTs had an average outer diameter of 15 nm and, length of up to ~2 microns and specific surface area of 182 m^2/g. Alumina coating on these CNTs by Nanodynamics Inc. was done by elsewhere by atomic layer deposition method (Hakim et al., 2007). Two coating cycles were followed. 27 cycles of ALD resulted in 25 nm of coating, whereas 54 cycles resulted in 50 nm (Fig. 1b).

2.2 Composite powder preparation and Spark Plasma Sintering (SPS)

Alumina + 11.2 vol% (5 mass%) MWNTs composites were prepared. Details of the processing can be found elsewhere (Inam et al., 2008). Briefly, MWNTs were dispersed in DMF via high power bath-sonication for 2 hours and then hand-mixed with alumina nanopowder for another 5 minutes. The liquid mixture was transferred to another jar filled with zirconia balls (milling media) of two different sizes (10 and 5 mm, mass ratio: 3:2). The jar was sealed and rotation ball milled for 8 hours at ~200 RPM. The milled slurry mixture was dried at 75°C for 12 hours on a heating plate and then transferred to a vacuum oven (100°C) for 3 days for complete removal of dispersant. A solvent-trap (filled with ice) was connected between the vacuum pump and the oven. The dried mixture was ground and sieved using a 250 mesh and then returned to the vacuum oven for another 4 days at the same

temperature for thorough extraction of the solvent. This lengthy drying procedure was followed because any residual solvent has a detrimental effect on the properties of CNTs reinforced nanocomposites (Lau et al., 2005; Moniruzzaman et al., 2006). Dried composite powder (~ 2 grams) was poured into a carbon die and cold pressed at 0.62 MPa for 5 seconds before sintering. Nanocomposite disks (thickness 2 mm and diameter 20 mm) were prepared by Spark Plasma Sintering (SPS) in a SPS 2040 furnace (Sumitomo Coal Mining Co, Japan). A pressure of 100 MPa was applied concurrently with the heating (rate 300°C/min) and released at the end of the sintering time, which was 3 minutes for all of the samples. The furnace has a pyrometer focused on a hole close to the sample in the upper punch to measure the processing temperature. Details of the SPS technique are reported elsewhere (Omori, 2000). The sintering temperatures were 1200°C and 1800°C. All of the samples were slowly cooled to avoid fracture due to thermal shocks and differential contractions. The densities for both sintered samples were 100% of the theoretical density.

2.3 Characterisations

Thermogravimetric analyses (TGA) were performed on a TA Instruments SDT Q600 TGA thermogravimetric analyzer. Samples were analyzed in platinum pans at a heating rate of 30°C/min to 1000°C in air flowing at 180 ml/min. Powder sample masses ranged from 30 to 40 mg, whereas sintered sample masses ranged from 30-50 mg. SPSed samples were fractured in order to observe the

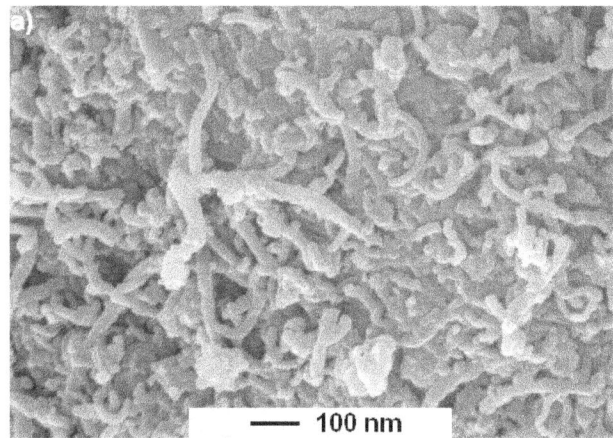

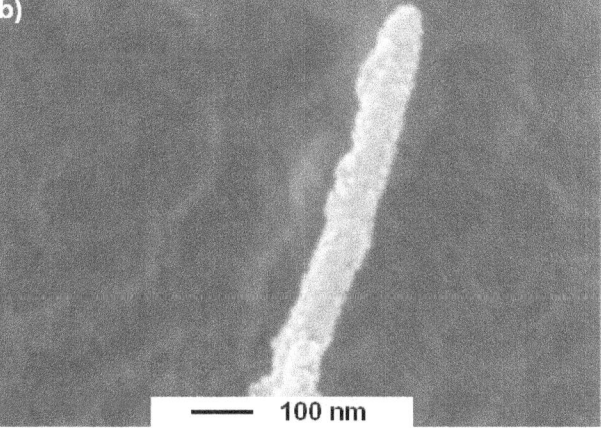

Figure 1. Multiwall CNTs: a) uncoated agglomerates; and b) single coated (54 ALD cycles) CNT.

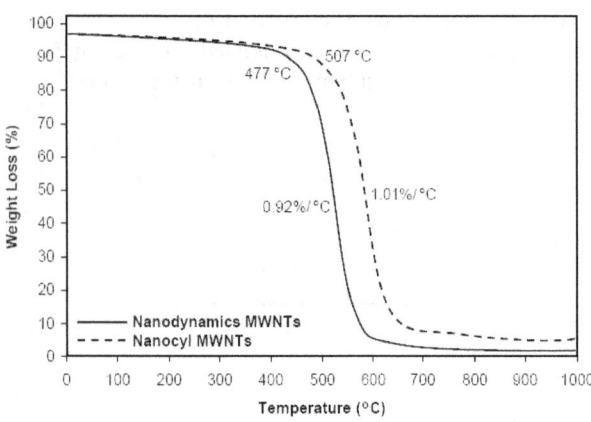

Figure 2. TGA of raw CNTs

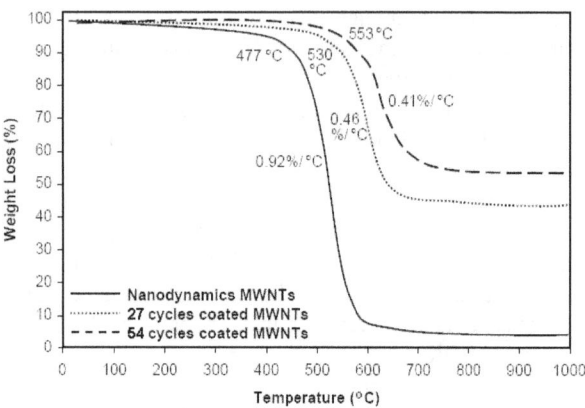

Figure 4. TGA of uncoated CNTs and alumina coated CNTs.

agglomeration and dispersion of CNTs. Fractured surfaces and powder samples were gold coated and observed in a FE-SEM (FEI, Inspect F, 20 kV, working distance 8-10 mm).

3. Results and discussion

Fig. 2 shows the oxidation behaviour of CNTs obtained from different suppliers. CNTs provided by Nanodynamics Inc, have larger average diameter as compared to the CNTs provided by the other source. The oxidative stability of CNTs is also influenced by defects (Li et al., 2008) and nanotube diameter (Yao et al., 1998; Li et al., 2008). Oxygen molecules react easily with larger surface areas, resulting in decreased oxidative stability of Nanodynamics CNTs. The oxidation of CNTs is not rapid and acute like combustion, which is also evident in other report (Tian et al., 2007). Because the kinetic energy of oxygen varies with temperature, there is not a critical temperature when the oxidation of CNTs starts (Tian et al., 2007) as

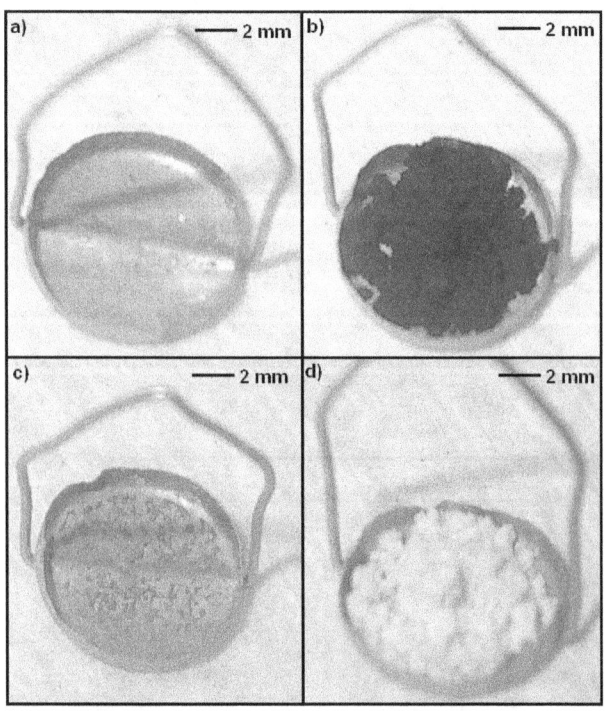

Figure 3. Platinum pan used for TGA: a) empty pan; b) CNTs before oxidation; c) impurities left after oxidation of uncoated CNTs; and d) alumina nanotubes left after oxidation of coated CNTs.

shown in Fig. 2. During the initial stage of TGA, all samples showed a slight mass loss due to the presence of amorphous carbon, as reported in other papers (Chen et al., 1998; Li et al., 2008). In the second stage of TGA, the curve slope is maintained almost the same in the definite temperature range for both types of CNTs. In the third stage of TGA, there was no weight gain observed during thermal treatment, since no oxidation of the impurities occurred. The weight loss for both types of CNTs was not 100 % due to the presence of impurities, which were found after the furnace pan (Fig. 3a) was cooled down (Fig. 3b and 3c).

To improve the oxidation resistance and chemical stability of CNTs, a protective film or coating was necessary to shield the CNTs against thermal or environment damage. The oxidative stability of CNTs was distinctly improved due to the protective alumina coating (Fig. 4). The onset oxidisation temperature for sample coated for 54 ALD cycles is now as high as 553°C in air atmosphere, which is 76°C (16%) higher than that of uncoated CNTs. Once the oxidation started, the degradation rate was also reduced to 0.41%/°C, which is 55% less than that of uncoated CNTs. The degradation process was delayed because it became more difficult for oxygen molecule to approach CNTs after coating. By analysing the third stage (after 700°C) of the TGA (Fig. 4), it is possible to quantify the mass content of CNTs in the coated nanocomposite. 27 ALD cycles resulted in 56.4 mass% of CNTs and 54 ALD cycles produced 44.9 mass% of CNTs in the coated nanocomposites. A thicker alumina coating could further inhibit the oxidisation of CNTs but it may decrease the mechanical properties of CNTs by making the coated CNTs brittle. After cooling down the TGA furnace, white coloured alumina nanotubes (Fig. 3d) were left in the platinum pan, which were previously surrounding CNTs. This could be one of the ways to mass-produce these alumina nanotubes (Fig. 5).

Dense CNT-dispersed alumina nanocomposites with different grain sizes were fabricated using SPS. In figure 6, the onset oxidisation temperature for sample coated for 54 ALD cycles is now as high as 588°C in air atmosphere, which is 81°C (16%) higher than that of raw CNTs. Once the oxidation started, the degradation rate was also reduced to 0.026%/°C, which is 97% less than that of raw CNTs. Representative images of the fractured surfaces of the SPSed nanocomposites were selected for studying the grain sizes (Fig. 7). Oxidative reactivity in

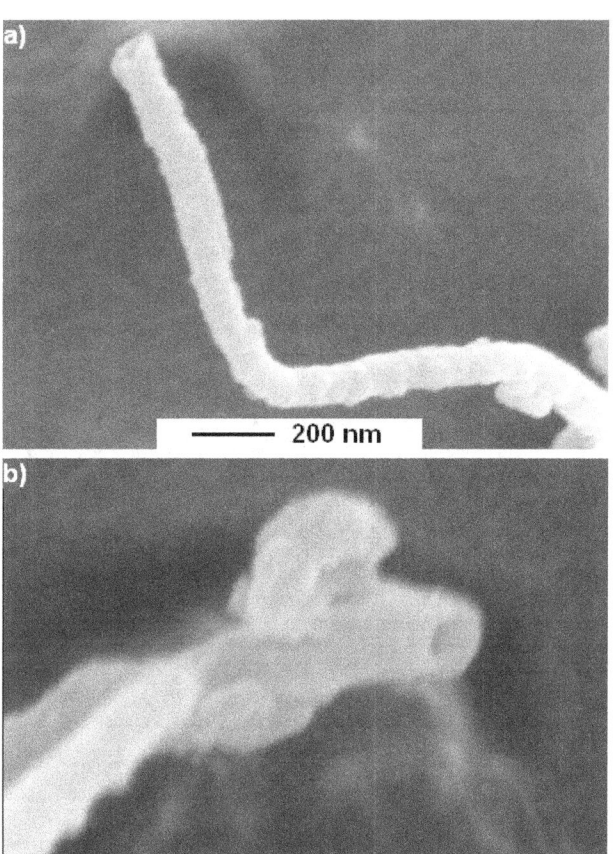

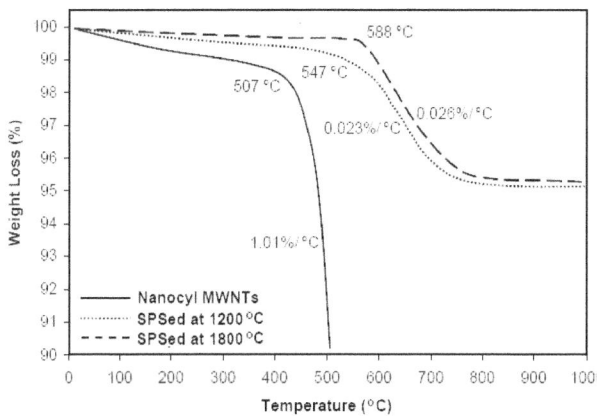

Figure 5. Alumina nanotube left after the oxidation of coated CNTs:
a) at lower magnification; and b) at higher magnification.

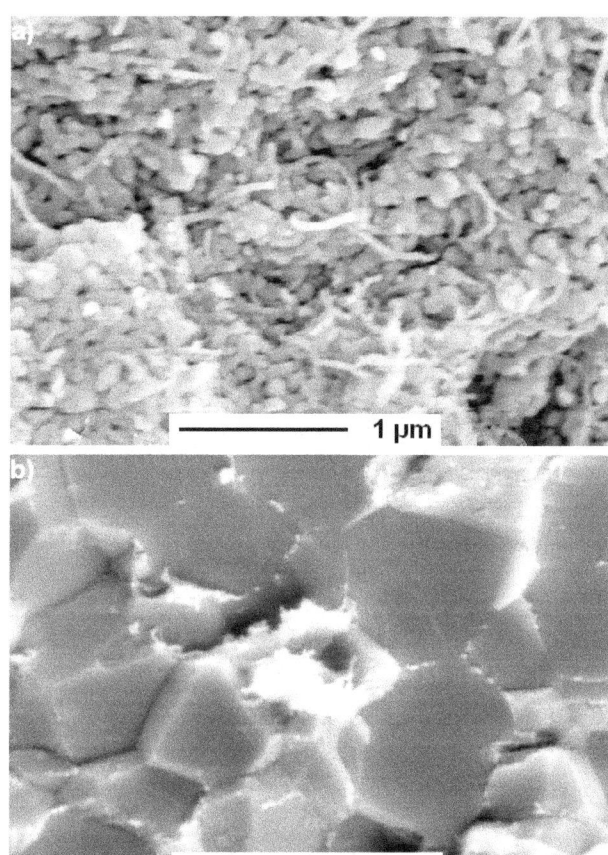

Figure 7. Representative fractured surfaces of the SPSed samples:
a) SPSed at 1200°C; and b) SPSed at 1800°C. Bright area represents CNTs.

these nanocomposites is dominated by the alumina grain size. Sample sintered at 1200°C produced finer grains (Fig. 7a). This produced large area of grain boundaries or easy entry path for oxidation reaction. However, in coarse-grained nanocomposite sintered at 1800°C (Fig. 7b), fewer grain boundaries made difficult for oxygen to approach CNTs, leading to a better oxidative stability.

4. Conclusion

It is necessary to preserve the chicken wire hexagonal structure of CNTs in ceramics for high-temperature applications. Fabricating ceramic-CNT nanocomposite by ALD method and SPS provide shielding that improves oxidative stability of the encapsulated CNTs. In this study, 16% improvements in the oxidation resistance were

observed for alumina-CNT nanocomposites that could be further improved by varying processing conditions. In ALD, a thicker alumina coating could further inhibit the oxidation of CNTs and enhance the thermal stability of CNTs. Oxidising CNTs after coating them with ceramics could be one of the ways for mass-production of ceramic nanotubes. In SPS, coarser grains protect CNTs more efficiently as compared to the finer ones, due to the presence of fewer grain boundaries. It was found that TGA is a good tool to evaluate the mass content of CNTs in the coated CNT and SPSed nanocomposites.

References

Bachtold, A., Hadley, P., Nakanishi, T. and Dekker, C. (2001). Logic circuits with carbon nanotube transistors. *Science*, Vol. 294, No. 5545, pp. 1317-1320.

Balazsi, C., Shen, Z., Konya, Z., Kasztovszky, Z., Weber, F., Vertesy, Z., Biro, L.P., Kiricsi, I. and Arato, P. (2005). Processing of carbon nanotube reinforced silicon nitride composites by spark plasma sintering. *Composites Science and Technology*, Vol. 65, No. 5, pp. 727-733.

Barsoum, M.W., Brodkin, D. and Raghy, T.E. (1997). Layered machinable ceramics for high temperature applications. *Scripta Materialia*, Vol. 36, No. 5, pp. 535-541.

Bocchini, S., Frache, A., Camino, G. and Claes, M. (2007). Polyethylene thermal oxidative stabilisation in carbon nanotubes based nanocomposites. *European Polymer Journal*, Vol. 43, No. 8, pp. 3222-3235.

Cao, J., Sun, J.Z., Hong, J., Li, H.Y., Chen, H.Z. and Wang, M. (2004). Carbon nanotube/ CdS core-shell nanowires prepared by a simple room-temperature chemical reduction method. *Advanced Materials*, Vol. 16, No. 1, pp. 84-87.

Chakrabarti, S., Pan, L., Tanaka, H., Hokushin, S. and Nakayama, Y. (2007). Stable field emission property of patterned MgO coated car-

Figure 6. TGA of SPSed nanocomposites and raw CNTs.

bon nanotube arrays. *Japanese Journal of Applied Physics*, Vol. 46, No. 7A, pp. 4364-4369.

Chen, C.M., Chen, M., Leu, F.C., Hsu, S.Y., Wang, S.C., Shi, S.C. and Chen, C.F. (1998). Purification of multi-walled carbon nanotubes by microwave digestion method. *Diamond and Related Materials*, Vol. 13, No. 4-8, pp. 1182-1186.

Costache, M.C., Wang, D.Y., Heidecker, M.J., Manias, E. and Wilkie, C.A. (2006). The thermal degradation of poly(methyl methacrylate) nanocomposites with montmorillonite, layered double hydroxides and carbon nanotubes. *Polymers Advanced Technologies*, Vol. 17, No. 4, pp. 272-280.

Dou, S.X., Yeoh, W.K., Shcherbakova, O., Wexler, D., Li, Y., Ren, Z.M., Munroe, P., Chen, S., Tan, K., Glowacki, B.A. and Driscoll, J.L.M. (2006). Alignment of carbon nanotube additives for improved performance of magnesium diboride superconductors. *Advanced Materials*, Vol. 18, No. 6, pp. 785-788.

Farmer, D.B. and Gordon, R.G. (2006). Atomic layer deposition on suspended single-walled carbon nanotubes via gas-phase noncovalent functionalization. *Nano Letters*, Vol. 6, No. 4, pp. 699-703.

Flahaut, E., Peigney, A., Laurent, C., Marliere, C., Chastel, F. and Rousset, A. (2000). Carbon nanotube–metal–oxide nanocomposites: microstructure, electrical conductivity and mechanical properties. *Acta Materialia*, Vol. 48, No. 14, pp. 3803-3812.

Fu, L., Liu, Y., Liu, Z. *et al.* (2006). Carbon nanotubes coated with alumina as gate dielectrics of field-effect transistors. Advanced Materials, Vol. 18, No. 2, pp. 181-185.

Hakim, L.F., King, D.M., Zhou, Y., Gump, C.J., George, S.M. and Weimer, A.W. (2007). Nanoparticle coating for advanced optical, mechanical and rheological properties. *Advanced Functional Materials*, Vol. 17, No. 16, pp. 3175-3181.

Heer, W.A.D. and Ugarte, D. (1993). Carbon onions produced by heat treatment of carbon soot and their relation to the 217.5 nm interstellar absorption feature. *Chemical Physics Letters*, Vol. 207, No. 4-6, pp. 480-486.

Heo, J.N., Kim, W.S., Jeong, T.W., Yu, S., Lee, J.H., Lee, C.S., Yi, W.K., Lee, Y.H., Yoo, J.B. and Kim, J.M. (2002). Effect of MgO film thickness on secondary electron emission from MgO-coated carbon nanotubes. *Physica B*, Vol. 323, No. 1-4, pp. 174-176.

Iijima, S. (1991). Helical microtubules of graphitic carbon. *Nature*, Vol. 354, No. 6348, pp. 56-58.

Inam, F., Yan, H., Reece, M.J. and Peijs, T. (2008). Dimethylformamide: an effective dispersant for making ceramic–carbon nanotube composites. *Nanotechnology*, Vol. 19, No. 19, pp. 195710.

Javey, A., Guo, J., Farmer, D.B., Wang, Q., Wang, D., Gordon, R.G., Lundstrom, M. and Dai, H. (2004). Carbon nanotube field-effect transistors with integrated ohmic contacts and high-k gate dielectrics. Nano Letters, Vol. 4, No. 3, pp. 447-450.

Jiang, L. and Gao, L. (2008). Densified multiwalled carbon nanotubes–titanium nitride composites with enhanced thermal properties. *Ceramics International*, Vol. 34, No. 1, pp. 231-235.

Kashiwagi, T., Grulke, E., Hilding, J., Harris, R.H., Awad, W. and Douglas, J.F. (2002). Thermal degradation and flammability properties of polypropylene-carbon nanotube composites. *Macromolecules*, Vol. 23, No. 13, pp. 761-765.

Kawasaki, S., Catalan, G., Fan, H.J., Saad, M.M., Gregg, J.M., Duarte, M.A.C., Rybczynski, J., Morrison, F.D., Tatsuta, T., Tsuji, O. and Scott, J.F. (2008). Conformal oxide coating of carbon nanotubes. *Applied Physics Letters*, Vol. 92, No. 5, pp. 053109.

Kong, Q.H. and Zhang, J.H. (2008). Synthesis of carbon nanotubes, and the effect on thermal stability in high-impact polystyrene. *Australian Journal of Chemistry*, Vol. 61, No. 1, pp. 72-76.

Laha, T., Agarwal, A., McKechnie, T. and Seal, S. (2004). Synthesis and characterization of plasma spray formed carbon nanotube reinforced aluminum composite. *Materials Science and Engineering A*, Vol. 381, No. 1-2, pp. 249-258.

Lau, K., Lu, M., Lam, C., Cheung, H., Sheng, F.L and Li, H.L. (2005). Thermal and mechanical properties of single-walled carbon nanotube bundle-reinforced epoxy nanocomposites: the role of solvent for nanotube dispersion. *Composites Science and Technology*, Vol. 65, No. 5, pp. 719-725.

Li, H., Zhao, N., He, C., Shi, C., Du, X. and Li, J. (2008). Thermogravimetric analysis and TEM characterization of the oxidation and

defect sites of carbon nanotubes synthesized by CVD of methane. *Materials Science and Engineering A*, Vol. 473, No. 1-2, pp. 355-359.

Ma, R.Z., Wu, J., Wei, B.Q., Liang, J. and Wu, D.H. (1998). Processing and properties of carbon nanotubes–nano-SiC ceramic. *Journal of Materials Science*. Vol. 33, No. 21, pp. 5243-5246.

Moniruzzaman, M., Du, F., Romero, N. and Winey, K.I. (2006). Increased flexural modulus and strength in SWNT/epoxy composites by a new fabrication method. *Polymer*, Vol. 47, No. 1, pp. 293-298.

Monthioux, M. and Kuznetsov, V.L. (2006). Who should be given the credit for the discovery of carbon nanotubes? *Carbon*, Vol. 169, No. 9, pp. 1621-1623.

Omori, M. (2000). Sintering, consolidation, reaction and crystal growth by the spark plasma system (SPS). *Materials Science and Engineering A*, Vol. 287, No. 2, pp. 183-188.

Pan, L., Konishi, Y., Tanaka, H., Chakrabarti, S., Hokushin, S., Akita, S. and Nakayama, Y. (2007). Effect of MgO coating on field emission of a stand-alone carbon nanotube. *Journal of Vacuum Science and Technology B*, Vol. 25, No. 5, pp. 1581-1583.

Son, Y.W., Han, S.W. and Ihm, J. (2003). Electronic structure and the field emission mechanism of MgO-coated carbon nanotubes. *New Journal of Physics*, Vol. 5, No. 1, p. 152.

Thostenson, E.T., Karandikar, P.G. and Chou, T.W. (2005). Fabrication and characterization of reaction bonded silicon carbide/carbon nanotube composites. *Journal of Physics D: Applied Physics*, Vol. 38, No. 21, pp. 3962-3965.

Tian, C.H., Ren, H.F., Geng, D.T. and Tian, X.X. (2007). Thermal stability of carbon nanotubes, in *Proceedings of the 7th International Symposium on Test and Measurement, ISTM'07*, August 5-8, 2007, Beijing, China, Vols. 1-7, pp. 4880-4882.

Wang, Y.H., Li, Y.N., Lu, J., Zang, J.B. and Huang, H. (2006). Microstructure and thermal characteristic of Si-coated multi-walled carbon nanotubes. *Nanotechnology*, Vol. 17, No. 15, pp. 3817-3821.

Wei, B.Q., Vajtai, R. and Ajayan, P.M. (2001). Reliability and current carrying capacity of carbon nanotubes. *Applied Physics Letters*, Vol. 79, No. 8, pp. 785-788.

Wind, S.J., Appenzeller, J., Martel, R. *et al.* (2002). Vertical scaling of carbon nanotube field-effect transistors using top gate electrodes. *Applied Physics Letters*, Vol. 80, No. 20, pp. 3817-3819.

Wood, A. (2003). Using carbon nanotubes to reinforce ceramics. *Chemical Week*, Vol. 165, p. 32.

Yamamoto, G., Sato, Y., Takahashi, T., Omori, M., Hashida, T., Okubo, A. and Tohji, K. (2006). Single-walled carbon nanotube-derived novel structural material. *Advanced Functional Materials*, Vol. 21, No. 6, pp. 1537-1542.

Yang, J., Lin, Y.H., Wang, J.F., Lai, M., Li, J., Liu, J., Tong, X. and Cheng, H. (2005). Morphology, thermal stability, and dynamic mechanical properties of atactic polypropylene/carbon nanotube composites. *Journal of Applied Polymers Science*, Vol. 98, No. 3, pp. 1087-1091.

Yao, N., Lordi, V., Ma, S.X.C., Dujardin, E., Krishnan, A., Treacy, M.M.J. and Ebbesen, T.W. (1998). Structure and oxidation patterns of carbon nanotubes. *Journal of Materials Research*, Vol. 13, No. 9, pp. 2432-2437.

Yuen, S.M., Ma, C.C.M., Chuang, C.Y., Hsiao, Y., Chiang, C. and Yu, A. (2008). Preparation, morphology, mechanical and electrical properties of TiO2 coated multiwalled carbon nanotube/epoxy composites. *Composites A: Applied Science and Manufacturing*, Vol. 39, No. 1, pp. 119-125.

Zhan, G.D., Du, X., King, D.M., Hakim, L.F., Liang, X., McCormick, J.A. and Weimer, A.W. (2008). Atomic layer deposition on bulk quantities of surfactant-modified single-walled carbon nanotubes. *Journal of the American Ceramics Society*, Vol. 91, No. 3, pp. 831-835.

Zhan, G.D., Kuntz, J.D., Wan, J. and Mukherjee, A.K. (2003). Single-wall carbon nanotubes as attractive toughening agents in alumina-based nanocomposites. *Nature Materials*, Vol. 2, No. 1, pp. 38-42.

Zhang, H.L., Lia, J.F., Yao, K.F. and Chen, L.D. (2005). Spark plasma sintering and thermal conductivity of carbon nanotube bulk materials. *Journal of Applied Physics*, Vol. 97, No. 11, pp. 114310.

Detection and Location of Impact Damage using Acoustic Emission

Mark Eaton, Matthew Pearson, Karen Holford, Carol Featherston, Rhys Pullin

Cardiff School of Engineering, Cardiff University, Queen's Buildings, The Parade, Newport Road, Cardiff, CF24 3AA, UK

Abstract: Impact damage is one of the greatest weaknesses of composite materials and efforts taken to improve its understanding, analysis and prevention have slowed the introduction of composite materials to large-scale and safety critical structures. In order to make this step, operators must be able to ensure the structural integrity and safe operation of composite structures during long service lives. One such approach to this is the use of Structural Health Monitoring (SHM) to continuously monitor the condition of a structure. In this paper a novel mapping based acoustic emission (AE) source location technique is used to detect and locate impact events on a carbon fibre composite panel.

Key Words: Impact damage, Acoustic emission, Source location, Composite materials.

1. Introduction

Composite materials offer many advantages as an aircraft material, not least is the potential for weight saving. For this reason their adoption by aircraft manufacturers has become extensive, in response to pressure to reduce both environmental impact and running costs of commercial aircraft. However, composite materials also have some disadvantages: susceptibility to impact damage being one of the greatest limitations. The efforts required to improve understanding, prevention and analysis of impact damage have slowed the introduction of composite materials to large-scale and safety critical primary structures. The potential for structures to sustain barely visible impact damage (BVID), where substantial damage can exist below the material surface with only a minor surface indent present externally, is of particular concern. Operators must, therefore, ensure the structural integrity and safe operation of such structures throughout there long service lives.

Commonly this problem is addressed through regular inspection of aircraft structure, however this approach has two limitations; firstly unique events such as an impact can happen at any time and are therefore not likely to be detected until the next inspection while an aircraft continues to fly. Secondly, down time for aircraft is very costly to an operator so taking an aircraft out of service for inspection is an expensive activity. An effective Structural Health Monitoring (SHM) tool would allow operators to continuously monitor the condition of an aircraft and therefore increase inspection intervals whilst increasing safety. The acoustic emission (AE) technique has great potential as an SHM tool; allowing continuous global monitoring of large structures and is capable of detecting and locating damage.

In this paper a novel "Delta T Mapping" algorithm for improved AE source location in complex structures and materials and an advanced statistical approach to signal arrival time estimation are described. The combined techniques were applied to the detection and location of a series of impact events on a wide carbon fibre composite tensile specimen. Comparison is made with the traditional Time of Arrival (TOA) location algorithm and improvements in detection and location accuracy are observed.

2. Background

Traditionally AE source location within a structure is resolved using the TOA approach; discussed in detail in the NDT Handbook (Miller, 2005) and by Rindorf (1981). User-defined inputs of sensor positions and a propagation velocity are required and source location is then resolved using the difference in arrival times (delta t, Δt) of a given signal at different sensor pairs. The accuracy of this calculation is dependent on two key areas: those relating to signal arrival time determination and those relating to processing.

Commonly the signal arrival time is determined at a given sensor when the sensor output reaches a certain user defined value, or crosses a threshold and is known as the "first threshold crossing" method. It is often the case that the true arrival time of the signal will not be detected; with a number of signal peaks occurring before the signal amplitude crosses the threshold. Previous attempts to improve signal arrival time estimation have included the use of filtering (Ding et al., 2004), cross-correlation (Ziola and Gorman, 1991) and wavelet transformation (Aljets et al., 2010; Hamstad et al., 2002; Jeong, 2001), however limitations still exist, with these approaches often still relying on a threshold at some stage and their performance has been seen to reduce in more complex structures. More recently threshold independent statistical approaches have been applied to the estimation of signal arrival time. Lokajicek and Klima (2006) utilised a 6th order statistical parameter to determine signal onset time, detecting signal arrivals to within ± 2 samples for 95% of analysed signals. Kurz et al. (2005) and Hensman et al. (2010) have utilised an approach based on the Akaike Information Criterion (AIC) to determine signal arrival times. The approach looks for a change in variance between the uncorrelated noise prior to signal onset and highly correlated signal after signal arrival. This is the approach that has been utilised in this work for signal arrival time determination and is discussed in more detail below.

The calculation of source position using the TOA algorithm relies on the assumption that the signal path from source to sensor is uninterrupted and that the signal velocity is constant in all directions and all areas of the structure. However this is rarely the case in real structures where holes, thickness changes, lugs and other features exist and is certainly not the case in composite materials where propagation velocity can vary greatly with material direction.

Numerous authors have attempted to address the variation in signal propagation velocity experienced in composite materials. Paget et al. (2003) developed a closed form solution for source position calculation based on the assumption of an elliptical wave front. However, the propagation wave front is only elliptical in composites with specific layups and closed form solutions are rarely stable in the presence of the uncertainties experienced in the measurement system. Others have utilised special sensor configurations to in an effort to achieve better source location. Aljets et al. (2010) used a closely arranged triangular sensor array to determine the angle of wave incidence upon the array and the propagation distance along this direction, to give a source position. Whereas Ciampa and Meo (2010) used a triangular array of three closely spaced sensor pairs (six sensors in total), allowing the source position to be described by six non-linear equations. Solving the equations with an iterative Newton method provides a source position without the need for prior knowledge of the wave velocity behaviour in the material. However, processing times are high, at around 2 seconds per event, and accuracy is likely to reduce in complex geometries.

An alternative approach to AE source location in complex materials and geometries is that of mapping. In this type of process a structure is mapped in such a way that a relationship can be formed between known physical positions upon a structure and the Δts that would result from a signal originating at these positions. Such a methodology has the potential to facilitate accurate source location even in situations where varying wave speeds, holes and geometry are present. Scholey et al. (2009) approached this by analytically calculating the expected Δts for an anisotropic composite panel from an array of points, or source positions. They describe the best-matched point search method which compares measured Δts with the analytical map to find the array point at which the difference is minimised and hence give the location. The accuracy is affected by the resolution of the mapping array, so small spacings of 1-2mm are used, and it is also important that accurate wave velocities are known for a given material. The approach is also not well suited to dealing with complex geometries, in which the calculation of arrival times becomes far more problematic. Baxter et al. (2007), Pullin et al. (2007) and Hensman et al. (2010) instead used artificial AE sources to determine Δts from known grid positions and therefore generate a map of Δts to aid source location in complex metallic structures. Baxter et al. (2007) and Pullin et al. (2007) used H-N sources to generate contour maps of constant Δt for each sensor pair, linearly interpolating between grid points to improve resolution. Mapped contours corresponding to measured Δts from real AE test data can then be selected for each sensor pair and overlaid to find a crossing point and hence a prediction of source location. Hensman et al. (2010) followed a similar methodology, but chose to represent the relationship between the Δts and the spatial grid using

Gaussian processes. Both mapping approaches were shown to improve source location in metallic structures with complex geometries and inherently compensate for variations in wave speed and any obstructions in the wave propagation path. This makes the mapping approach ideally suited to use in composite materials where propagation velocity can vary greatly before the effects of geometry are considered. In this paper the "Delta T Mapping" methodology is utilised to improve AE source location accuracy of impact events in composite materials.

3. Arrival time estimation

The accurate arrival time estimation of an AE signal is paramount to ensuring accurate location calculation. In this work the AIC based approach, discussed above, is used and involves the minimisation of equation (1) below.

$$AIC(t) = t\log_{10}\left(\mathrm{var}\left(x[1;t]\right)\right) + \\ + (T - t - 1)\log_{10}\left(\mathrm{var}\left(x[t;T]\right)\right) \quad (1)$$

The signal is split into two parts, that from time 0 to time "t" and that from time "t" until the end of the signal. Equation (1) describes the similarity in entropy between two parts of the signal, for every time "t" throughout the signal duration. When "t" becomes aligned with the onset of the signal, the minimum similarity is observed between the high-entropy uncorrelated noise prior to signal onset and the low-entropy waveform showing marked correlation after signal onset. Hence the minimum of the AIC function corresponds to the signal onset time, as can be seen in Fig. 1.

4. Delta T Mapping source location

The "Delta T Mapping" methodology discussed above (Baxter et al., 2007; Pullin et al., 2007) has 5 associated steps, which are outlined briefly below:

Determine area of interest – Delta-T source location can provide complete coverage of a part or structure, or it can be employed as a tool to improve source location around specific areas of expected fracture, which could potentially be identified via finite element modelling.

Construct a Map System – A grid is placed over the area of interest within which AE events will be located. It should be noted that sources are located with reference to the grid and not the sensors and it is not required that sensors be placed within the grid.

Obtain time of arrival data from an artificial source – An artificial source (nominally a H-N source (ASTM,

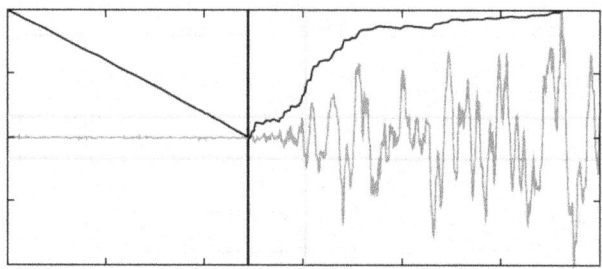

Figure 1. Arrival time estimation using AIC based approach. The vertical grey line indicates the threshold based arrival time. The black trend is the AIC function and the vertical black line indicates the estimated arrival time at its minimum.

2010; Hsu and Breckenridge, 1981) is generated at the nodes of the grid to provide AE data for each sensor. An average result of several sources is used for each node. Missing data points can be interpolated from surrounding nodes.

Calculate Delta T map – Each artificial source results in a difference in arrival time or Delta T for each sensor pair (an array of four sensors has six sensor pairs). The average Delta T at each node is stored in a map for each sensor pair. The resulting maps can be visualised as contours of constant Delta T.

Locating real AE data – The Delta T values from a real AE event are calculated for each sensor pair. A line of constant Delta T equivalent to that of the real AE event can then be identified on the map of each sensor pair. By overlaying the resulting contours, a convergence point can be found that indicates the source location. As with time of arrival, a minimum of three sensors is required to provide a point location and more sensors will improve the location. In theory all the lines should intersect at one location, however in practice this is not the case. Thus in order to estimate a location all convergence points are calculated and a cluster analysis provides the most likely location.

5. Experimental Procedure

A carbon fibre composite panel was manufactured from Advanced Composite Group (ACG) MTM28-1/HS-135-34%RW uni-directional pre-preg using a $((0,90)_4)_s$ layup. Following the lay-up process, the panel was placed in a vacuum bag and cured in an oven at 120°C. After trimming the overall dimensions were 200mm wide and 370mm long, as shown in Fig. 2. A 2cm grid was applied to the central area of interest of 200mm x 160mm for the purposes of training the Delta T Maps. Training data from five H-N sources were collected at each grid point to allow average Δts to be calculated and arrival times were measured using the AIC method discussed above. The specimen was instrumented with 4 Pancom Pico-Z AE sensors, placed at the corners of the area of interest at positions (10mm, 150mm), (190mm, 150mm), (10mm, 10mm) and (190mm, 10mm), presented in Fig. 2 by white circles. The sensors were attached using a cyanoacrylate adhesive that also provides an appropriate acoustic couplant.

The specimen was impacted at position of (30mm, 30mm) in the area of interest, shown in Fig. 2 by a grey cross. A circular clamp was placed on the specimen in such a way as to leave a 100mm diameter unsupported area centred about the impact location. Five repeated low velocity impacts of 5, 7, 8, 8 and 10J were performed at the impact location using an Instrom Dynatup 9250HV impact test machine with a 20mm diameter hemispherical impact tup and an impactor mass of 5.81kg. Following the final impact event a c-scan inspection of the panel was conducted to assess the position and extent of the damage sustained.

AE monitoring was conducted throughout each impact event using a Physical Acoustic Limited (PAL) PCI-2 acquisition system.

Sensor outputs were pre-amplified by 40dB prior to recording and signal acquisition and arrival time measure-

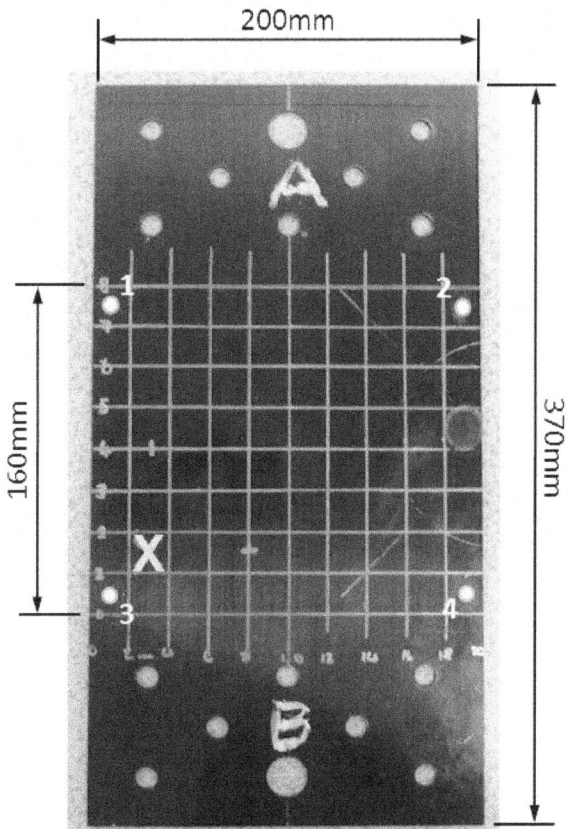

Figure 2. Carbon fibre specimen and dimensions – white circles indicate sensor positions and the grey cross indicates impact location.

ment were triggered using a 50dB threshold. A sampled version of all detected signals was also recorded at a sample rate of 5MHz. The sampled waveforms were processed post test using the AIC function to determine the true signal arrival times, which were in turn used in the calculation of source position by the "Delta T Mapping" algorithm. For comparison source positions were also calculated by the traditional TOA location algorithm using the threshold based arrival time estimations.

6. Results and Discussion

Fig. 3 shows the c-scan results obtained from post impact test inspection of the specimen. A grid representing the monitored area of interest is superimposed over the image and damage resulting from the impact events is highlighted. The specimen bolt holes are visible along the top and bottom edge of the image, outside of the area of interest. Additional areas of signal loss can be seen in the four corners of the area of interest and in the middle of the right hand edge of the specimen, resulting from the residue of the adhesive used for sensor coupling and are not caused by damage to the specimen. A 272mm² area of delamination was induced by the impact events, however only a barely visible indent was detectable on the impacted surface and only minor splitting was visible on the back face.

Fig. 4 presents the calculated location of the impact events, within the area of interest, using the traditional TOA algorithm (squares) and the Delta T Mapping algorithm (diamonds). It is immediately apparent that not only is the TOA location algorithm less accurate but that it has

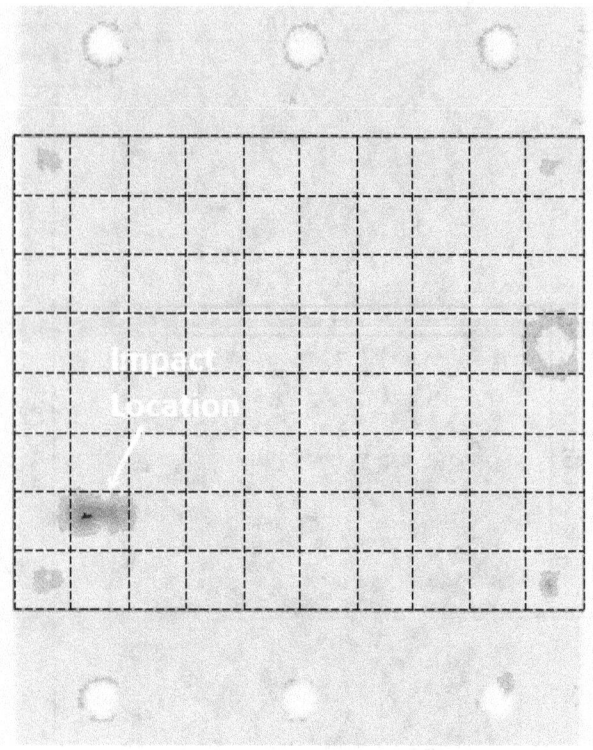

Figure 3. C-scan image from post impact testing inspection, overlaid with a grid representing the area of interest.

Table 1
Location Errors

Impact No.	Euclidean Error (mm)	
	TOA	Delta T Mapping
1	20.6	8.5
2	-	6.3
3	39.4	10.4
4	-	5.3
5	-	3.5

"Delta T Mapping" methodology detecting and locating all five impact events, compared with only two detected using the TOA algorithm. Furthermore the combined approach of "Delta T Mapping" and AIC arrival time estimation demonstrated a significant improvement in location accuracy; offering a 58% and 73% improvement in accuracy when compared with the only two events located by the TOA approach.

These results highlight the limitations of using traditional location algorithms in complex materials; limitations that are only likely to increase with application to composite parts with complex geometry, such as holes, curvature and ply drops. The results also demonstrate how such limitations can be overcome through accurate arrival time determination and the use of advanced location techniques such as "Delta T Mapping".

In order for the "Delta T Mapping" technique to continue working effectively there must be minimal change to a given structure. If a structural change occurs that affects wave propagation behaviour, such as damage, a change in material properties or structural repairs, then subsequent location accuracy may be affected. Such changes to a structure are also likely to affect the accuracy of other advanced location techniques discussed above. In the event of such a change, a structure could be remapped in order to provide consistently accurate location results following a structural change.

Further work will be undertaken to assess the detection of impact damage under subsequent loading, hence offering a second opportunity to detect impact damage, if an initial impact event was not detected. Additional work will aim to demonstrate the "Delta T Mapping" technique in more complex composite geometries.

also located fewer events than Delta T Mapping. In fact, due to the errors arising from a threshold based signal arrival time measurement and the lack of account for wave speed variation the TOA algorithm was only able to resolve a location for 2 (impact 1 and 3) out of the 5 impact events. The Delta T Mapping technique however, has located all 5 impact events close to the impact location and importantly within the observed area of damage.

Table 1 presents the euclidean location error measured relative to the impact location at (30, 30). It should be noted however that signals can be assumed to have originated from any position within the observed area of damage. In the two cases where the TOA algorithm was able to resolve a location the Delta T Mapping algorithm offers a 12.1mm and 29mm improvement in location accuracy for impacts 1 and 3, corresponding to a 58% and 73% improvement in location accuracy, respectively. This demonstrates how techniques for improving AE source location, such as Delta T Mapping, can provide considerable improvements in location accuracy over traditional methods and in doing so improve detection sensitivity.

7. Conclusions

A series of five impact tests were conducted on a wide carbon fibre composite tensile specimen and monitored using AE. Subsequent visual inspection of the specimen revealed only a barely visible indent at the impact site; however C-scan inspection revealed a significant delamination area of $272mm^2$.

AE source location for each impact event was calculated using the traditional time of arrival technique and the novel "Delta T Mapping" approach combined with AIC based arrival time estimation. A significant improvement in detection sensitivity was observed, with the

References

Aljets, D., Chong, A., Wilcox, S. and Holford, K. (2010). Acoustic emission source location in plate like structures using a closely arranged triangular sensor array, *Journal of Acoustic Emission*, Vol. 28, pp. 85-98.

ASTM. (2010). A standard guide for determining the reproducibility of acoustic emission sensor response, *American Society for Testing and Materials*, Vol. E976.

Baxter, M.G., Pullin, R., Holford, K.M. and Evans, S.L. (2007). Delta T source location for acoustic emission, *Mechanical Systems and Signal Processing*, Vol. 21, No. 3, pp. 1512-1520.

Ciampa, F. and, Meo, M. (2010). A new algorithm for acoustic emission localisation and flexural group velocity determination in anisotropic structures, *Composites Part A: Applied Science and Manufacturing*, Vol. 41, No. 12, pp. 1777-1786.

Ding, Y., Reuben, R.L. and Steel, J.A. (2004). A new method for waveform analysis for estimating AE wave arrival times using wavelet decomposition, *NDT & E International*, Vol. 37, pp. 279-290.

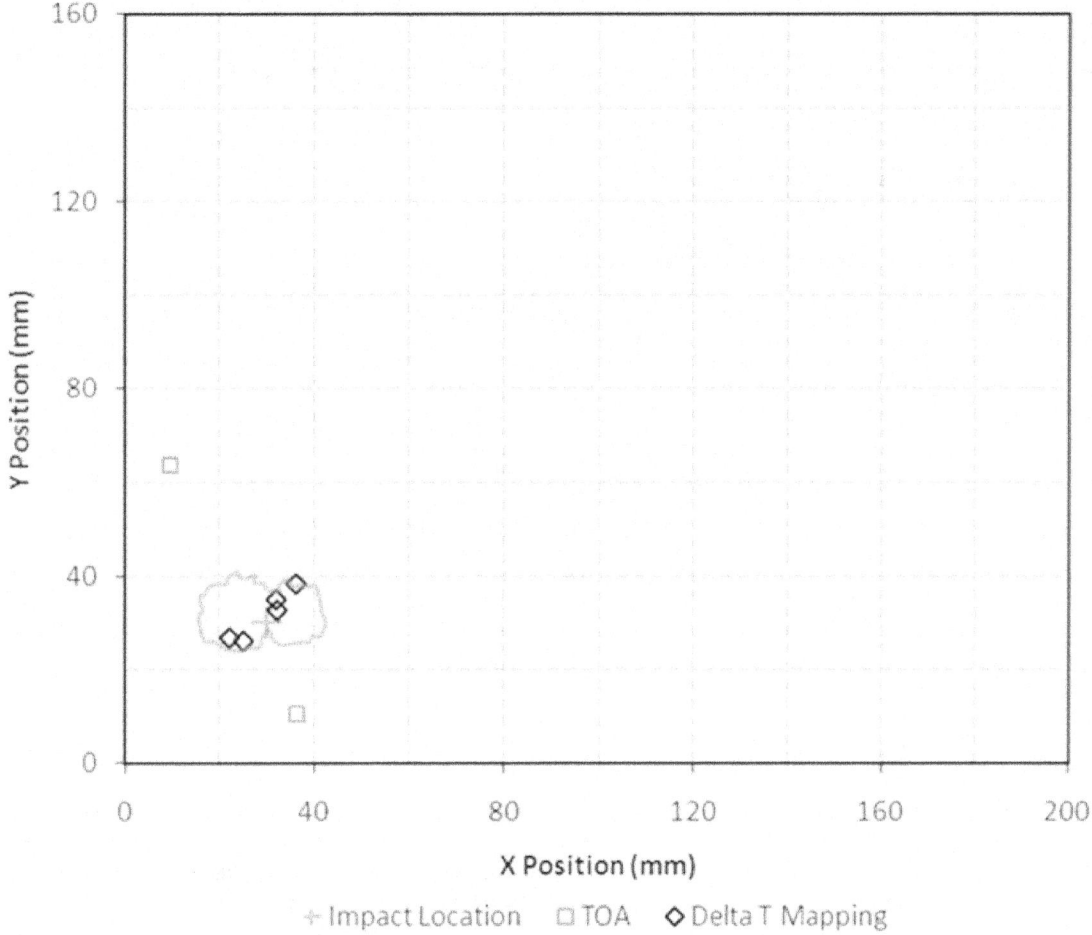

Figure 4. AE location of impact events using both TOA (squares) and Delta T Mapping (diamonds) algorithms. The impact location is represented by a grey cross and an outline of the damaged region is overlaid for reference.

Hamstad, M.A., O'Gallagher, A. and Gary, J. (2002). A wavelet transform applied to acoustic emission signals: Part 2: Source location, *Journal of Acoustic Emission*, Vol. 20, pp. 62-82.

Hensman, J.J., Mills, R., Pierce, S.G., Worden, K. and Eaton, M. (2010). Locating acoustic emission sources in complex structures using gaussian processes, *Mechanical Systems and Signal Processing*, Vol. 24, pp. 211-223.

Hsu, N.N. and Breckenridge, F.R. (1981). Characterization and calibration of acoustic emission sensors, *Materials Evaluation*, Vol. 39, No. 1, pp. 60-68.

Jeong, H. (2001). Analysis of plate wave propagation in anisotropic laminates using a wavelet transform, *NDT & E International*, Vol. 34, pp. 185-190.

Kurz, J.H., Grosse, C.U. and Reinhardt, H-W. (2005). Strategies for reliable automatic onset time picking of acoustic emission and of ultrasound signals in concrete, *Ultrasonics*, Vol. 43, pp. 538-546.

Lokajicek, T. and Klima, K. (2006). A first arrival identification system of acoustic emission (AE) signals by means of a higher-order statistics approach, *Measurement Science and Technology*, Vol. 17, pp. 2461-2466.

Miller, R.K., Carlos, M.F., Findlay, R.D., Godinez-Azcuaga, V., Rhodes, M.R., Shu, F. and Wang, W.D. (2005). Acoustic emission testing, in: Miller, R.K., Hill, E.V.K. and Moore, P.O. (eds.) *NDT Handbook*, 3rd ed., Vol. 6ASNT.

Paget, C.A., Atherton, K. and O'Brien, E.W. (2003). Triangulation algorithm for damage location in aeronautical composite structures, in: *Proceedings of IWSHM-4*, Stanford, CA, USA.

Pullin, R., Baxter, M.G., Eaton, M.J., Holford, K.M. and Evans, S.L. (2007). Novel acoustic emission source location, in: *6th International Conference on Acoustic Emission*, Lake Tahoe-Nevada, USA.

Rindorf, H.J. (1981). Acoustic emission source location in theory and in practice, *Bruel and Kjaer Technical Review*, Vol. 2, pp. 3-44.

Scholey, J.J., Wilcox, P.D., Wisnom, M.R., Friswell, M.I., Pavier, M., and Aliha, M.R. (2009). A generic technique for acoustic emission source location, *Journal of Acoustic Emission*, Vol. 27, pp. 291-298.

Ziola, S.M. and Gorman, M.R. (1991). Source location in thin plate using cross-correlation., *Journal of the Acoustical Society of America*, Vol. 90, No. 5, pp. 2551-2556.

An Experimental Method for Determination of Characteristic Time Points in Nanosuspensions Processing under Influence of Ultrasonic Field

Vladimir Tarasov[1], Nikolay Stepanishev[1], Raisa Boyarskaya[2]

[1] Rocket Engineering Department, Faculty of Special Machinery, Bauman Moscow State Technical University, 5 2nd Baumanskaya Street, Moscow, 105005, Russia
[2] Department of Mechanical Manufacturing, Faculty of Manufacturing Engineering, Bauman Moscow State Technical University, 5 2nd Baumanskaya Street, Moscow, 105005, Russia

Abstract: Regularities of nanosuspensions preparation process in an ultrasonic field have been established. The technique for experimental determination of characteristic aspects of nanosuspensions preparation technological process has been proposed.

Key Words: Nanosuspensions, Carbon nanotubes, Ultrasonic field influence, Agglomerate, Processing time, Concentration of nanotubes.

1. Introduction

Producing of high quality nanosuspensions is an important technical objective for various fields of Engineering. Modifying the binder with nanomaterials for improving the quality of polymer composites matrix is among the applications. It has been experimentally observed (Tarasov and Stepanishev, 2010a; Tarasov and Stepanishev, 2010b) that addition of carbon nanotubes into the matrix increases the strength and improves the processing characteristics of the composite matrix.

Various investigations where nanoparticles have been added into the polymer matrix (Veedu et al., 2006; Bekyarova et al., 2007; Zhua et al., 2007) demonstrate the importance of uniform distribution of the nanoparticles over the entire volume, thus maximizing the main advantage of nanoparticles, which is providing strong connections with the molecules of the binder.

Electronic microscope researches showed that carbon nanotubes (CNTs) have a complex dimensional geometry (Fig. 1a), which is necessary for composite material matrix reinforcement. Analysis of the physical properties CNTs explains their tendency to form agglomerates (Fig. 1b) under the influence of Van der Waals intermolecular interaction forces. Agglomerates can be formed either during the synthesis or later during operations with CNTs (cleaning, an introduction to polymers, etc.).

Ultrasonic methods are particular prospect here. However, during the design of the processing route it is important to assign a reasonable duration of ultrasonic treatment.

In this paper we propose a method of experimental determination of the specific processing time required for preparing nanosuspensions using ultrasonic treatment. This method allows their determination of a reasonable duration of ultrasonic treatment.

2. Model of preparing nanosuspensions using ultrasonic treatment

The required concentration k of nanotubes if they are evenly distributed in the volume is:

$$k = \frac{m_{TP}}{\rho \vartheta} \quad (1)$$

where m_{TP} is the total mass of the nanotubes; ρ, ϑ are density and volume of the binder, respectively.

Describing the process of ultrasonic matrix nanomodification, it has been assumed that in the initial stage of introducing nanotubes into the binder a nanotube hyper-

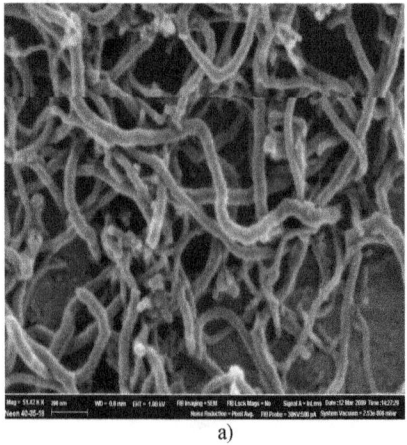
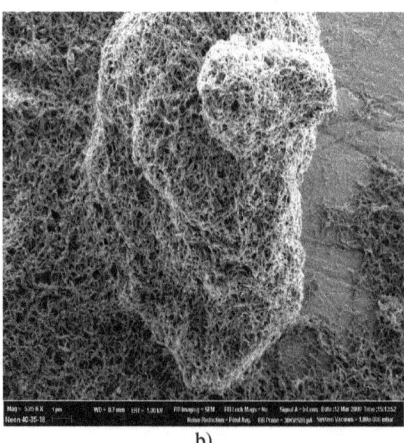
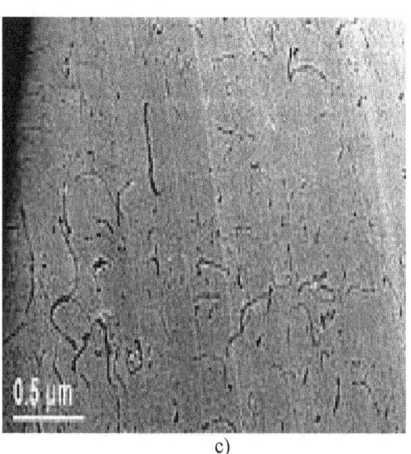

a)   b)   c)

Figure 1. State of nanotubes ensemble in the binder: a) Photo of the nanotube ensemble with the electronic microscope "NEON 40-35-18"; b) CNTs agglomerates; c) electronic microscope JEM-200 CX photo of nanotube distribution in nanosuspension after ultrasonic field influence.

agglomerate is formed. Each nanotube has mass, m_0 and length, l^*. Attractive forces arise between the nanotubes that create tensile stresses in the nanotube hyperagglomerate

$$P_r = K_r \frac{a^2}{x^2} \qquad (2)$$

where K_r is the constant of expansion process, a^2 is the constant proportional to nanotube surface .

Repulsive forces appear in hyperagglomerate with a quasi-regular structure, which is characterized by the distance between the nanotubes $l < l^*$. Nanotubes can be elastically deformed, which can cause reduction of the distance between particles, y, where $y = l - x$. The resulting repulsion pressure is described by:

$$P_{ot} = \begin{cases} K_{ot} \dfrac{l-x}{l} & \text{when } x \leq l \\ \\ 0 & \text{when } x > l \end{cases} \qquad (3)$$

Resulting stress is:

$$P = -K_{ot} \frac{y}{l} + K_r \frac{a^2}{(l-y)^2} \qquad (4)$$

There is a value y_0, when $P = 0$:

$$y_0 = \frac{l}{4}\left(1 - \sqrt{1 - 8\frac{K_r}{K_{ot}}\frac{a^2}{l^2}}\right) \approx \frac{K_r}{K_{ot}}\frac{a^2}{l^2}l \qquad (5)$$

The value of the resulting tension $|P|$ in the agglomerate will increase under the compression and expansion proportionally to y. When expansion reaches the value of $y = y_0$ nanotubes in agglomerate will reach maximum distance from each other and nanotubes will separate from the agglomerate.

The dynamics of the particle separation from agglomerate during ultrasonic treatment of suspension is described as follows. The agglomerate spot dimension is denoted as R_0, and it will be determined in accordance with photodetection data obtained during the nanomodification process. Considering that the number of nanoparticles in agglomerate is:

$$N = \frac{m_{TP}}{m_0} \qquad (6)$$

the thickness of agglomerate is:

$$b = l\frac{m_{TP}}{m_0}\left(\frac{l}{R_0}\right)^2 \qquad (7)$$

In this paper the stiffness of the nanotube frame is assumed less than stiffness of the binder. Thus, vibrations in the frame will be initiated by liquid.

The wave motion initiated by the generator in the binder is assumed to be plane and is described by the equation:

$$P(t,Y) = P_0 \cos\left[2\pi f\left(t - Y/C\right)\right] \qquad (8)$$

where P_0, f are amplitude and frequency of pressure fluctuations in the acoustic wave; t, Y are time and spatial coordinate, respectively.

The mass velocity of the binder is:

$$V(t,Y) = V_0 \cos\left[2\pi f\left(t - Y/C\right)\right] \qquad (9)$$

where

$$V_0 = \frac{P_0}{\rho C} \qquad (10)$$

The maximum displacement of particles during vibration is:

$$z_m = \int_0^{1/2f} V dt = \frac{V_0}{\pi f} \qquad (11)$$

Since the amplitude of the ultrasonic transducer's vibration is known from equipment manufacturer's data, the values V_0, P_0 are easily determined.

In the vicinity of $Y = 0$ the relation (9) can be written as

$$V = V_0 \sin 2\pi ft + V_0 \cos 2\pi ft \frac{2\pi fY}{C} \qquad (12)$$

The first term of this equation describes vibration of agglomerate without changing its size, and the second term describes the expansion of agglomerate. For the analysis of agglomerate destruction in an ultrasonic field the increasing movement of the binder is represented as:

$$\Delta V = V_0 \sin 2\pi ft \frac{2\pi fY}{C} \qquad (13)$$

The relative motion of nanotubes in an ultrasonic field is represented as:

$$m_0 \frac{dU}{dt} = S\rho C \Delta V = S\rho 2\pi fYV_0 \sin 2\pi ft \qquad (14)$$

where U is a mass velocity of the relative motion of the nanotubes, S is the nanotubes cross section.

The solution of the equation has the form

$$U = \frac{S\rho YV_0}{m_0}\left(1 - \cos 2\pi ft\right) \qquad (15)$$

Relative movement of the particles ΔY is determined by Y coordinate out of the center of agglomerate, and is equal to:

$$\Delta Y = \frac{S\rho YV_0}{m_0}\left(t - \frac{\sin 2\pi ft}{2\pi f}\right) \qquad (16)$$

During the time of one pressure pulse that is equal to the half of the oscillation period $1/2f$, the particle will shift on the value of:

$$\Delta Y = \frac{\pi S\rho Yz_m}{2m_0} \qquad (17)$$

As the number of influencing pulsations ft grows the relative displacement of nanotubes in the agglomerate will increase

$$\Delta Y = ft\frac{\pi S\rho Yz_m}{2m_0} \qquad (18)$$

and at a certain time will become equal to the limiting magnitude y_0. At this moment, separation of nanotubes from the agglomerate and their diffusion into the binder will occur. This relation may be used for obtaining a formula describing the change of coordinates of nanotubes separation front Y_Φ per time

$$Y_\Phi(t) = \frac{y_0 2m_0}{\pi S\rho z_m}\frac{1}{ft} = \frac{\Omega}{ft} \qquad (19)$$

where $\Omega = \dfrac{y_0 2m_0}{\pi S\rho z_m}$.

The coefficient Ω can be determined experimentally. Values m_0, S may be theoretically estimated. Thus the relation (19) allows to estimate y_0.

The mass of nanotubes diffusing into a binder is equal to

$$m_{dif}(t) = m_{TP}\frac{b - Y_\Phi(t)}{b} \qquad (20)$$

Uniform distribution of nanotubes in the working space of the processing chamber depends on the processes of diffusion with the basic relation:

$$\rho V_n k = D\frac{\partial k}{\partial n} \qquad (21)$$

The left side $\rho V_n k$ describes the nanotubes mass flow rate per unit area in n direction, V_n is mass velocity in the direction n, k is concentration of nanotubes in the binder, D is diffusion coefficient.

Accounting for the mass balance we obtain the diffusion equation in general form:

$$\frac{\partial k_t}{\partial t} = \frac{D}{\rho}\Delta k_t \qquad (22)$$

where Δ is the Laplace operator, k_t is the current concentration of nanotubes at an arbitrary time.

As a particular case we can assume that the diffusion process occurs mainly in the direction Y, from the ultrasonic generator to the bottom of the volume. Then the diffusion process is described by the differential equation of parabolic type,

$$\frac{\partial k_t}{\partial t} = \frac{D}{\rho}\frac{\partial^2 k_t}{\partial Y^2} \qquad (23)$$

which has an analytical solution. The character of the diffusion process may be different depending on the ratio of the velocities of nanotubes mass separation from agglomeration and diffusion.

If a suspension viscosity is low, the concentration of nanotubes will be uniform within the volume of the working space and will monotonically vary with time. In this case the change of the current concentration k_t of nanotubes with time is described by the following equation:

$$\frac{k_t(t)}{k} = \frac{m_{dif}(t)}{k\rho L^3} = 1 - \frac{\Omega^*}{t} \qquad (24)$$

where $\Omega^* = \dfrac{\Omega}{bf}$.

If the suspension viscosity is high, equation (23) can be transformed to:

$$\frac{\partial k_t}{\partial t} = \omega^*(k - k_t) \qquad (25)$$

which has the solution

$$k_t = k\left[1 - \exp(-\omega^* t)\right] \quad when \ k_t \leq k \qquad (26)$$

where $\omega^* = \dfrac{\omega}{\eta\rho L}$ is an empirical coefficient; η, ρ are the viscosity and density of the binder, and L is the dimension of the working capacity for nanosuspensions preparation.

3. Experimental

To monitor the effectiveness of ultrasonic dispersion and to explore the nature of the diffusion process of nanoparticles experimental studies with laboratory ultrasonic disperser LUZD-1.5/21-3.0 were carried out under the following technological regimes:

1) the power of the generator was 1.5 kW;
2) the frequency was $f = 21\ 000$ Hz;
3) the emitter's end amplitude was $z_m = 20\mu m$.

In this paper an optical method for determining the homogeneity of the suspension is proposed by fixing the change in its colour gamma in accordance with the scheme of the tests shown in Fig. 2.

Fig. 3 shows the photographs of the separate stages of ultrasonic dispersion process made by digital camera. Change of colour intensity is characteristic for the presented photos.

The sequence of frames, which characterizes nanosuspensions colour intensity change is shown in Fig. 4.

Analysis of the photographs shows that the distribution of carbon nanotubes (CNTs) in the matrix volume is uniform. As agglomerates separate into individual nanoparticles under the influence of ultrasonic treatment and fill the working volume, the intensity of nanosuspensions' color changes with time, and reach a maximum value.

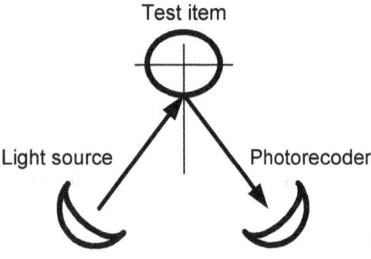

Figure 2. The scheme of homogeneity control of suspension with nanoparticles per its colour.

  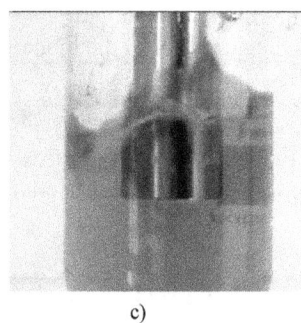 

a)   b)   c)   d)

Figure 3. Photos of the stages of nanoparticles dispersion in liquid under the influence of ultrasonic field: a) introduction of nanoparticles; b)-d) stages of ultrasonic treatment

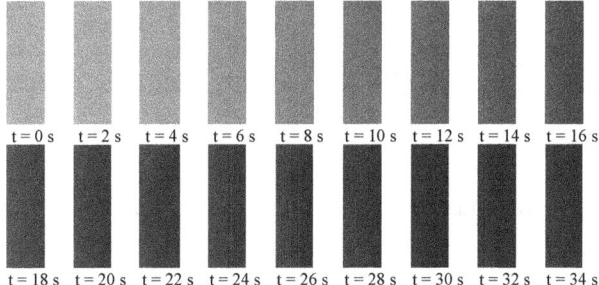

Figure 4. Photographs of binder filling process with nanotubes in an ultrasonic field. The elapsed time is shown under the image.

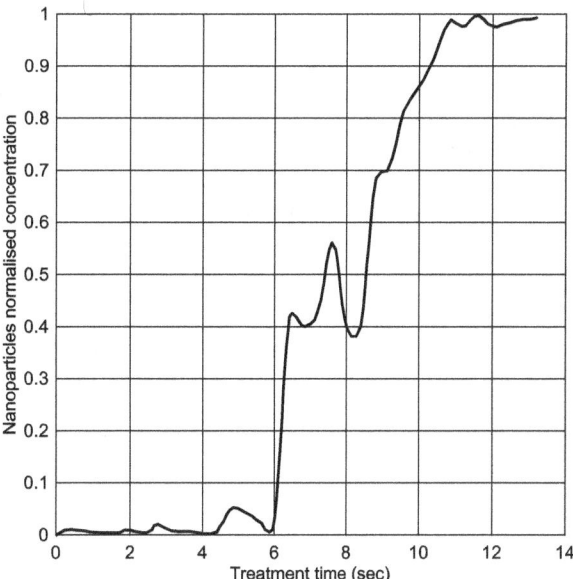

Figure 5. Graphic dependence of the intensity of the colour of the nanosuspension during ultrasonic treatment.

No colour change was detected after a year of observations which could be attributed to the strong bond of carbon nanotubes with molecules of the binder.

Processing of the colour intensity of the pictures was performed with the help of a computer program "Image Analysis – Media Cybernetics – Image Pro Plus 6.0" (Meyer Instruments, 2006). The results of the data are presented in Fig. 4.

After normalization of the colour intensity level, the dependence of nanotube concentration increase with time was obtained (Fig. 5). This allows definition of the beginning of the destruction of agglomerates during ultrasonic treatment. Statistical analysis of the experimental data gives a value of constant $\Omega^* = 6.4$ sec.

Additional analysis of experimental data in the interval of time from 6.4 sec. shows the necessity to transform equation (26) to the form

$$k_t = k\left\{1 - \exp\left[-\omega^*(t - t_0)\right]\right\} \quad \text{when } k_t \leq k \quad (27)$$

where t_0 is the time of beginning of agglomerates destruction.

The fit to the experimental data ($\omega^* = 0.67$ sec^{-1}) is shown in Fig. 6.

Analysis of experimental data confirmed the mechanism of agglomerate destruction and showed for the destruction of the agglomerates is required ultrasonic treatment over time t_0. Process constants, were obtained which allow deeper understanding of the essence of the physical processes taking place.

The dependences obtained provide a basis for reasonable choice of the intensity of the ultrasonic field in accordance with the requirements for the treatment process duration. To do so, specifying the tolerance ε of the nominal concentration values, we obtain the dependence for designation of the suspension ultrasonic treatment time.

$$t_k = t_0 + \frac{1}{\omega^*} \ln \frac{1}{\varepsilon} \quad (28)$$

4. Conclusion

The proposed method allows determination of the time to agglomerate destruction beginning and the total time of ultrasonic treatment of suspension, using the experimental data.

To maximize the reinforcement effect of the carbon nanotubes the optimal mass fraction of particular carbon nanotubes must be obtained by any available method, for example, by the method described in (Tarasov and Stepanishev, 2010b).

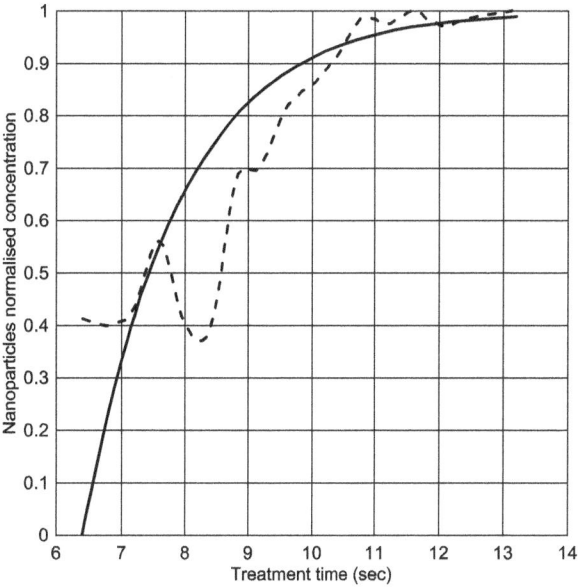

Figure 6. Nanotubes normalized concentration change in time (dotted line are experimental data, solid line model).

References

Tarasov, V.A., Stepanishev, N.A. (2010a). Nanotechnology applications for hardening polyester matrix, *The Bulletin BMSTU, Ser. Engineering, Spec. Issue of Actual Problems of Rocketry Systems*, pp. 207-217

Tarasov, V.A., Stepanishev, N.A. (2010b). Hardening of the polyester matrix with carbon nanotubes, *The Bulletin BMSTU, Ser. Instrument Making, Spec. Issue Nanoengineering*, pp. 53-65.

Veedu, V. P., Cao, A., Li, X., Ma, K., Soldano, C., Kar, S., Ajayan, P. M. and Ghasemi-Nejhad, M. N. (2006). Multifunctional composites using reinforced laminae with carbon-nanotube forests, *Nature Materials*, No 5. pp. 457-462.

Bekyarova, E., Thostenson, E. T., Yu, A., Kim, H., Gao, J., Tang, J., Hahn, H. T., Chou, T.-W., Itkis, M. E. and Haddon, R. C. (2007). Multiscale carbon nanotube–carbon fiber reinforcement for advanced epoxy composites, *Langmuir*, Vol. 23, No. 7, pp. 3970-3974.

Zhua, J., Imamb, A., Cranec, R., Lozanod, K., Khabasheskue, V. N. and Barreraf, E. V. (2007). Processing a glass fiber reinforced vinyl ester composite with nanotube enhancement of interlaminar shear strength, *Composites Science and Technology*, Vol. 67, No. 7-8, pp. 1509-1517.

Meyer Instruments (2008). *Media Cybernetics Products from Meyer Instruments*. [Online]. Available at: http://www.meyerinst.com/ html/ mediacy/imagepro.htm (Accessed: 29 January 2012).

Advances in Bonded Repair of CFRP Aircraft Structures by Surface Inspection

Kerstin Albinsky, Kai Brune, Stefan Dieckhoff, Olaf Hesebeck, Uwe Lommatzsch, Susanne Markus

Fraunhofer Institute for Manufacturing Technology and Applied Materials Research - IFAM, 12 Wiener Street, Bremen, D-28359, Germany

Abstract: Due to the increasing amount of CFRP materials in aircraft structures it is necessary to have a reliable repair process. Bonded repair has several advantages in comparison to bolted repair regarding aerodynamic effects or stress concentration in the structure. In case of a damage of CFRP structure contaminants like hydraulic oil or de-icing fluids often appear on adherent surfaces. Due to the fact that these contaminants have a huge influence on the bond strength it is necessary to have a reliable measuring device to monitor the adherent surface before bonding. There are no implemented measuring techniques to monitor the adhesive surface in field of bonded repair available up to now. The two measuring techniques X-ray fluorescence analyses as well as the FTIR-spectroscopy have been tested regarding the detection of contamination on CFRP adherent surfaces in field of bonded repair.

Key Words: Aircraft structures, Bonded repair, CFRP, Infrared-spectroscopy, Surface monitoring, X-ray fluorescence.

1. Introduction

Due to the demand of reduced fuel consumption and CO_2 emission it is necessary to implement light-weight design in combination with carbon fibre reinforced plastics (CFRP) in aircraft structures. During the entire live cycle of an aircraft, damages of structures occur that are caused by foreign objects collision (like tool drop during maintenance or ground service vehicle), lightning and hail storm.

It is compulsory to have a useful, reliable and fast in-field repair solution for CFRP aircraft structures. State of the art for the repair of CFRP structure is the use of bolting or adhesive bonding (or a combination of both) (Baker et al., 2002), (Ahn and Springer, 2000), (Composite, 2002).

There are several advantages of adhesive bonding technology over bolted repair (see Tab. 1). Adhesive bonding does not disrupt the CFRP structure by hole-drilling with a resulting stress concentration in the structure. Additionally, properly designed repair patches do have an excellent strength-to-weight ratio and equivalent or superior load levels to mechanically fastened joints. By using a flash bonding repair, no aerodynamic penalty is put on the repaired structure.

Up to now bonded repair of primary structure is only certified for small and military aircrafts or cosmetic repair. In order to increase the acceptance of the adhesive bonding technology for the repair of large civil aircraft structures, it is important to prove the reliability of structural adhesive joints (Wetzel et al., 2011).

A defined surface state (e.g. free of contaminants) is important to achieve a reliable adhesive bonded joint. Because of missing NDT-tools for the assessment of adhesive bonds, it is compulsory to have a surface inspection tool for the detection of contamination of CFRP surfaces before bonding.

The inspection tool needs to be sensitive enough to detect the relevant amount of contamination on CFRP components. In addition, those tools need to be suitable for operation in field of repair (weight, size, health and safety). Further on, it is important that the measuring techniques have no impact on the part to be bonded afterwards.

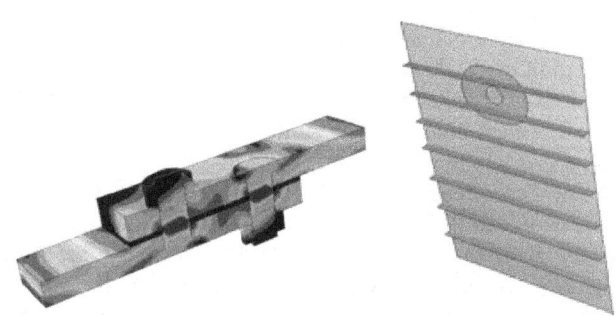

Figure 1. Bolted repair (left) and bonded repair (right)

Table 1
Overview of advantages and disadvantages of bolted and bonded repair

	Bolted Repair	Bonded repair
Advantages	• Long experience • Heating not required • Tolerant to surface contaminations	• Can sustain load levels equivalent or even superior to bolted repair • Favourable strength-to-weight ratio • Flush repair for aerodynamic smoothness
Disadvantages	• Drilling holes required → disrupts integrity → stress concentration • Non smooth surface (aerodynamic req. for control surface difficult to achieve) • Corrosion issues • Ti fasteners required • Effect on radar signature	• Requires adequate surface preparation • Not allowed/certified for primary a/c structures (exception for military, small a/c, cosmetic repair)

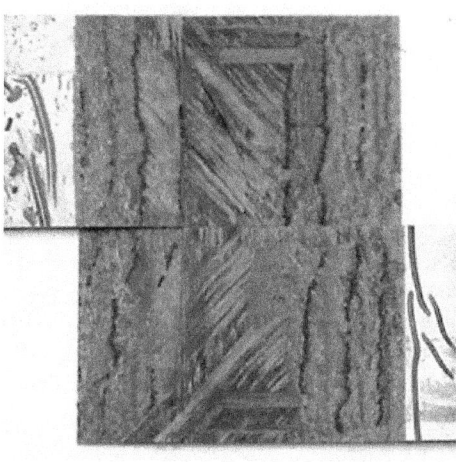

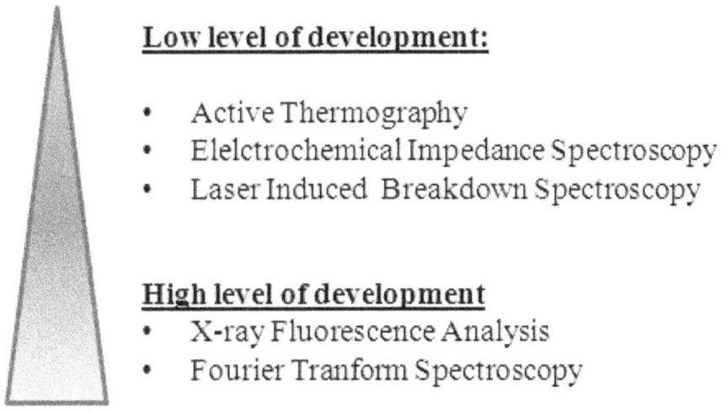

a)

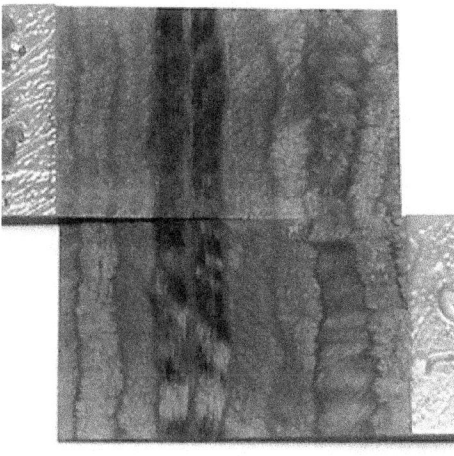

b)

Low level of development:

- Active Thermography
- Elelctrochemical Impedance Spectroscopy
- Laser Induced Breakdown Spectroscopy

High level of development
- X-ray Fluorescence Analysis
- Fourier Tranform Spectroscopy

Figure 2. Level of development of different measuring techniques for the monitoring of adherent surfaces in field of bonded repair

Regarding the state-of-the-art for evaluation of adherent surfaces in field of bonded repair only one method is established at the moment. In order to ensure that there is no contamination on CFRP surface the water break test is applied (Cremers and Radziemski, 2006), (Vadillo et al, 1996). In case of this method the surface state is evaluated by its wetting behaviour. The water break test is not quantifiable and limited for surfaces with a high roughness. Today there is no measuring technique implemented to monitor the state of adherent surfaces before bonding in field of repair. There are innovative measuring techniques that might be useful for surface inspection in field of bonded repair. The active thermography, electronic impedance spectroscopy as well as the laser induced breakdown spectroscopy allow the monitor of surfaces, but still have a low level of development. The X-ray fluorescence analysis and Fourier transform IR spectroscopy are techniques that already have a high level of development and are directly useful for testing on CFRP adherent surfaces (see Fig. 2).

That's why these techniques are chosen for further testing.

2. Influence of contaminants on bond strength

In order to study the influence of different contaminants, scarfed joints were prepared of contaminated CFRP laminates. They were tested under tensile load before and after ageing (1000 hours in water at 70°C, tested at 80°C). All specimens were exposed in the contaminant for three days, except for the hand cream which was applied to the

surfaces directly. Before bonding, all surfaces have been cleaned extensively by a paper towel. All of the selected contaminants had an influence on the bond strength (see Fig. 3). The shear strength of a reference sample without surface contamination was about 22 N/mm² with a cohesive failure in CFRP and the adhesive (see Fig. 4a). The hydraulic oil contaminated sample had a joint strength that was 24% lower than the initial strength. The failure mode was cohesive failure of the adhesive (see Fig. 4b).

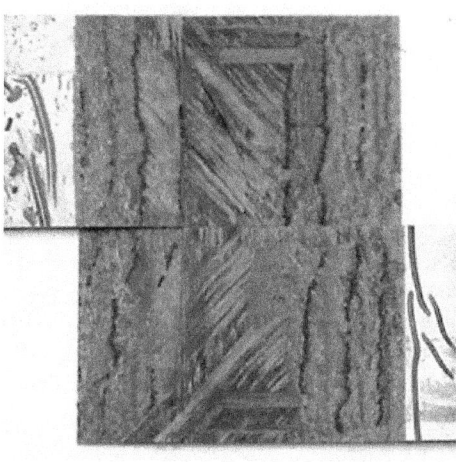

a)

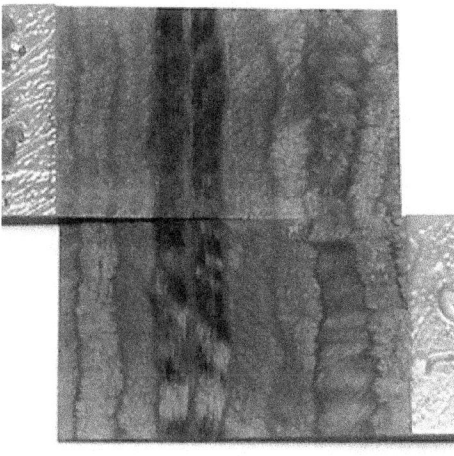

b)

c)

Figure 4. Examples of fracture surfaces of scarfed joints after ageing and tensile test: a) clean sample, b) runway de-icing fluid contaminated sample, c) hydraulic oil contaminated sample.

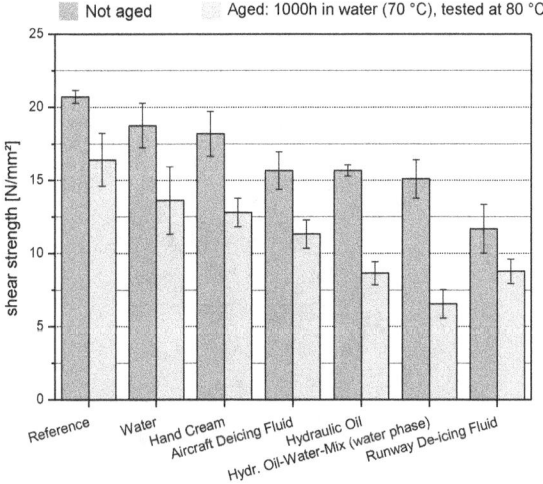

Figure 3. Influence of contaminants on bond strength

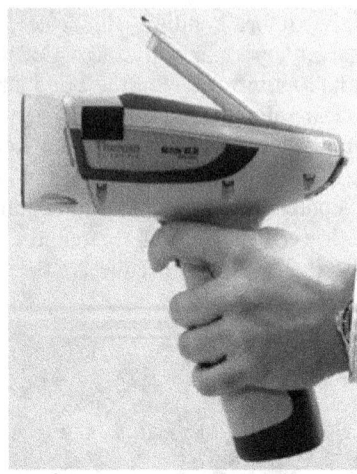

Figure 5. Handheld XRF analyzer

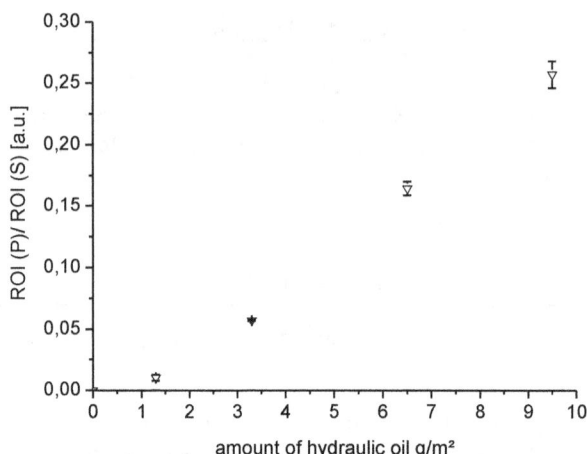

Figure 7. Relation of phosphorus and sulphur signal measured by means of XRF over amount of contaminant on CFRP surface.

The samples contaminated with runway de-icing fluid had joint strength of only 56% of initial strength with adhesive failure (see Fig. 4c). Because these contaminants (hydraulic oil, de-icing fluid) have a huge influence on bond strength they have been chosen for further testing of measuring techniques.

3. Surface Monitoring of CFRP surfaces

3.1 Detection of Skydrol by means of X-ray fluorescence analysis (XRF)

In order to detect hydraulic oil contamination on CFRP surfaces a handheld X-ray fluorescence analyzer was applied (see Fig. 5).

An X-ray excitation source is used to irradiate the sample with resulting fluorescence of the atoms in it. The characteristic X-ray emission is detected and analyzed.

Because the emitted X-rays are characteristic for an atom, a qualitative identification of elements is provided. In order to have a basis for quantitative analysis of elements, the intensities of X-rays from an unknown to those of known standard need to be compared (Kalnicky and Signghvi, 2001).

In the acquired X-ray spectra the intensity is plotted versus the energy of the emitted photons. The position of the peak leads to the identification of the elements present in the sample.

XRF is a fast measuring technique and non-destructive to samples. In order to prevent a high influence of ambient air in the measuring head, nitrogen is used as a floating gas.

In Fig. 6 a spectrum of a clean CFRP reference sample and a hydraulic oil contaminated sample is shown. The element peaks of sulfur and chlorine result from the CFRP resin. The argon peak is caused by residual ambient air between sample and measuring head. As the hydraulic oil is composed of phosphoric acid esters, phosphorus is the element that represents the hydraulic oil. It is possible to differentiate between the reference and the hydraulic oil contaminated sample by means of the intensity of the phosphor peak.

As XRF has a very high information depth of several millimeters it was tested if this method is able to detect contaminants, that only form a thin film on the surface. Selected amounts of surface contaminations were prepared by wiping the hydraulic oil on the surface. The amount of contamination on surface was evaluated by gravimetric measurements. Each surface state was meas-

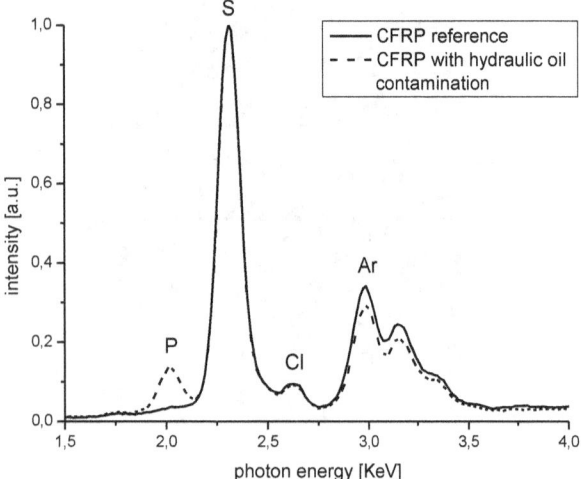

Figure 6. XRF spectra of reference and contaminated samples, measured by XRF

Figure 8. Handheld FTIR spectrometer

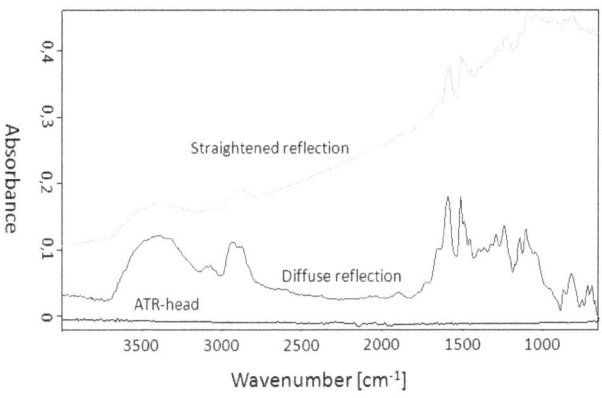

Figure 9. IR spectra of CFRP surfaces measured with different measuring heads

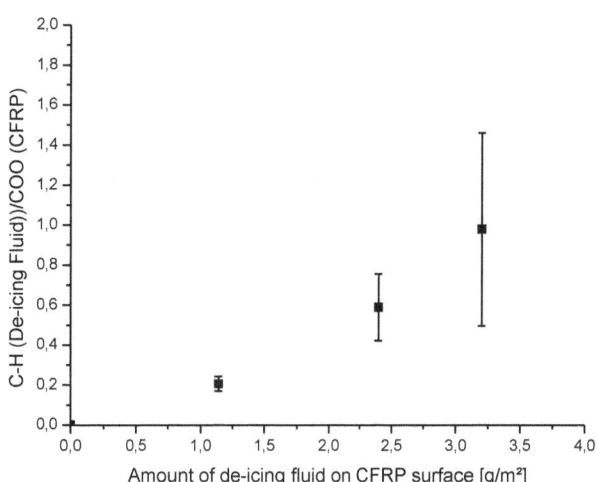

Figure 11. Ratio of hydro carbonyl band (de-icing fluid) and carbonyl band (CFRP) measured by FTIR spectroscopy

ured by XRF at 5 different positions with a measuring time of 30 seconds.

In order to quantify the amount of contaminant on the surface, the relation between the signals of the sulphur from the CFRP was compared to the signal of the phosphor (hydraulic oil). The intensity of the two peaks was in each case integrated over a defined region of interest (ROI). The ratio of the integrated signal intensities of phosphor and sulphur in the ROI were plotted against the amount of hydraulic oil on the specimens in order to calibrate the measuring technique (see Fig. 7). The ratio of phosphor to sulphur increased with the amount of contaminant on the surface.

It is possible to detect a thin film of contaminant on the CFRP surfaces by means of XRF. The detection limit for the detection of hydraulic oil by means of XRF was estimated to be 500 mg/m² (45 mg/m² phosphorus) assumed a signal to noise ratio of 3:1 as significant. Additional tests have shown that this amount of contaminant does not have any influence on the bond strength.

3.2 Detection of Runway-De-icing Fluid by means of FTIR

A handheld Fourier Transform Infrared Spectrometer (FTIR) (see Fig. 8) was used in order to detect runway de-icing fluid on CFRP surfaces.

An infrared light source in the measuring device illuminates the material surface. Infrared radiation inserts the material and is absorbed only at frequencies, corresponding to molecules modes of vibration. For the material an absorption spectra is determined by a sensor and plotted versus the frequency (Koenig, 1992). By evaluation of

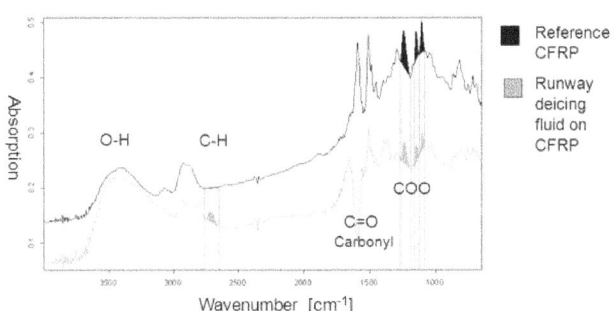

Figure 10. IR spectra of a clean and a runway de-icing fluid contaminated CFRP surface

absorption spectra, the molecule composition of material can be determined.

In order to detect de-icing fluid on CFRP surfaces it was necessary to define the most suitable FTIR measuring principle for the analysis of intense absorbing CFRP material. An internal reflectance (ATR), an external reflectance (RA) as well as a diffuse reflectance (DRIFT) measuring head were tested. Fig. 9 shows the spectra of CFRP surfaces measured by different measuring heads.

It is obvious, that only by diffuse reflection useful spectra from CFRP surface are obtained. Therefore the diffuse reflectance measuring head was used for further detection of contaminations on CFRP surfaces. The scarfed surfaces are too rough for external and internal reflectance measurements.

In Fig. 10 spectra of a clean and a CFRP surface contaminated by runway de-icing fluid are shown. The contaminated sample shows a distinct increase of hydrocarbon band centered at 2710 cm[-1]. The runway de-icing fluid is a mixture of potassium formate and water. The increase of the hydrocarbon band results from potassium format that builds a solid film on the adhesive surface. In order to quantify the amount of deicing fluid, different defined surface with different amounts of deicing fluid on CFRP surfaces were measured. Each sample was measured at 5 different positions with 512 spectra at a resolution of 8 cm[-1]. The surface states were prepared by wiping the contaminant on the surface with gravimetric evaluation and drying for 1 hour at 80°C in oven before measuring. In order to quantify the amount of contaminant on surface the hydrocarbonyl band (de-icing fluid) and the carbonyl-band centered at 1239 cm[-1] (CFRP) were integrated. The ratio of both integrated band was plotted against the amount of contaminant (see Fig. 11).

It is obvious that there is an increase of bands of contaminant and CFRP resin. Additionally for higher amounts of contamination, an increase of standard deviation was determined. This effect results from the bad wetting behaviour of de-icing fluid on the surface. With higher amounts, an inhomogeneous distribution of contaminant on the surface was achieved. The detection limit of runway de-icing fluid on CFRP surface was estimated to be 450 mg/m² (150 mg/m² Potassiumformate).

Bond tests have shown that this amount of contaminant still shows a huge reduction of bond strength.

4. Summary

Due to the implementation of CFRP structures in aircrafts it is important to have a reliable repair techniques. Although bonded repair has several advantages, it is necessary to monitor adherent surfaces before bonding. Contaminants might reduce the bond strength and degrade the bonded structure. It was shown that contaminants like hydraulic oil as well as de-icing fluid result in a huge reduction of bond strength. That´s why it is necessary to control the properties of the adhesive surfaces by surface monitoring. There are several innovating measuring techniques which might be useful for the monitoring of surfaces in field of repair. The two handheld measuring techniques (X-ray fluorescence analysis, FTIR-spectroscopy) provide a high level of development and have been tested regarding the detection of contaminants.

The detection of hydraulic oil was evaluated by means of X-ray fluorescence analysis. Different concentrations of contaminants on the surface were measured and a detection limit was determined at 500 mg/m² of hydraulic oil on the surface. Additional tests have shown that this amount of contaminant do not have an influence on bond strength.

The detection of the contamination de-icing fluid was tested by FTIR-spectroscopy. In the first step the most suitable measuring head was chosen to get a useful spectrum of scarfed CFRP surfaces. Different concentrations of contaminants were measured on the CFRP surfaces by means of diffuse reflectance FTIR measuring head and a detection limit of 450 mg/m² has been determined. Additional bond tests have shown that this amount of contaminant leads to a significant reduction of bond strength.

In contrast the X-ray fluorescence analysis is a technique to detect the contaminant hydraulic oil Skydrol on the CFRP surface and is promising for the implementation in repair process of CFRP structures. The IR-spectroscopy is not sensitive enough to detect de-icing fluid and is therefore not useful to detect this kind of contaminant in field of bonded repair.

5. Acknowledgments

The research leading to these results has received partial funding from the European Union's Seventh Framework Programme (FP7/2007-2013) under grant agreement n° ACP0-GA-2010-266226 (ENCOMB, Extended Non-Destructive Testing of Composite Bonds).

References

Ahn S. and Springer, G. (2000). *Repair of Composite Laminates*. Technical Report. DOT/FAA/AR-00/46.

Baker A. A., Rose, L. R. F. and Jones, R. (2002). *Advances in the Bonded Composite Repair of Metallic Aircraft Structures*. Oxford: Elsevier.

Composite Materials Handbook-MIL 17 (2002). Vol. 3, US Department of Defence.

Cremers, D. A. and Radziemski, L. J. (2006). *Handbook of Laser-Induced Breakdown Spectroscopy*. New York: Wiley.

Kalnicky, D. J. and Singhvi, R. (2001). Field portable XRF analysis of environment samples, *Journal of Hazardous Materials*, Vol. 83, No 1 -2, pp. 93-122.

Koenig, J. L. (1992). *Spectroscopy of Polymers*. ACS Professional Reference Book. Washington: American Chemical Society.

Vadillo, J. M., Palanco, S., Romero, M. D. and Laserna, J. J. (1996). Applications of laser-induced breakdown spectrometry (LIBS) in surface analysis, *Fresenius Journal of Analytical Chemistry*, Vol. 355, No 7-8, pp. 909-912.

Wetzel, M., Rieck, T. and Holtmannsptter, J. (2011). Contamination in adhesive bonding for aviation applications. Detection of effect of adhesion limiting contaminations, *Adhesion: Adhesive and Sealants*, Vol. 3. pp. 29-33.

A Qualitative Visual Analysis of the Fractured Surfaces of Epoxy/Carbon Fibre Composite Prepared by the Melt and the Solution Technologies

Islam Alexandrov, Galina Malysheva, Tatiana Guzeva

Rocket and Spacecraft Composite Structures Department, Faculty of Special Machinery, Bauman Moscow State Technical University, 5 2nd Baumanskaya Street, Moscow, 105005, Russia

Abstract: This study is the investigation of the adhesion properties of epoxy matrix to carbon fibre for quasi-static loadings. The shear adhesion strength can be served as a measure of the adhesion. Two technologies of impregnation were employed: the solution technology (matrix was introduced into the solvent) and the melt technology (the composition was heated to reduce the matrix viscosity). Results of microstructural analysis of fracture surface in micro-carbon-fibre-reinforced polymers have been presented. It was found that the damage varies according to the matrix thickness. The effect of the thickness of the matrix on the elastic-strength characteristics of carbon fibre was determined. They were compared with the obtained theoretical and experimental values. Examples of real construction products made from carbon-reinforced plastics using melt and the solution technologies are presented.

Key Words: Polymer-matrix composites, Melt technology, Solution technology, Fracture surfaces.

1. Introduction

Polymer composites are widely used for lightweight structures (aircrafts, sporting goods, wheel chairs, etc.), in addition to vibration damping, electronic enclosures, asphalt (composite with pitch, a polymer, as the matrix), solder replacement, etc. Polymer composites can be tailored for various properties by appropriately choosing their components, proportions, distributions, morphologies, degrees of crystallinity, crystallographic textures, as well as the structure and composition of the interface between components (Deborah et al., 2010; Bratukhin et al., 2007; Komkov et al., 2011).

At present, due to the ever-increasing loads on new design equipment, the research on fracture mechanisms of polymer composites are quite significant and have great practical importance. The purpose of this work is to investigate structure of carbon fibres samples impregnated with epoxy matrix after tensile tests.

2. Methods of research

Carbon fibre (UKN-2500) was impregnated with room temperature curing epoxy matrix. Epoxy resin with epoxy groups 22 and hardener – hexamethylenediamine was used as the matrix. These samples are the simplest type of polymer composite materials samples. Samples were unidirectional composite with continuous fibres.

In the present work we have used two technologies for impregnation. They are widely used in Russia in the prepregs manufacturing because they are easy to process and very cheap. The first technology is conventionally called the solution technology and the other is known as the melt technology. The important step in solution technology is to inject solvent (consisted of ethanol and acetone mixture) in the matrix after resin and hardener mixing. The concentration of the matrix was 55% by weight. The resulting mixture was stirred until a homogeneous solution viscosity was achieved. In the melt technology, no solvent

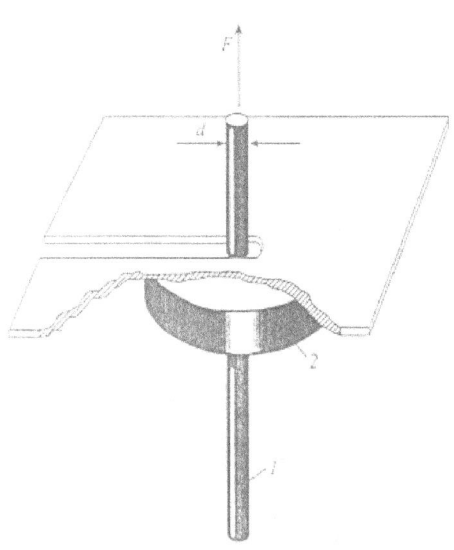

Figure 1. Schematics of determining the shear adhesive strength in compounds: 1 – fibre, 2 – epoxy matrix.

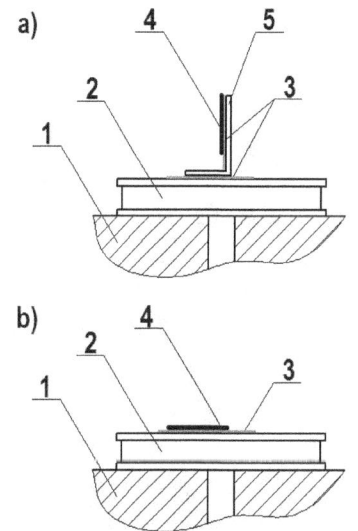

Figure 2. Schematics of samples fixation on microscope substrate:
a) vertical; b) horizontal.
1 – cartridge box, 2 – substrate, 3 – carbon double side sticky tape,
4 – sample, 5 – device for vertical fixing

was used. After mixing hardener with the resin, it was heated to 70°C. In the melt technology the solvent was not used. After hardener's injection to the matrix it was warmed by using special equipment up to the temperature +70°C.

After manufacturing, samples were tested in static tensile testing machine "TEXTECHNO" STATIGRAPH M. The displacement rate was set to 2.5 mm/min. The stretching was carried out along the fibre. Special samples were produced to evaluate the matrix adhesion (Fig. 1). The shear adhesion strength's which determined by pulling fibres from the cured resin layer (the pull-out method). It was used to be served as a measure of the adhesion (Bazhenov, et al., 2010; Gorbatkina, et al., 1987).

The distinctive feature of this method is the large scatter of the thickness of the layer matrix on fibre which is connected with large differences in theirs viscosity. The thickness of the matrix layer which was produced by us-

ing the melt technology was 1-2mm; by using the solution technology was 0.1-0.3 mm.

The value of failure load was determined on the adhesiometer (micro-tensile testing machine). It includes a system of loading, force-measuring device, and samples-fixing device. Load measurement accuracy is ±1%. All tests were carried out at room temperature.

Adhesive strength of each sample was calculated by the formula:

$$\tau_i = \frac{F_i}{S_i}$$

were F_i is load required for pulling out the fibre from the cured matrix layer; S_i is area of the compound which is equal to $S_i = \pi d_i l_i$, d_i is diameter of the fibre, l_i is thickness of the matrix layer on the fibre.

The surface of the fractured samples were analysed by electron microscope Phenom which captures images in the resolution less than 50 nm. The samples were fixed

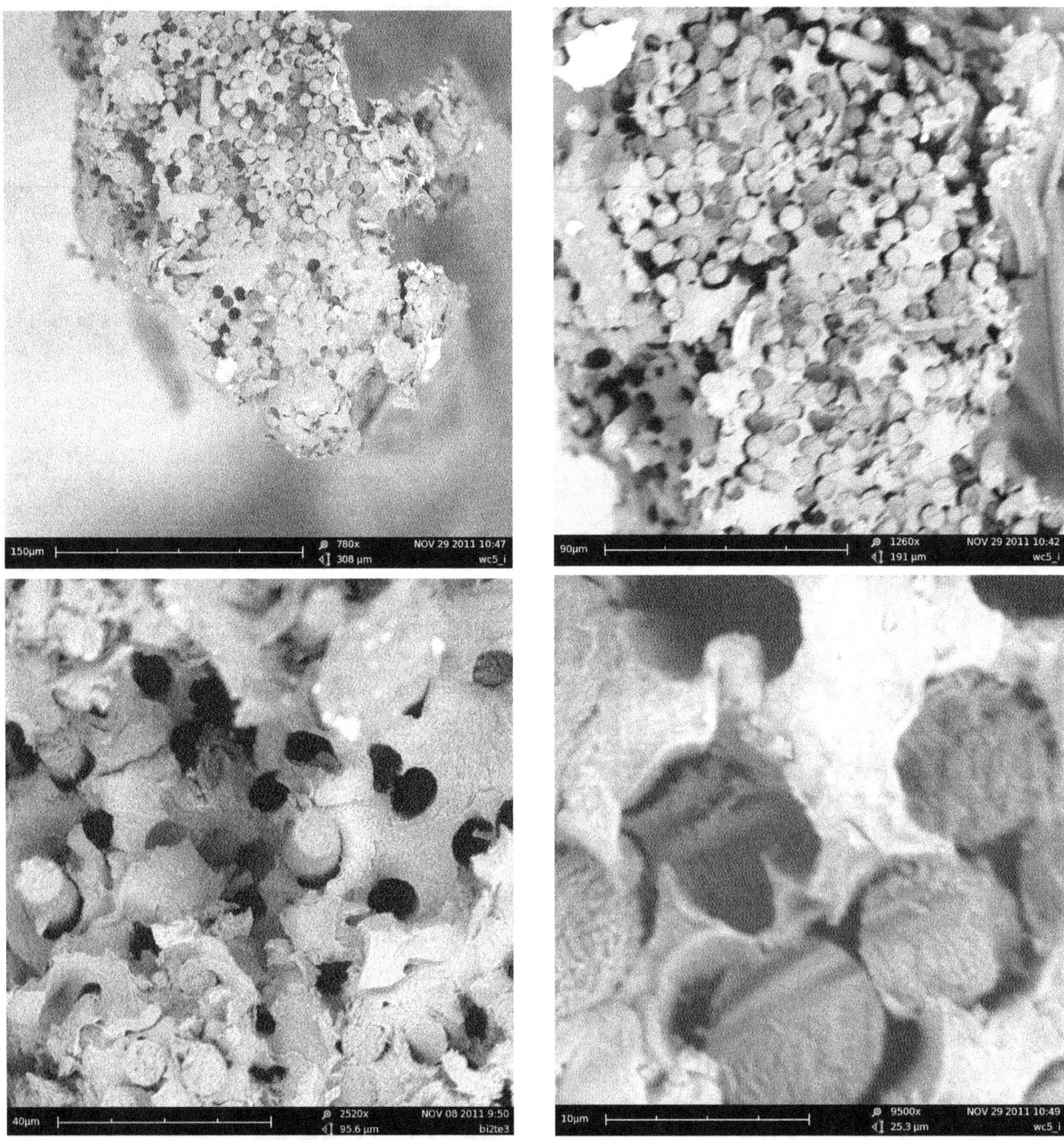

Figure 3. Fractured surfaces of the carbon-fibre-reinforced polymer produced by melt technology.

horizontally and vertically on a special substrate (Fig. 2). It allowed studying the structure in two dimensions.

The samples were fixed on the substrate surface with double-sided carbon sticky tape. Afterwards, substrates were sequentially mounted in a special cartridge-box which then was placed under a microscope.

3. Theoretical analysis

The deformation of the fibres ε_f and matrix ε_m takes place during the deformation of micro-plastic ε

$$\varepsilon = \varepsilon_f = \varepsilon_m \qquad (1)$$

However, equation (1) is only valid if the matrix has corresponding deformation characteristics with available thicknesses. The dependence of the adhesion strength on the matrix's thickness is shown in Table 1. The thinner (up to certain limits) the matrix's layer thickness, the higher the value of adhesion strength between the compo-

Table 1
The results of measurements of adhesive strength
by the pull-out method

Data	The matrix preparation technology	
	The melt technology	The solution technology
The value of adhesion strength, MPa, depending on the area of contact matrix with the fibre, mm^2		
0.2	82	67
0.4	65	58
0.6	56	51
0.8	47	45
1.0	38	36

nents and vice versa. During the manufacturing of PMC products by computation method with using the prepregs, the consistency of matrix in PMC usually is about 30-35%. If a micro-plastic sample does not satisfy the equa-

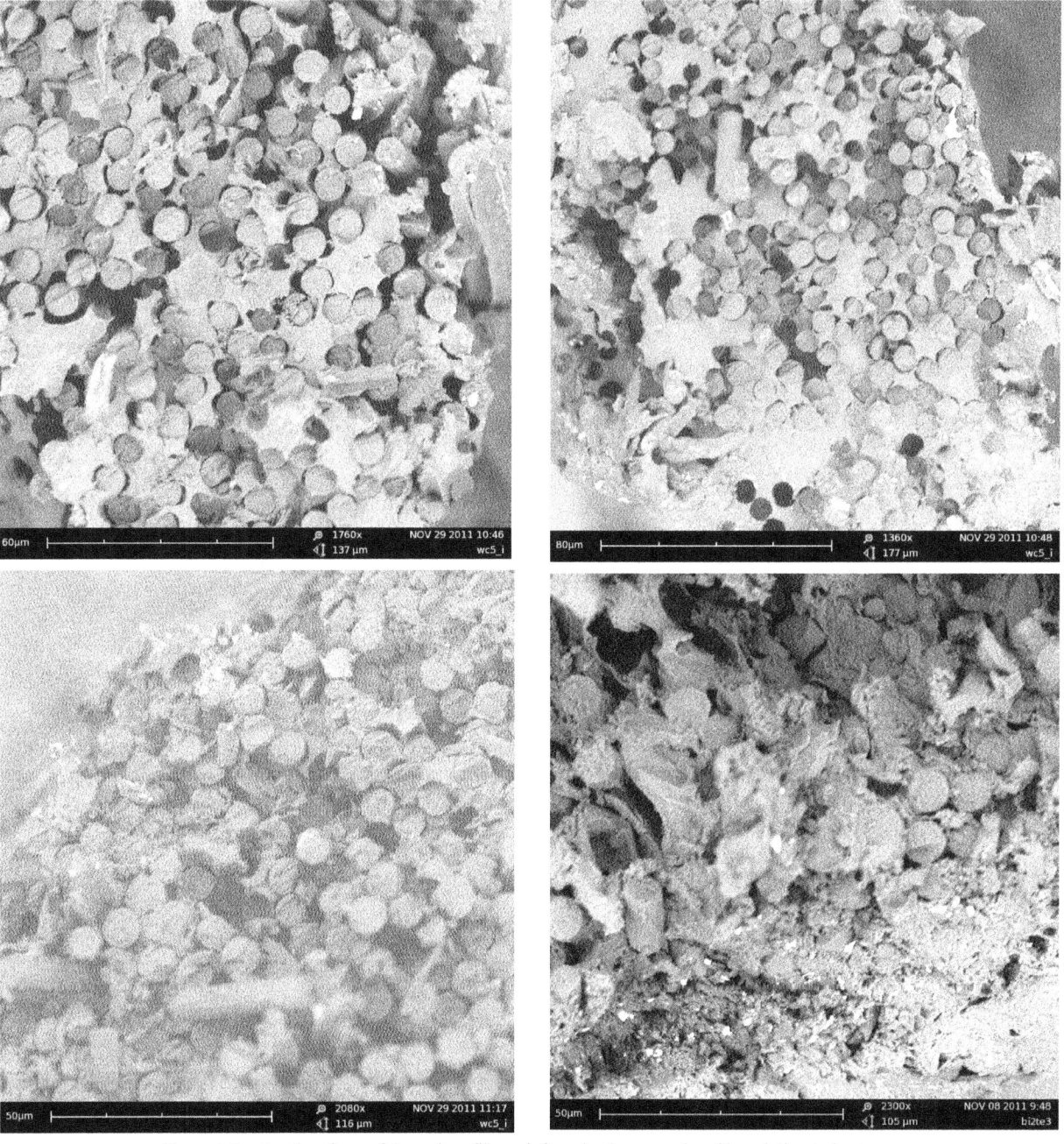

Figure 4. Fractured surfaces of the carbon-fibre-reinforced polymer produced by solution technology.

tion (1), it is necessary to use different technologies for manufacturing, which would reduce the thickness of the matrix.

The load applied to the micro-plastic will be distributed between its components according to equation

$$\sigma V = \sigma_f V_f + \sigma_m V_m \qquad (2)$$

where σ, σ_f, σ_m – stress in micro-plastic, fibre and matrix; V, V_f, V_m – micro-plastic, fibre and matrix volume fraction.

The elastic modules of fibre E_f and matrix E_m will be

$$\frac{\sigma_m}{\sigma_f} = \frac{E_m}{E_f} \qquad (3)$$

Substituting (3) in (2)

$$\sigma V = \sigma_f \left(V_f + \frac{E_m}{E_f} V_f \right) \qquad (4)$$

In the direction of micro-plastic reinforcement modulus of elasticity E can be determined by the additivity rule as follows:

$$E = E_f V_f + E_m V_m = E_f V_f + E_m \left(1 - V_f\right) \qquad (5)$$

Equation (5) is valid if the tension of alternating components (i.e., the matrix and mono-filament) is loaded sequentially. The equations obtained allow to determine approximate elastic characteristics of the micro-plastic. However, the stress of the matrix and the fibres are not equal. Therefore, there will be a slippage between the mono-filaments and the matrix at practice.

4. Results and discussions

Production of samples for the study of adhesion strength in the matrix-fibre is quite laborious. However, carefully selected mode of compounds formation allows receiving well-formed interface and providing reproducibility of measurements of the adhesive strength results. The results are shown in Table 1.

Table 2 shows the calculated and experimental data on the mechanical characteristics of the obtained carbon plastics.

Processing at different structural levels (molecular, submicro-, micro-and macro-levels) is the main issue for studying fracture of any material.

Results of structural analysis show failure and allow us to estimate complexity of factors which are quite random in nature. It includes variations for the factors like laying monofilaments in yarn, the thickness of matrix between monofilament, monofilaments' and matrix's defects (Fig. 3, Fig. 4). Therefore, the obtained results of structural analysis suggest the following conclusions:

a) Monofilaments of carbon fibre had almost no damaged after being used (i.e., ripped). Thus, a high quality product can be obtained .

b) All monofilaments have regular (round) shapes, their diameters are close to each other i.e. 8 μm. Such form of monofilaments are required when using plastics for the evaluation of the stress-strain state of PMC structure.

c) Amount of matrix located between monofilament varies considerably.

Table 2
The mechanical characteristics of carbon plastics

Data	Value
The initial values of the composite material components	
Modulus of elasticity, GPa - carbon fibre - epoxy matrix	210 6
Tensile strength, GPa - carbon fibre - epoxy matrix	2.5 0.05
The theoretical values of the composite material characteristics	
Modulus of elasticity, GPa, at the content of the matrix,% 50 40 30	108.0 128.4 148.8
Tensile strength, GPa, at the content of the matrix,% 50 40 30	1.275 1.520 1.765
The experimental values of the composite material characteristics	
Modulus of elasticity, GPa, the fibre content of 69.6% (the melt technology)	185
Tensile strength, GPA, with 76% fibre content (the solution technology)	0.5

Quantitatively, the matrix content in the samples was not determined. However, it should be noted that in melt technology, content of the matrix in PMC is more as compared to that in the solution technology.

Epoxy matrix is brittle (relative elongation of 0.5%) and carbon fibres has a good ductility (relative elongation of 2.5%). Therefore, the micro-plastic will break at the time when the deformation of unidirectional composite becomes equal to the failure strain of the monofilament. It is brittle fracture (there is no localization of micro-damage that would occur at a higher deformability of matrix). Fractured monofilaments and the surfaces with breaks are quite visible on the images of fracture surfaces microstructures. These surfaces of discontinuity have clear and straight edges. It can be seen from the images that the polymer matrix is crushed during the failure.

Accordingly, there are three main types of micro-plastic fractures, i.e. in between individual monofilaments, failure at the interface, and the failure of the matrix itself. The fourth type of failures includes all three types together. Our presented results were analysed for micro-plastics manufactured from three different processes: without solvent, with solvent and the melt technologies.

References

Bazhenov, S.L., Berlin, A.A. and Kulkov, A.A. (2010). *Polymer Composite Materials*. Dolgoprudny: Intellect.

Bratukhin, A.G., Pogosyan, M.A., Surov, V.I. and Tarasenko L.V. (2007). *Construction and Functional Materials of Modern Aircraft Building*. Moscow: Moscow Aviation Institute.

Chung, D.D.L. (2010). *Composite Materials: Science and Applications*. London: Springer-Verlag.

Gorbatkina, Y.A. (1987). *Adhesive Strength in Polymer-Fibre Systems*. Moscow: Khimiya.

Komkov, M.A. and Tarasov, V.A. (2011). *Winding Technology of Composite Structures, and Missile Weapons: Textbook*. Moscow: Bauman Moscow State Technical University.

Effect of Grits Size on Machining Quality in Drilling Carbon Fiber Reinforced Plastics with Brazed Diamond Core Drill

Yan Chen, Yucan Fu, Jiuhua Xu, Juan Mu, Weifeng Wang

College of Mechanical and Electrical Engineering, Nanjing University of Aeronautics and Astronautics, 29 Yudao Street, Nanjing, 210016, P.R.China

Abstract: To solve the drilling-induced problems, such as poor machining quality and easy tool wear during drilling carbon fiber reinforced plastics (CFRP), the brazed diamond grits core drill has been developed and tested in this paper. The influences of grits size and set style on thrust force, the dimensional tolerance of drilled holes are investigated. The results showed that the maximum of thrust force existed when the diamond grits were set disorderly. When the grits size of diamond with set orderly minimized, the thrust force first kept stable and then increased. The dimensional tolerance, roundness and roughness decreased with the decrease of grits size.

Key Words: Brazed diamond core drill, CFRP, Machining quality, Thrust force.

1. Introduction

In 2011, China has stepped forward to convince airlines that it can make safe aircraft after Boeing and Airbus. Zhou (2011) argued that the large passenger aircraft programs are being implemented and composites on the C919 (Chinese trunk liner) account for 20 percent of the aircraft's structural weight. Furthermore, this is the first time that composites have been used in primary structures in Chinese commercial aircraft to change the structures to be lighter and more efficient like the Boeing B787 and the Airbus A380.

Most composite parts in aircraft are generally made as near-net-shape parts, but a composite part needs to be assembled to other parts in most cases. So, drilling is often necessary to assemble different parts. In conventional machining, the drilling by twist drill is the most applied method though as much as 40% for all materials removal processes (Brinksmeier, 1990). Twist drill with a long, noncutting chisel edge and the rake angle decreases toward the drill's center compromises that limit its performance in generating hole for composite materials (Tsao, 2008). Thus, during the process of drilling the composite laminates, there always exist some problems, such as delamination, spalling, edge chipping and so on. Especially, the delamination would decrease the strength and stiffness of the composite laminates, so it reduces the load carrying capacity. In order to solve the drilling-induced delamination, most of researches are concentrated on geometry and material of drill, drilling parameters (Lin, 1996; Chen, 1997; Enemuoh, 2001; Hocheng and Tsao, 2003; Hocheng and Tsao, 2005; Davim, 2007; Gaitonde, 2008; Rawat, 2009). The research by Hocheng and Tsao (2006) concluded that core drill allows for the largest drilling feed rate among varied drills to avoid delamination. The cutting end of conventional core drill is made of diamond grits by electroplated process. It also can utilize the best abrasive resistance of diamond to prolong the tool's life. However, little study was devoted to core drill by other manufacturing process in machining composites.

Compared with the electroplated tools, the brazed diamond tools have higher crystal exposure as shown in Fig. 1 (Hwang, 2000; Shi, 2003; Webster, 2004) and better grit retention because of the joining of chemical nature

among the diamond grits, brazing alloy and matrix. The diamond grits of conventional diamond grinding tools are disorderly set. However, the diamond grits of brazed diamond tool can be orderly set according to the demands of machining, i.e. the distance of diamond grits keeps constant. Thus, the brazed diamond tool has more chip clearance space and higher resistance of grits falling-off during drilling composites so that it has the advantages of high sharpness, grains utilization ratio and longevity. And it has already been proved by A. Trenker and H. Seidemann (2002) as shown in Fig.1 (c).

The main objective of the present work focuses on the relationship between machining quality and the diamond grits size of brazed diamond core drills. In the present study, the influences of grits size and set style on thrust force, the dimensional tolerance of drilled holes are investigated in order to drill damage-free and precise holes.

2. Experimental procedure

The CFRP laminates are drilled in this study. The laminates were fabricated from the flat woven carbon/epoxy of T300 fibers. The laminates were cut into 60mm×100mm using diamond-edged saw to fit the clamp. 15 plies make the plate thickness 2.4 mm. The fiber volume fraction is 60%. As shown in Fig. 2, the ends of the core drills were brazed with diamond grits, and the three kinds of size of 40/45 mesh (355~425μm), 50/60 mesh (250~300μm) and 80/100 mesh (150~180μm) were used, respectively. To validate the influence of grits set styles on the drilling CFRP, two set styles (orderly-set and disorderly-set) were adopted for the brazed core drill with 80/100 mesh diamond grits. The orderly-set diamond grits realized by the self-developing jigs and fixtures that the diamond grits could still keep orderly set after brazing. The brazing alloy was Ag-Cu-Ti and the diameter of core drills is 6 mm. Brazing operation was carried out in vacuum brazing. The system was maintained at a vacuum below 1×10^{-2} Pa at the brazing temperature (900~920°C) and held for certain time (8~10min). Then the specimen was furnace-cooled after brazing completed.

The drilling test were carried out on a HG410J engraving and milling machine as shown in Fig. 3. The engraving and milling machine has maximum spindle speed of

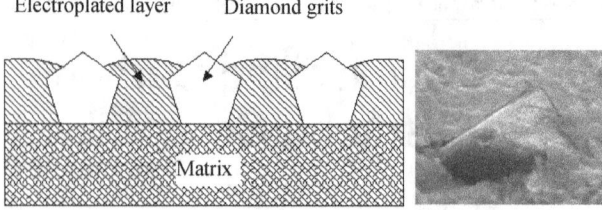

Electroplated layer   Diamond grits

Matrix

a) Topography of electroplated diamond grits tool

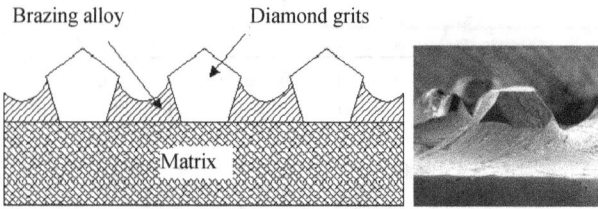

Brazing alloy   Diamond grits

Matrix

b) Topography of brazed diamond grits tool

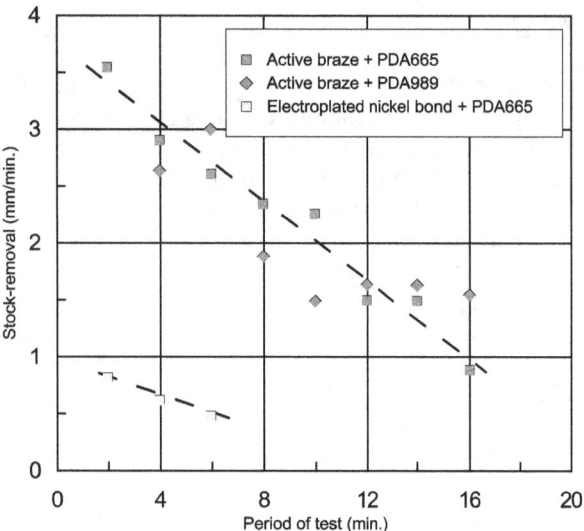

Active braze + PDA665
Active braze + PDA989
Electroplated nickel bond + PDA665

(c) Comparison of stock-removal with electroplated and brazed grinders

Figure 1. Topography of two kinds of diamond grits tools

24 000rpm and maximum feed speed of 5 m/min. To utilize the dynamic sharp of diamond grits, all tests were run without coolant at spindle speeds of 15,000 rpm and feed speed of 20, 60, and 150 mm/min, respectively. The test of different parameter combinations were replicated three times. The thrust force during drilling process was meas-

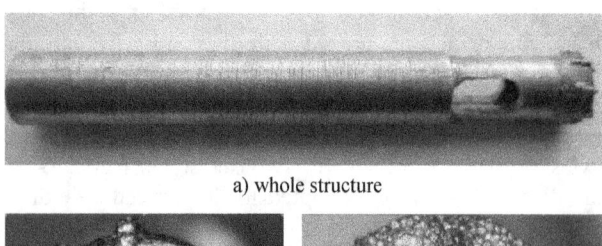

a) whole structure

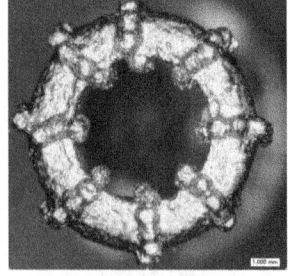

b) orderly-set core drill end   c) disorderly-set core drill end

Figure 2. Photograph for brazed diamond core drill.

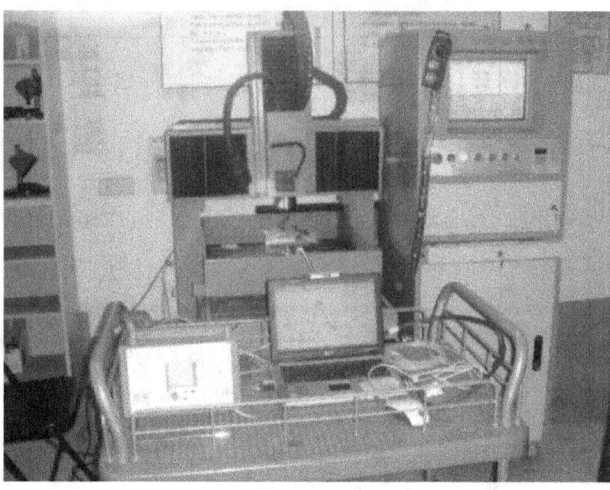

Figure 3. Experimental setup.

ured with a data acquisition system composed of a 9272 Kistler dynamometer and a 5070 A Kistler amplifier.

After drilling, the diameters and roundness of the drilled holes were measured by a numerical type all-powerful tool microscope JGW-S. The images of entry and exit of holes were observed by a digital microscope HIROX KH-7700 and roughness of the hole wall was measured by a roughness measure instrument Perthometer M1, while the surface quality of the hole wall was observed through scanning electron microscope (SEM).

3. Experimental results and discussion

3.1 Thrust force in the drilling process

Due to the signal variations during drill rotation, the thrust force values were averaged over one spindle revolution. Also, to reduce the influence of outlier values, the final results used were the average of three experiments run under identical conditions.

The influences of the diamond grits size and set styles on thrust force were illustrated in Fig. 4. A clear trend was found regarding the effect of feed speed independently of the grits size or set style as same as most researches. The set style of diamond grits has an important influence on the thrust force. At any feed speed, the disorderly-set diamond grits core drill has the maximum thrust force. The thrust force of brazed diamond core drill with

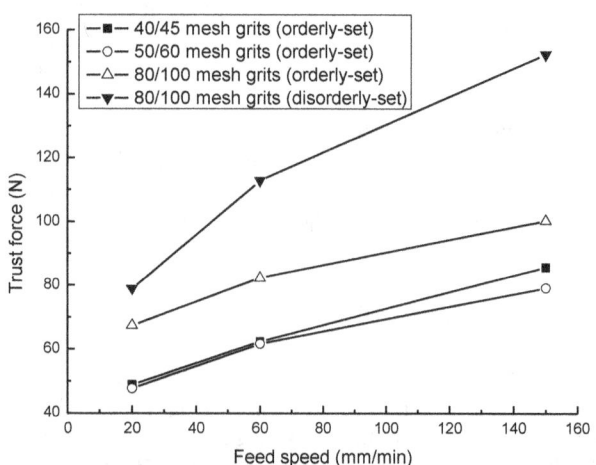

40/45 mesh grits (orderly-set)
50/60 mesh grits (orderly-set)
80/100 mesh grits (orderly-set)
80/100 mesh grits (disorderly-set)

Figure 4. The main effects of grits size and set style on thrust force.

80/100 mesh grits is larger than those with 50/60 and 40/45 mesh grits when the grits are set orderly. Although 50/60 mesh grits is finer than 40/45 mesh grits, the former thrust force is roughly equivalent to latter thrust force.

During the drilling process with diamond core drill, chips are accumulated in the intergrit space. They are composed of small CFRP aggregates. Indeed, the smaller the grit size is, the lower the intergrit space is. And when the grits are set disorderly, the intergrit space is lower because of higher grits density. So, chips do not have enough space between grits to lodge before being evacuated. Mechanism of cut is deteriorated and the thrust force increases, especially used the disorderly-set grits core drill. When the diamond grits are orderly set, the CFRP chip can flow through the intergrit space. So, the friction between the chip and tool could be greatly reduced. Consequently, grist size must be chosen considering intergrit space in order to avoid tool saturation. Considering experimental results, there is enough space to lodge chips in the brazed diamond core drills with 80/100 mesh grits. This value depends also on other factors such as material bulking.

In fact, the traditional electroplated diamond core drill is seldom used in factory. This is partly due to the higher thrust force than that used twist drill. But more important is the disorderly-set grits of electroplated diamond core drill whose intergrit space is so low that results in strong friction among chip, tool and workpiece material. Even worse, the heat by the friction may burn matrix.

According to the experimental result, when the diamond grits were set orderly, the intergrit space was enough to lodge chips. Therefore the thrust force decreased so greatly that avoid matrix burn due to reduction of friction among chip, tool and workpiece material. So, the use of brazed diamond core drill in factory may become easy.

3.2 Topography of the hole at the entry and exit

Delamination is the typical damage encountered during the drilling of CFRP laminates due to the relatively poor strength in the thickness direction. Such damage is generally initiated when the thrust force exceeds a threshold value at critical stages, e.g., at the entry and exit of the drill bit. The delamination at the entry side of the tool was peel-up style and delamination at the exit side was push-out style (as shown in Fig. 5) used twist drill (Schulze et al., 2011). But the direction of axial force was down, so

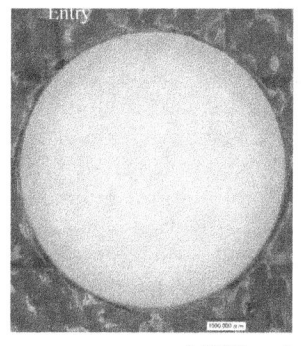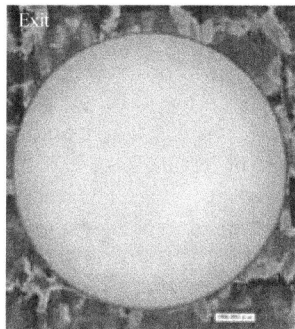

a) 40/45 mesh grits (orderly-set)

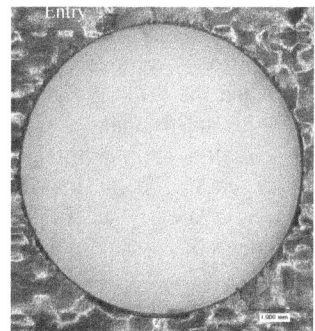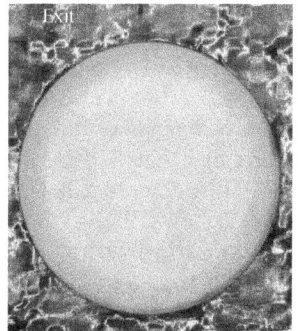

b) 50/60 mesh grits (orderly-set)

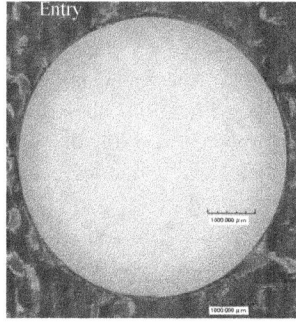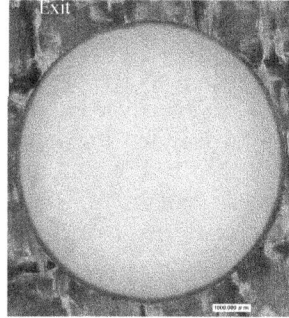

c) 80/100 mesh grits (orderly-set)

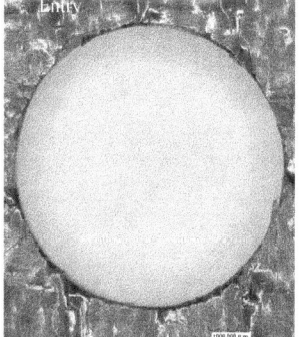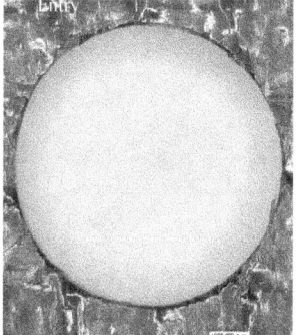

d) 80/100 mesh grits (disorderly-set)

Figure 7. Topography of hole at the entry and exits (spindle speed: 15000 rpm, feed speed: 150 mm/min).

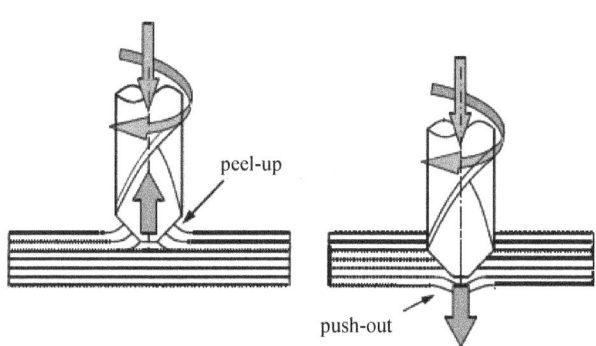

Figure 5. Schematic of delamination mechanisms with twist drill.

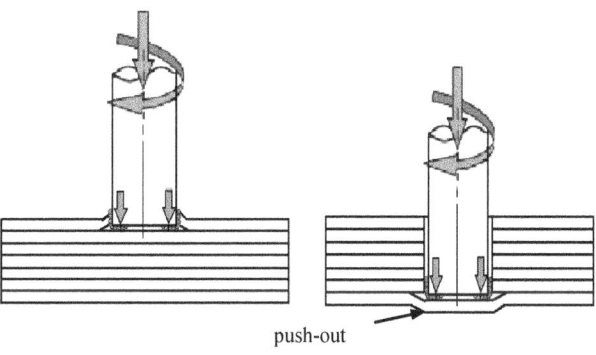

Figure 6. Schematic of delamination mechanisms with core drill.

the peel-up delamintion used core drill at the entry side seldom occurred (as shown in Fig. 6). Therefore, the delamination factor was smaller.

On the other side, at the exit side, the direction of axial force was also down, so the push-out delamintion used core drill at the exit side occurred like twist drill. Due to the differences of cutting mechanism and structure between twist drill and core drill, the delamintion at the entry hardly occurs. Delamination at the exit didn't occur by examination through ultrasonic C-Scan except the disorderly-set core drills used, so the delamination factor wasn't discussed in the present study.

To examine the other typical defects such as spalling, burr and chipping at the entrance and exit, the topography of hole was observed. The drilling quality mainly depends on the thrust force and thrust force increases with the increase of feed speed. So, the topographies with the varied core drills under the highest feed speed were mainly illustrated in Fig. 7. The qualities of holes drilled by brazed diamond core drill with orderly-set grits were better than that with disorderly-set 80/100 mesh grits under same cutting parameters. And from the topography, the former holes at the entry and the exit had no evident defects. But there existed spalling, matrix burn and burr at the entry and the exit of hole drilled by brazed diamond core drill with disorderly-set 80/100 mesh grits.

As above mentioned, the thrust force was biggest used the disorderly-set grits core drill, and the drilling quality mainly depends on the thrust force. So, the quality was worse than that used orderly-set grits core drills. Although the thrust force used the orderly-set 80/100 mesh grits core drill was bigger than that used orderly-set 40/45, 50/60 mesh grits core drill, the topography was similar to the latter topography. This reason may be that the thrust force used the orderly-set 80/100 mesh grits core drill didn't exceed the threshold value.

3.3 Other quality measurement

It is important to achieve dimensional tolerances and surface quality required by technical specifications because it may reduce the composite performance without desired assembly precision. So, the diameter, roundness and roughness of drilled hole were measured in order to attain damage-free and precise holes as shown in Tab. 1. As above mentioned, the final results used also were the average of three experiments run under identical conditions.

The diameters of the brazed diamond core drills were 6 mm, and the diameters of drilled holes were smaller than that of drills, i.e. the holes feature undersize. The diameter deviation increases with the increase of grits size of the

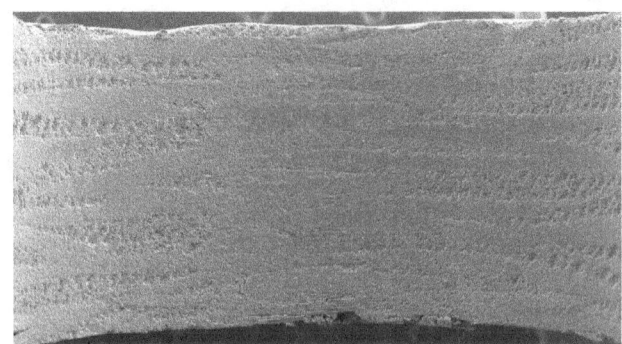

Figure 8. Image of hole wall used brazed core drill.

tool and the decrease of feed speed. This reason is that the cutting process used core drill is shifted from workpiece material separation used twist drill towards material deformation and elastic deformation of material increases compared with material separation. When the tool is drawn back the workpiece material relaxes and returns to its original position. Furthermore, the exposing heights among grits increase with the increase of grits size, so it is not identical in the brazed diamond core drill. To improve the quality, the diamond grits must be carefully chosen or use finer diamond grits such as 80/100 mesh. When the feed speed increase, the drilling time could be reduces. So, the friction-induced heat could be relieved that results in the decrease of diameter deviation. But the measurement indicated that the accuracy to size of drilling holes was able to be guaranteed according to the demand of parts. And with the increase of hole number, the diameter of hole with the brazed core drill hardly changed because of good wear resistance of diamond grits.

It showed the roundness and roughness value of hole wall used brazed diamond core drill increased with the increase of grits size and feed speed; it can not be compared with twist drill applications in metallic material where surface roughness depends essentially on feed. Indeed, each grit creates a mark in the drilled surface that contributes to the machining. The surface topography used brazed diamond core drill was generated by the traces of the individual abrasive particles on the machined surface as shown in Fig. 8. Therefore, dimensions of this mark depend mainly on grits size and feed speed. And each abrasive grit on the brazed diamond core drill acted as a single point cutting edge with a large negative rake angle and a wide edge angle, so the carbon fiber was compressed, i.e. the single fibers are exposed to compressive stress rather than shear stress. After drilled, the elastic recovery of carbon fiber, so the torn fibers are protruding out of the hole surface. Furthermore, the smaller the grits size, the stronger the compressing and rubbing function

Table 1
Measurement values of the drilled holes

Grits size	Set style	Diameter minimum (mm)			Diameter maximum (mm)			Roundness (μm)			Roughness Ra (μm)		
		Feed speed (mm/min)			Feed speed (mm/min)			Feed speed (mm/min)			Feed speed (mm/min)		
		20	60	150	20	60	150	20	60	150	20	60	150
40/45	orderly	5.956	5.962	5.980	5.965	5.993	5.999	35.4	37.1	38.3	2.98	3.56	4.53
50/60	orderly	5.975	5.976	5.979	5.979	5.980	5.986	22.4	28.6	31.7	2.55	3.13	3.97
80/100	orderly	5.982	5.984	5.987	5.985	5.989	5.995	21.2	25.4	28.6	2.24	2.85	3.45
80/100	disorderly	5.985	5.987	5.992	5.987	5.991	5.997	20.4	24.1	27.2	2.25	2.82	3.25

are. To reduce the value of roughness and roundness, the finer diamond grits should be applied and the feed speed should be decreased. That is why grits size must be chosen according to the surface roughness requirements. Moreover, checking tools before machining is important. Indeed, impact of the grits quality on the generated surface is very important.

Although the roundness and roughness was smaller used the disorderly-set 80/100 mesh grits core drill, the matrix burn was found on the surface of whole hole. That might result in the decrease of CFRP performance. Concerning choice of grit size and set style, cautions must be taken in order to check that glass transition temperature is not reached in order to avoid matrix burn. Under the enough intergrit space, the finer the diamond grits are, the better drilling quality is.

4. Conclusions

Based on the experimental results obtained from the machining quality after drilling CFRP, the following conclusions can be extracted.

(1) The thrust force depended mainly on grits set style during the process of drilling CFRP. The thrust force used the disorderly-set grits core drill was larger than that used orderly-set grits core drill. When the grits size of diamond with set orderly minimized, the thrust force first kept stable and then increased.

(2) There were not found delamination, spalling and burr at the entry and the exit through topography observed when the holes were drilled used orderly-set grits core drill. However, spalling, burr and matrix burn were found at the entry and the exit used the disorderly-set grits core drill.

(3) The diameters of drilled holes were smaller than that of drills, i.e. the holes feature undersize. The diameter tolerance value of hole decreased with the decrease of grits size. And the roundness and roughness value of hole wall increased with the increase of grits size and feed speed.

5. Acknowledgements

This research was supported by the National Nature Science Foundation of China (No. 51075210), State Major Science and Technology Special Projects (No. 2012ZX04003031), and National Basic Research Program of China 973 Program (2009 CB724403).

References

Brinksmeier, E. (1990). Prediction of tool fracture in drilling, *CIRP Annals – Manufacturing Technology*, Vol. 39, No 1, pp. 97-100.

Chen, W.-C. (1997). Some experimental investigations in the drilling of carbon fiber-reinforced plastic(CFRP) composite laminates, *International Journal of Machine Tools and Manufacture*, Vol. 37, No 8, pp. 1097-1108

Davim, J.P., Rubio, J.C. and Abrao, A.M. (2007). A novel approach based on digital image analysis to evaluate the delamination factor after drilling composite laminates, *Composites Science and Technology*, Vol. 67, No 9, pp. 1939-1945

Enemuoh, E.U., El-Gizawy, A.S. and Okafor, A.C. (2001). An approach for development of damage-free drilling of carbon fiber reinforced thermosets, *International Journal of Machine Tools and Manufacture*, Vol. 41, No 12, pp.1795-1814.

Gaitonde, V.N., Karnik, S.R. and Rubio, J.C. (2008). Analysis of parametric influence on delamination in high-speed drilling of carbon fiber reinforced plastic composites, *Journal of Materials Processing Technology*, Vol. 203, No 1-3, pp. 431-438.

Hocheng, H. and Tsao, C.C. (2003). Comprehensive analysis of delamination in drilling of composite materials with various drill bits, *Journal of Materials Processing Technology*, Vol. 140, No 1-3, pp. 335-339.

Hocheng, H. and Tsao, C.C. (2005). The path towards delamination-free drilling of composite materials, *Journal of Materials Processing Technology*, Vol. 167, No 2-3, pp. 251-264.

Hocheng, H., Tsao, C.C. (2006). Effects of special drill bits on drilling-induced delamination of composite materials, *International Journal of Machine Tools and Manufacture*, Vol. 46, No 12-13, pp.1403-1416.

Hwang, T.W., Evans,C.J., Malkin, S. et al. (2000). High speed grinding of silicon nitride with electroplated diamond wheels,Part.1:wear and wheel life, *Journal of Manufacturing Science and Engineering*, Vol. 122, No 2, pp. 32-41.

Lin, S.C. and Chen, I.K. (1996). Drilling carbon fiber-reinforced composite material at high speed, *Wear*, Vol. 194, No 1-2, pp. 56-162.

Rawat, Sanjay, Attia, Helmi. (2009). Characterization of the dry high speed drilling process of woven composites using machinability maps approach. *Manufacturing Technology*, Vol. 58, No 1, pp. 105-108.

Schulze, V., Becke, C., Weidenmann, K. and Dietrich, S. (2011). Machining strategies for hole making in composites with minimal workpiece damage by directing the process forces inwards, *Journal of Materials Processing Technology*, Vol. 211, No 3, pp 329-338.

Shi, Z. and Malkin, S. (2003). An investigation of grinding with electroplated CBN wheels, *Annals of the CIRP*, Vol. 52, No 1, pp. 267-270.

Trenker, A. and Seidemann, H. (2002). High-vacuum brazing of diamond tools, *Industry Diamond Review*, Vol. 62, pp 49-51.

Tsao, C.C. (2008). Investigation into the effects of drilling parameters on delamination by various step-core drills. *Journal of Materials Processing Technology*, Vol. 206, No 1-3, pp. 405-411.

Webster, J. and Tricard, M. (2004). Innovation in abrasive products for precision grinding, *Annals of the CIRP*, Vol. 53, No 2, pp. 597-617.

Zhou, L.-D. (2011). Composite Material Application and Challenge for Commercial Aircraft, in *2011 International Forum on Composite Material Applications for Large Commercial Aircraft*, September 19-20, Shanghai, China.

Rheological Effects in Constrained Flows of Viscous Fluids and the Necessity of Their Modelling in Composite Technologies Optimization

Boris Semenov, Svetlana Rapokhina, Anna Sedykh, Ngo Thanh Binh

Rocket and Spacecraft Composite Structures Department, Faculty of Special Machinery, Bauman Moscow State Technical University, 5 2nd Baumanskaya Street, Moscow, 105005, Russia

Abstract: The fountain effect and die swell phenomena in viscous fluids flows were considered. These effects emerge spontaneously in industrial equipment used for production of composite materials. The experimental results which confirm the presence of these effects and show the importance of their consideration for controllable structure formation are presented. The research was conducted for such technologies as semisolid metal (SSM) billet pressing (or thixoforming), carbon fibre-reinforced rod pultrusion and powder injection molding (PIM) technology. An opportunity to model complex flow in shape-generating equipment is discussed.

Key Words: Composite materials, Die swell phenomena, Fountain effect, Rheology, Viscous fluid.

1. Introduction

Current methodology of composite materials items design and production includes and, is based on, two specific steps. The first one is identification of the microstructures which are theoretically predicted to lead to the desired properties or exploitation targets. The second is identification of the ways to produce the materials in order to obtain the appropriate microstructures. This approach implies the development of specialized composite materials in terms of knowledgebase, materials mechanics and physics, as contrasted to empirical laws from experimental measurements.

Constraint or widening of the cross-section area of a moving two-phase viscous flow are common techniques in construction of shape-generating equipment during composite materials production, moulds for die forming or powder injection moulding, shape-generating equipment for drawing (pultrusion), pressing, rolling and winding of parts reinforced by fibre bunches.

Constraint of flow by just one shape-generating surface is the basic feature of flow forming processes (Fig. 1) (Wyatt and Dew-Hughes, 1974; Kucherov, 2006). Individuality of different processes of flow forming is defined not by configuration of the material flow area (lighter part with the boundary CD on Fig. 1) but more significantly by the distinctive boundary conditions defined on the acting face of the shape-generating tool. In general, exhibition of other rheological effects is analyzed in conjunction with course of impregnation of areas between filaments and threads on different stages of processing package prepregs of fibrous composites.

The purpose of this paper is to analyze and discuss two important rheological effects and their ability to affect the quality of the real structure and the level of material properties in shaping products in terms of a few examples of typical composite manufacturing technologies. These effects are fountain effect of the viscous fluid in a narrow channel and die swell of the flow at the exit of the channel in an expanding cavity. They appear primarily during extrusion of highly-viscous liquids (various polymers in the processing are in this condition), and expressed the

necessity to consider the characterization and modelling of two new rheological effects fundamentally changing the material structure of the products produced.

2. Rheological effects in SSM billet pressing

Usually some natural changes of container construction and shape-generating die (compare Fig. 1a and 2) are considered as necessary structural solutions for both traditional materials and composites processing. Consider a case when a narrow channel is added to an existing process. In this case the medium conveying speed enhances according to implemented drawing ratio.

Often the model of medium movement is viscous incompressible Newtonian or Bingham fluid flow developing in isothermal conditions. Leakage at the channel entrance and exit and presence of slip near the channel wall

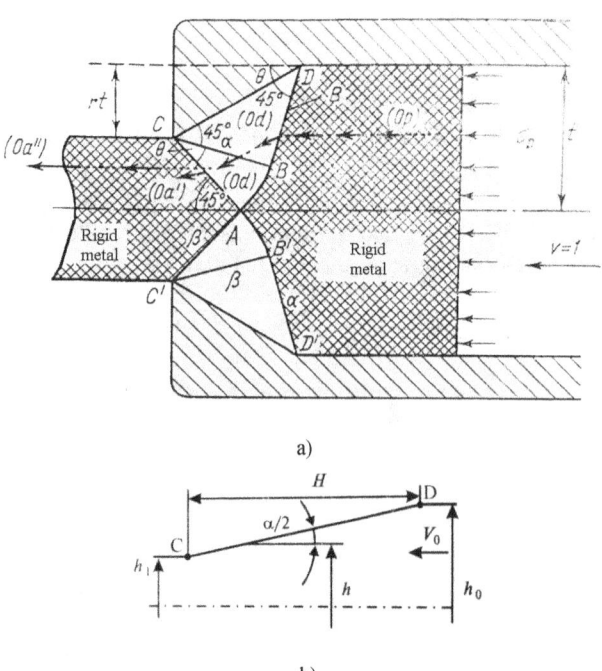

a)

b)

Figure 1. Pressing through the smooth die with wedge functional area.

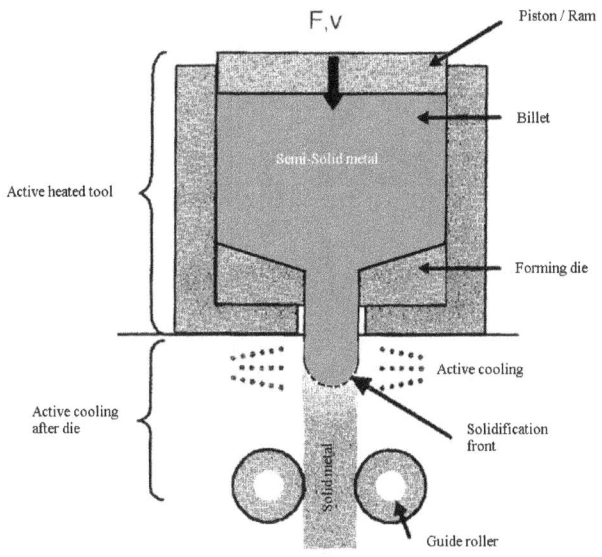

Figure 2. Practical realization of isothermal extrusion concept.

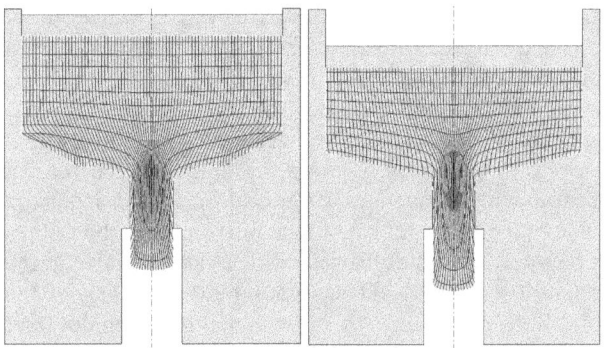

Figure 3. The results of SSM billet pressing modeling in software package Q-Form with different tool types and loading conditions.

are ignored. The results of modeling of pressing of SSM billets made of an alloy AK7 in the software package Q-Form are presented in Fig. 3. Different loading conditions and die types were explored but the same boundary conditions on the wedge surface and die channel eliminating slip were used.

The program predicts the die swell phenomena but cannot predict the principal morphological changes during this experiment (Fig. 4 and 5). It was assumed that in this experiment all macro- and micro- changes are induced by

the fountain effect. It begins at the moment of narrow die channel filling, which induces the fluid part of the suspension to reflux to the container (Fig. 6). This causes super plastic deformation of the solid particles in suspension and transforms them into fibres even before they are in the narrow die channel. As a result, there is fibrous composite material with the unique properties formed at the exit (Semenov et al., 2011). Constriction or expansion of the cross-sectional area of moving viscous fluid flow is a versatile processing method so form typical design elements of shaping equipment in the production of shaped objects of composite materials: molds for stamping and molding of particle- or fibre-reinforced binders. The flow forming processes' features noted above in these processes are implemented in the scheme shown in Fig. 7 in which a smooth die with a wedge-shaped or flat working area is complemented by a narrow channel of different lengths attached to the shaping cavity or without cavity (Paradies et al., 1996). Under these conditions the leading edge of the flow (on the free surface) inside the channel takes on a parabolic profile (Fig. 7) which can be preserved and hereafter passes into the solidification front during the curing (Fig. 2) or is modified during the passage through the channels. Disadvantages of this approach to the viscous fluid flow modelling has long been observed by Richardson (1970) and by Ditter and Hirt (1992).

3. Die swell phenomena in carbon fiber-reinforced pultruded rod

The term "die swell" is used to describe the enlargement of a fluid stream upon emerging from a nozzle. Die swell is usually characterized by the swelling ratio D/D_0 where D_0 is the nozzle diameter and D is the final swelled diameter of the fluid stream. Die swell has been observed with Newtonian fluids as well as non-Newtonian ones. Experimentally, it has been shown to be a function of Reynolds number, the shear dependence of the viscosity, the length to diameter ratio of the nozzle, and in the case of polymer the molecular weight of the fluid. This effect has been observed in relation to the carbon fibre-reinforced pultruded rod. Rod is made on a pultrusionline pulling Toho Tenax HTA 5131 carbon fibre through a bath with a solution of vinyl alcohol, the system dies and

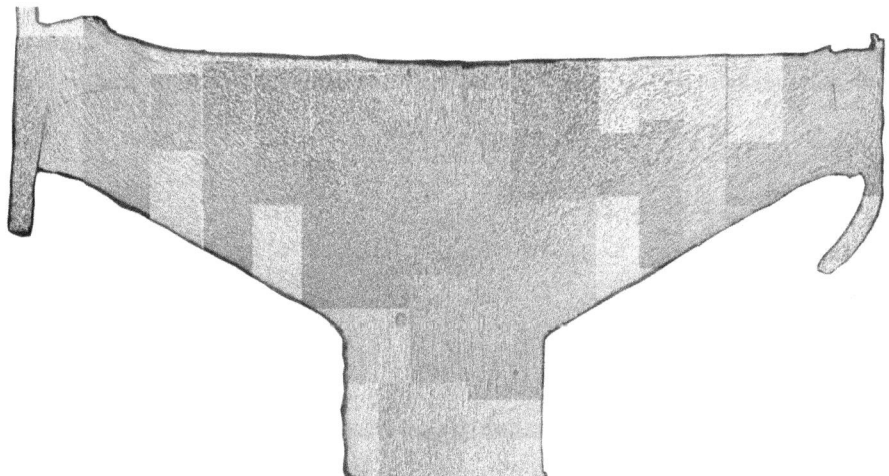

Figure 4. The real macro- and micro- structure of remainder after SSM billet pressing.

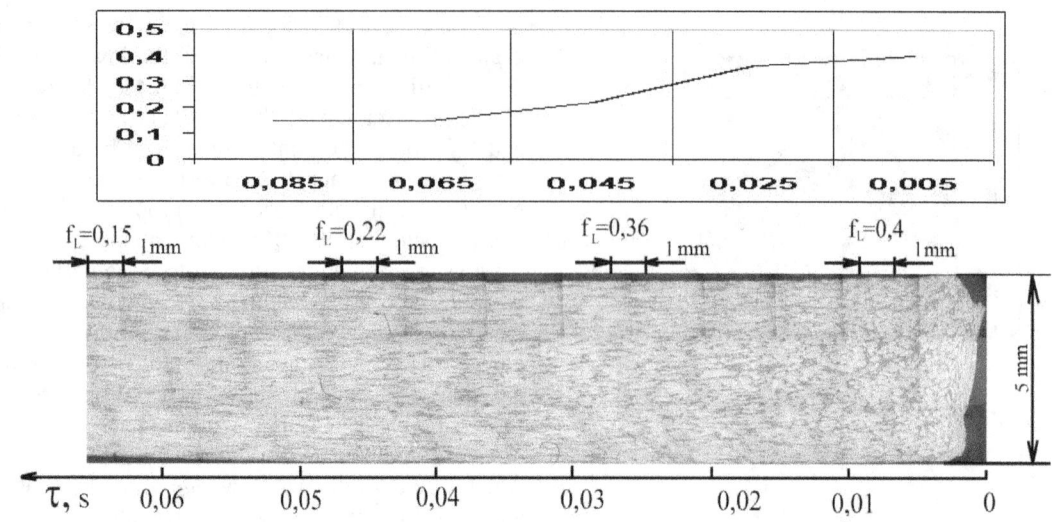

Figure 5. Microstructure and eutectic component of the rod of and alloy AK7 on at the initial flowing part.

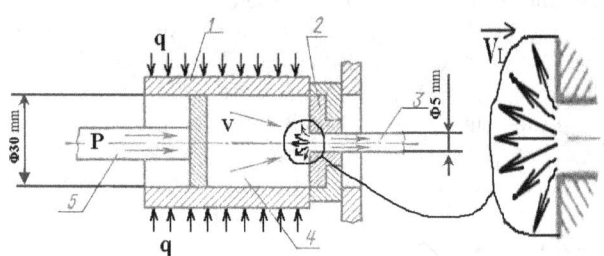

Figure 6. The supposed directions of the reverse relative movement (reflux) of liquid suspension part.

heating devices. The experimental results indicate the existence of the described die swell effect of carbon rod determined the ratio D/D_0 where D is the rod cross section diameter and $D_0 = 0.650$ mm – the diameter of the shaping expansion nozzle with an average rod fibre volume filling in excess of 70%. It was established that the rod of a hybrid composite material forms at the output of the ROD-line. This rod is the unidirectional carbon-vinyl fibre-reinforced plastic with the approximate (average) rate ratios of reinforcement 0.70 for carbon fibres and 0.19 for PVA. The rod can be further affected by internal rupture (Fig. 8). The table gives the results of measurements of the diameters, the total and reinforced cross-sectional area on the rod segments in length 250 mm selected from the continuous pulling of the thread through the 3.6-m path of the product.

For this purpose each rod was measured in cross section of the sample using the program Image Expert Pro 3 where 20 measurements of the diameter (up to three

points) were made. Based on this data the scatter of the diameter, its average value, the area of the die swell defects and the proportion of carbon fibre-reinforced parts in section of the rod were determined (Tabl. 1).

For the analyzed rods common defects at the micro level are the presence of non-reinforced sections, whose rate varies significantly along the length of the formed product with an average value of 19.2%, and a random value of the estimated rate of the reinforcing carbon fibres in non-defect parts of the rods which varies relative to the mean close to 79%. The viscous fluid jet filling of the solid fibres moving freely in the radial direction does not change the overall picture. As a result, in a moving volume of the solid-liquid medium the random process of structuring of the suspension liquid component develops and achieves a high degree of completion. In places where the die swell of a viscous medium leads to the formation of long unreinforced sections in rods there is growth of a new type of reinforcing element: rod or film-type polymer PVA.

Random fluctuations in the volume fraction of carbon filaments produced by die swell of the jet can significantly affect the strength and stiffness of the rod. The results of mechanical tests of a rod samples made from yarn Toho Tenax HTA 5131 are shown below. The test machine: Zwick; standard tests: tensile; load cell 10 N; loading rate: 60 mm/min. Deformation of the samples and the fracture mode are shown in Fig. 9. The data show that there was a low measured value of the coefficient of variation of failure load, which amounted to 4.31%, the

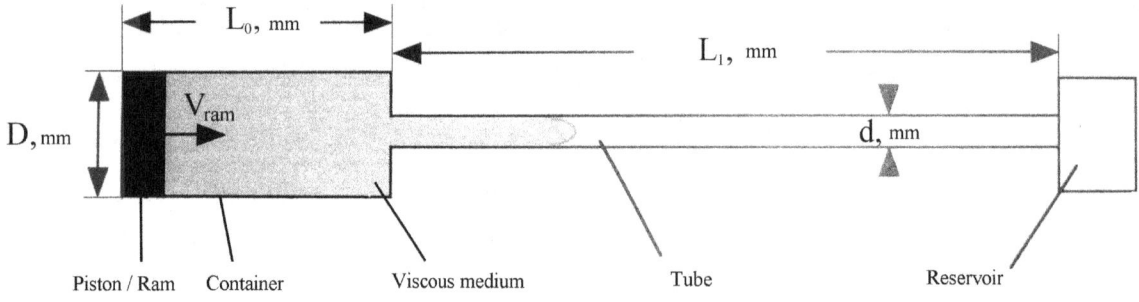

Figure 7. The typical experimental setup comprising an injector, a specialized blank (feedstock), a transport channel and a shaping cavity.

54

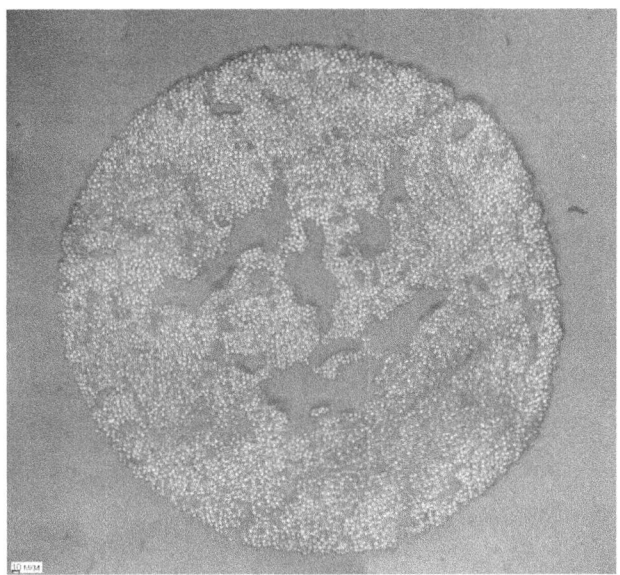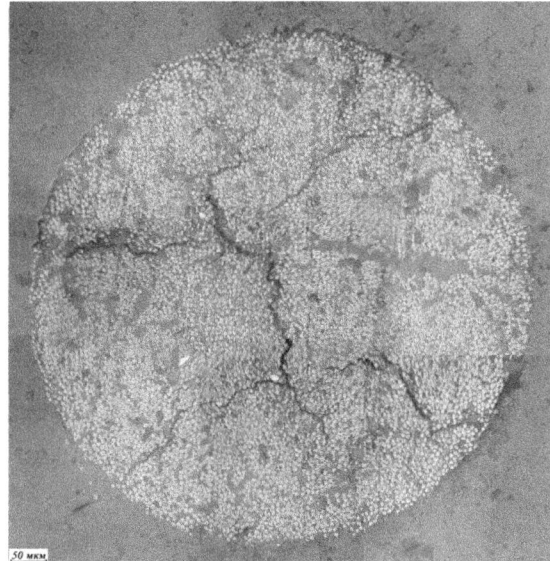

Figure 8. Typical defects in the rod macrostructure caused by the action of the radial tensile stress developed during the process of the cure of the pultruded composite.

coefficient of variation of the rod formed material strength was larger at 9.84% and the coefficient of variation of the elastic modulus of the composite was 11.49%.

The coefficient of variation of the material controlled parameter growth is a measure of reduction of the product quality which was caused by uncontrollable random fluctuations of the formed rod cross-sectional area. As has been established and quantified the volume fraction of rod carbon-reinforced components changed uncontrollably at the same time. Experimentally determined values of the strength of the reinforced sections σ^*_c increase linearly with increasing volume fraction of filler obeying the rule of mixtures (Fig. 10). Thus, the high significance of the effect of the die swell on the quality of composite technology has been confirmed experimentally.

4. Fountain effect

The term "fountain effect" was first used by Rose to describe the tendency of the leading edge of a very viscous fluid filling a channel to resemble a fountain. This effect is demonstrated in Fig. 11 provided by Ditter and Hirt (1992), where the fluid near the wall lags the fluid at the center of the channel due to the no-slip condition at the wall. Herewith elements of the molten polymeric fluid undergo complex shear and stretching motions as they

catch up with the free flow front and then move outwards to the cold walls.

This phenomena has clear implications with regard to fluid particle (polymer chains) orientation at the surface and thus the surface characteristics. Liquid particles in the centre of the channel align such a way that they move along streamlines which are parallel to the free flow surface. Those particles can change the orientation near the channel exit. Fig. 12a shows streamlines and fluid element deformation in fountain flow (Vlachopoulos and Strutt, 2003), and Fig. 12b shows the structural changes in semisolid SSM billet with 60% viscoelastic spherical particles during motion in a narrow channel and its exit. While molecular and particle orientation change is used in extrusion of SSM billets to improve the mechanical properties, in another similar process – powder injection molding - orientation is generally not desirable.

5. Rheological effects in PIM-technologies

The fountain effect occurs in composite materials production using PIM-technology (injection molding of suspensions of fluid polymers filled with solid free of distortion powders with subsequent debinding and sintering). It is visible during channel filling. The consequence of fountain effect is powder – binder separation (Fig. 13). This

Table 1
The results of measurements of the diameter and quantitative analysis of defects in the rod microstructure

Characteristic	Value					
Rod index number	1	2	3	4	5	6
Rod total area, mm^2	0.341	0.355	0.317	0.381	0.376	0.374
Average diameter D, mm	0.668	0.681	0.696	0.681	0.698	0.682
Rod fiber-reinforced area, mm^2	0.304	0.271	0.246	0.291	0.306	0.322
Die swell defect area, mm^2	0.0363	0.0841	0.0810	0.0902	0.0702	0.0520
Defect rate, F_D	0.107	0.237	0.245	0.237	0.187	0.139
Die swell ratio, D/D_0	1.028	1.048	1.070	1.048	1.073	1.049
Rod estimated diameter without die swell, mm	0.623	0.588	0.550	0.608	0.624	0.640
Estimated rate of carbon fibers, %	74.8	84.0	≈91.0	78.1	74.3	70.6

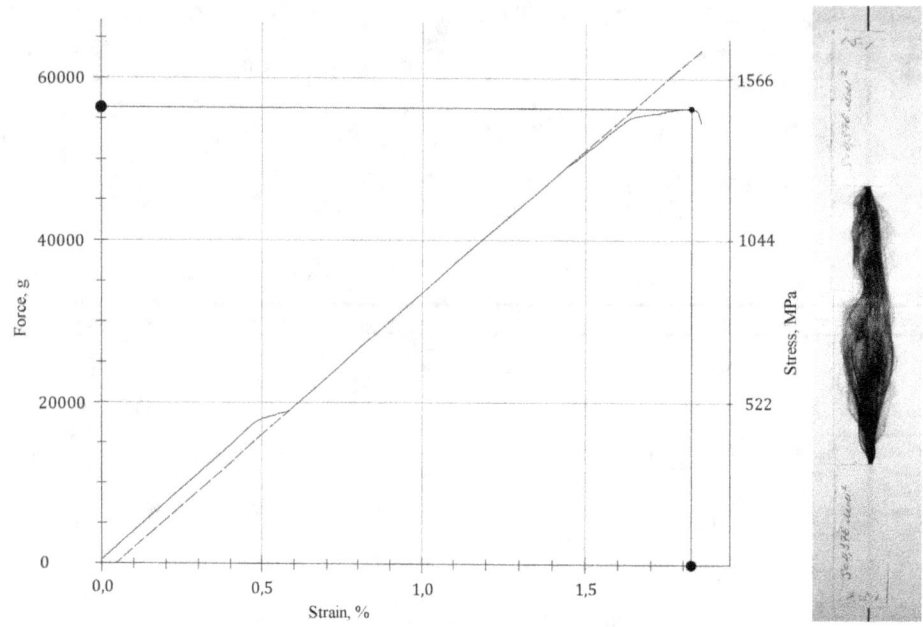

Figure 9. Stress-strain curve of one of the samples (group 5, sample 1-5-1) and the fracture mode ($D/D_0 = 1,073$; $F_D = 0,187$).

orientation is intensified in the process of packing. Further cooling can lead to warpage.

For PIM-technology the development of the fountain effect is more important with the increase of temperature and velocity. Due to the adhesion of the feedstock at the wall of the channel, temperatures of cavity and feedstock significantly influenced by the powder content. The binder is a liquid and flows easily; while powder follows it. The low influence of injection speed on powder content in the center of pipe nodal point can lead to separate movement of two phases in channels (Hausnerova, 2011). Phase separation is a term commonly used for an adhesive failure of powder and binder in PIM-technology.

Thus, a separation of these two ingredients causes quality issues such as visual defects, mechanical weak points, warpage and local voids. Modeling PIM-technologies is well established. A basic requirement for high quality production is a homogeneous mixing of powder and binder, adhesion and mutual interaction between them.

In this case, the history of the feedstock, which is travelling through injection machine, nozzle, runner system and cavity, has to be implemented in the software simulation of composite technologies to predict the phase separation area and forecast macro- and microstructures of the molded part. The examples discussed show the characteristic features distortion during thermal cycling of composite materials.

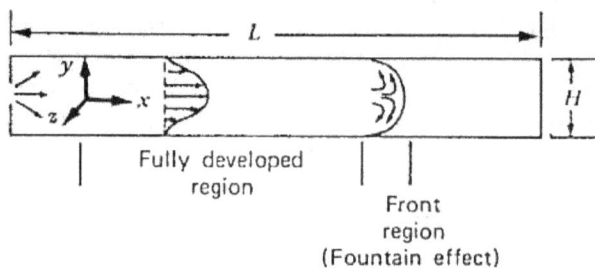

Figure 11. Sketch of a channel flow which exhibits the fountain effect.

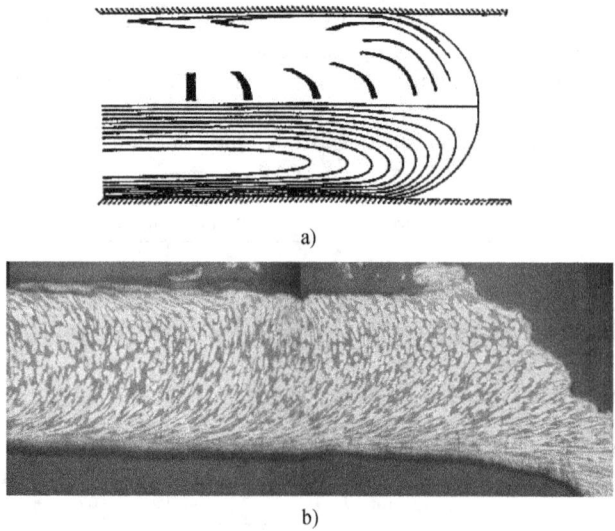

Figure 10. Calculated strength and the variation of the strength of the rods reinforced sections.

Figure 12. Fountain flow streamlines and fluid element deformation illustration: a – viscous fluid flow in narrow channel; b – micro- and macro- structural changes of suspended metal which contains viscoelastic spherical particles.

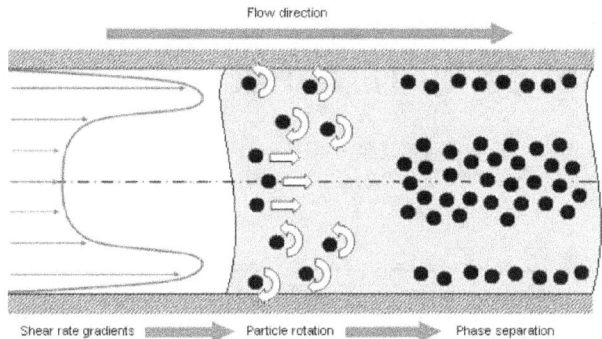

Figure 13. Local shear stress gradients causing
powder – binder separation.

6. Conclusion

Rheological effects such as die swell phenomena, fountain effect and phase separation have a great influence on the fluid flow and final parts quality in a number of current technologies and should be considered for proper materials processing.

In this paper the effects were shown in such applications as thixoforming, PIM-technologies and rod pultrusion. Unfortunately, study of these rheological effects resulting in a constraining or expanding viscous fluid flows, which blanks of composite materials always are, is not given due consideration in composite technologies.

References

Ditter, J.L. and Hirt, C.W. (1992). *The Fountain Effect and Die Swell Phenomena*, Technical Report FSI-92-TN35, Flow Science Technical Notes.

Hausnerova, B. (2011). Powder injection moulding – An alternative processing method for automotive items, in *New Trends and Developments in Automotive System Engineering*, ed. Chiaberge, M. Rijeka: InTech, pp.129-146.

Kucheryaev, B.V. (2006). *Continuum Mechanics (Theoretical Foundations of the Composite Metal Forming with Tasks and Solutions, Examples and Exercises)*. Moscow: MISiS.

Paradies, C.J., Rappaz, M., Imwinkelried, T. and Gabathuler, J.P. (1996). Simulation of the pressure die casting of a thixotropic aluminum alloy, in *Proceedings of the 4th International Conference on Semi-Solid Processings of Alloys and Composites,* eds. Kirkwood, D.H. and Kapranos, P. Sheffield, UK, pp. 115-119.

Richardson, S. (1970). The Die Swell Phenomenon, *Rheologica Acta*, Vol. 9, No 2, pp. 193-199.

Semenov, B.I., Kushtarov, K.M., Dzhindo, N.A. and Binh, N.T. (2011). Thixoforging and thixoextrusion of semisolid alloys slurries, *Blanking Productions in Mechanical Engineering (Forging and Stamping, Foundry and Other Productions)*, No 2, pp.21-23.

Semenov, B.I., Kushtarov, K.M., Dzhindo, N.A. and Binh, N.T. (2012). Thixoforging of cylinder SSM billet method in the mode of superplasticity of the solid phase. Patent № 2444412 , 10.03.2012.

Vlachopoulos, J. and Strutt, D. (2003). Overview polymer processing, *Materials Science and Technology*. Vol. 19, pp. 1161-1169.

Wyatt, O.H. and Dew-Hughes, D. (1974). *Metals, Ceramics and Polymers: An Introduction to the Structure and Properties of Engineering Materials*. Cambridge: Cambridge University Press.

A Review of Cutting-edge Techniques for Material Selection

Noori Brifcani[1], Richard Day[1], David Walker[1], Suzanne Hughes[2], Ken Ball[2], Dave Price[2]

[1] Glyndŵr University, Plas Coch, Mold Road, Wrexham, LL11 2AW, UK
[2] Qioptiq Ltd, Glascoed Road, St. Asaph, Denbighshire, LL17 0LL, UK

Abstract: Selecting the optimum material for a given application is a complex task for engineers and designers across all industrial fields. There are a huge number of materials now available with a range of different properties and behaviours and so it has become even more necessary to carry out a systematic process in order to screen and/or rank the materials to give a promising number of candidates. The output of the material selection process depends upon which method is used. In some methods, a chart can be used to identify promising candidates whereas in others a single 'optimum' material may be chosen or a ranked list of candidates identified. This paper aims to summarise the documented techniques for material selection, evaluating the methods that are currently available, and compare the methods for consistency and effectiveness.

Key Words: Materials selection, Material screening, Performance indices, MCDM, TOPSIS, VIKOR, ELECTRE.

1. Introduction

Choosing the optimum material for an engineering application is a difficult but very important task. The selection of a cheaper material may mean greater competitiveness and more sales, the selection of a lighter material may increase fuel economy and reduce emissions in an automobile or aircraft, and the selection of an inappropriate material for a task may result in critical failure or poor performance. More recent demands from customers and legislation from governments have made material selection even more important. Examples of this include reducing the mass of a car in order to reduce emissions to meet regulations which are predicted to become tighter in years to come, and reducing the burden on a soldier by reducing the mass of the equipment that is carried.

There are over 160,000 materials available (Ramalhete et al., 2010), which gives an insight to the scale of the material selection task. Materials can be grouped into several general categories: Metals & Alloys, Polymers, Ceramics and Composites, with the materials inside each group usually having several properties in common. Each material is defined by its properties which are usually measured in tests carried out in accordance to standards (for example ISO or ASTM). These properties can be grouped into Mechanical Properties, such as Young's Modulus and Tensile Strength; Physical Properties, such as Density; Electrical, such as resistivity; Thermal, such as melt-ing point; and others, such as Cost. Some material properties have a quantitative value, such as the Hardness of a material measured by the Vicker's Hardness Test. Other properties can have a qualitative value often described in a linguistic nature, such as Corrosion Resistance being 'Poor', 'Good', 'Very Good', etc. These material properties are the profile used to compare one material against others for a given application.

With the huge range of properties that describe a material, it would be very rare to find a material that has the absolute ideal values for a function – instead, a trade-off of properties is usually required based on the requirements (Ashby, 2011). Material Selection Techniques are systematic tools that can aid a designer or engineer in defining the material requirements for a required function, and then finding the material that would suit this function

best. Selection of a material should be investigated in parallel with initial design and product development, as the material selected will have individual properties that influence how it can and should be manufactured and therefore how it should be designed. Material Selection can also be used to identify alternative materials for an existing product, in order to reduce cost or mass or meet new legal requirements, for example.

The material selection techniques available vary in how they are used and the output of the method. In the method proposed by Ashby (2011), for example, the output is given on a chart with a calculated material performance index gradient that can be used to identify candidate materials. Others, such as the Multiple Criteria Decision Methods, are purely numerical and the output is often a screened, ranked shortlist of candidates which can then be investigated further.

This paper aims to research and review the documented material selection methods and their applications. In addition to this, the paper aims to consider other implications in the process, such as methods of identifying weightings of importance, and the material database resources available for the analyst.

2. Material selection methods

There are several documented methods that have been used for the selection of materials and these vary in function, from 'free-search' methods such as from Ashby (2011) to more quantitative methods such as the Weighted Property Method and the use of Multiple Criteria Decision Making techniques. In all methods, there is an importance in the first instance to fully understand the problem, so that the requirements and objectives can be selected carefully. Failure to understand the problem can result in a selection method giving unreasonable or even impossible solutions to a material selection problem.

Jahan et al. (2009) discovered that, at the time of research, the most popular methods documented for material selection were TOPSIS, ELECTRE and AHP, all techniques within the Multiple Criteria Decision Making (MCDM) methodology. There is a need to select a suitable method in accordance to the nature of the material selection problem (Cicek et al., 2010). In addition to the available methods, there are many academic papers that focus on modifying the traditional approaches or applying modified approaches to material selection problems. Some of the alternative approaches discovered will be discussed in this paper.

2.1 Ashby free-search

Ashby (2011) states that any given component desires a profile of material properties in order to function optimally. Clearly, however, it would be unrealistic to expect the exact profile of required properties to meet the property profile of a material. This means that some property trade-offs are required in order to find the overall most appropriate candidate material. Ashby (2011) defines a 'translation' step, where the requirements of a design are converted into constraints and objectives which can then be used to identify materials. Once these constraints are found, they can be applied to a material database (some example databases are identified later in this paper) in order to screen for potential candidates. These screened materials are then graphically shown according to a design objective or performance index, for example having the lowest cost or density, or the highest thermal conductivity, or a combination of a number of material properties (Ashby, 2011). No material selection technique can promise to give the perfect answer and so further research from documentation is required. This is important with aspects such as bi-metallic corrosion properties, manufacturing processes, availability, surface coatings, supplier relations, and other variables that are not assessed in the selection process. Fig. 1 shows a basic flow chart of the strategy proposed by Ashby (2011).

Parate and Gupta (2010) used Ashby's approach to choose a suitable material for an electrostatic actuator. Performance indices were developed for the component based on Ashby's methodology and material selection charts were used to find the best material candidates. It was noted in the paper that there is an ever-expanding database of materials available and the charts allow for new materials to be added. Parate and Gupta (2010) used selection charts with variables of Actuation Voltage vs. Speed and Fracture Strength vs. Displacement. They identified the best candidate materials for two actuator types - high actuation force and high actuation speed.

The method proposed by Ashby (2011) has the advantages of being intuitive and also relatively simple with a limited amount of calculations. CES Selector software from Granta, developed with Ashby, combines Performance Index generation and Material Selection Charts with a developed Material Database to allow a capability of carrying out the full selection technique efficiently in one piece of software.

Disadvantages of the Ashby method are that it requires a significant amount of work to calculate performance indices, select the required chart axes and then create the material selection charts. The procedure is not as systematic as some other methods, and it also does not give a ranked list of alternatives or assign a value of suitability. The CES Selector software from Granta does give a good solution to some of these issues as it contains several material databases, such as CAMPUS and Material Universe, as well as allowing for performance index calculations, creating selection charts and inputting gradients onto the charts based on performance indices. The output is still a chart, however, which can mean it is difficult to choose the optimum material(s) for a solution. Fig. 2 shows a material selection chart with gradients from performance indices overlaid. It is clear to see the material families and how the properties (Young's Modulus and Density in this case) are similar in each group.

2.2 Weighted Property Method

The Weighted Property Method is a very simple numerical decision-making technique. Firstly, the functions of

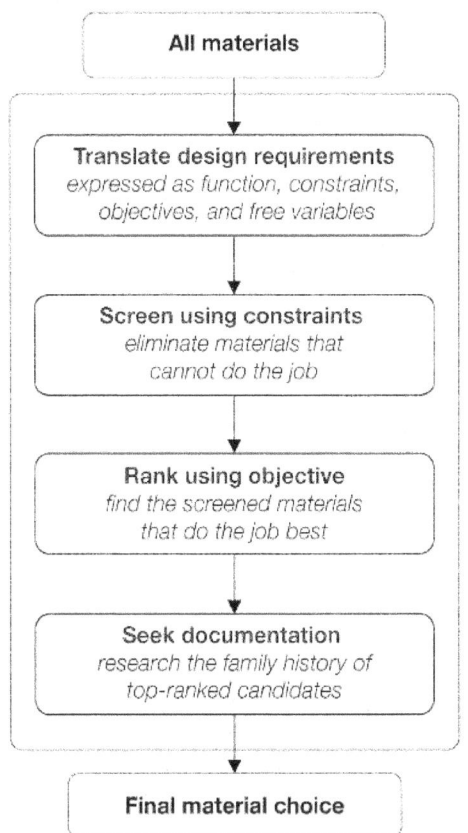

Figure 1. The strategy proposed by Ashby in four main steps (Ashby, 2001)

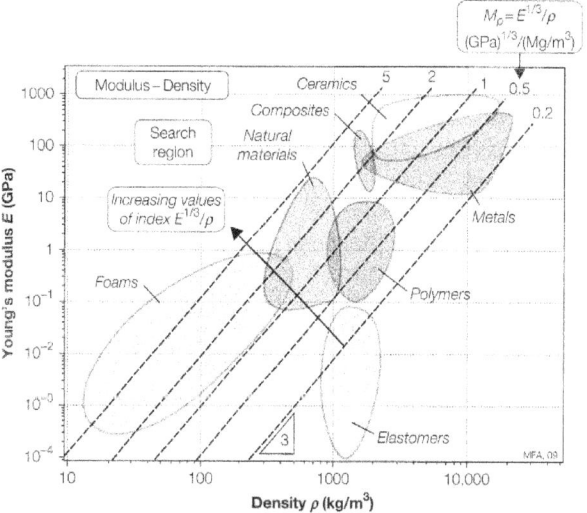

Figure 2. An example material selection chart showing Young's Modulus against Density, with performance index gradient lines shown (Ashby, 2001)

the application are assessed and each important attribute (material property) is then assigned a weighting of importance. These weightings are assigned by a designer/engineer/etc. and the sum of all of the weightings should be equal to 1. Each material property has a unique scale of measurement, such as Tensile Yield Strength in MPa and Young's Modulus in GPa (SI units), so it is necessary to scale the numbers to allow an overall comparison index to be calculated. In order to obtain these scaled property values, a simple calculation is done. In material properties where a larger value is favoured, the numerical value of each property is divided by the largest value of that property across all candidate materials, and multiplied by 100. For material properties where a smaller value is more favourable (for example density or cost), the lowest value is divided by each value and multiplied by 100 (Findik and Turan, 2012). In order to find the weighted property index for each material, the scaled property values are first multiplied by the assigned weighting factor, to give the weighted scaled values. The weighted property index is then the sum of all of the weighted scaled values for each candidate material. This index can be used to compare any number of materials for suitability in the application. In cases where a qualitative value is given for a property, this can be converted to a quantitative value by applying a scale (Findik and Turan, 2012). For example, a corrosion resistance of 'Excellent' could convert to a value of 5, 'Very Poor' to a value of 1 and other linguistic values in-between.

An advantage of this method is its simplicity – a spreadsheet can be created using this method in minutes and any number of materials can be evaluated. It can also take into account any number of material properties and does not involve difficult arithmetic or expensive software. The output of the method also gives numerical values and this allows a ranked shortlist to be created and also means that the suitability of each material can be compared.

The method is completely reliant on the weighting values, as these define the importance of each property for the function. This means that change in the weightings results in a change in the selected material and so there is the problem of bias and mistakes in the weighting values. Actually selecting the weightings, where there is no 'right answer' also gives a further question – how can we obtain weightings that truly portray the requirements of the application? Importance weighting methods are discussed later in this paper.

Findik and Turan (2012) used this technique, as well as design considerations and joining methods, to identify materials that would allow the reduction in weight of a train load wagon. Required properties for the function included high specific strength and stiffness, corrosion resistance and low cost. Aluminium, magnesium and titanium alloys were considered as substitutes for the steel wagons and Al-alloys were selected as the most suitable by using the Weighted Property Method.

2.3 Multiple Criteria Decision Making

Multiple-Criteria-Decision-Making (MCDM) processes were not initially created for material selection; however material selection does fit well in the methodology. The MCDM technique involves generating alternatives (e.g. from a material database or from gathering data), establishing the required criteria and evaluating the alternative materials using a set of criteria weights; the outcome is a ranked list of alternative solutions (Jahan et al. 2009). A number of MCDM methods are reviewed in this section.

2.3.1 TOPSIS

The Technique for Order Preference by Similarity to an Ideal Solution (TOPSIS) method is based on the factor that the chosen alternative (material) should have the shortest distance from the ideal solution and the longest distance from the negative-ideal solution (Opricovic and Tzeng, 2004). Shanian and Savadogo (2006) state that there are a number of features of TOPSIS which give it good potential for a material selection problem. The method allows for an unlimited number of alternatives (materials) and attributes (material properties), and it allows for trade-offs due to the fact that no attribute is considered alone – it is always seen as a trade-off with others. The output of TOPSIS is a ranked list with a numerical value for each alternative – allowing comparisons of suitability – whereas other techniques may only give the list. The method uses a pre-determined set of weighting criteria which are defined by the analyst/engineer. Pair-wise comparisons are avoided which means that the method is fast and allows for linking a database to the method, making it systematic and fast (Shanian and Savadogo, 2006).

The TOPSIS procedure starts with normalising the material property values to eliminate differences in units and applying weightings to create the weighted normalised decision matrix. Ideal and negative ideal solutions are then determined – in a case where a higher value is better, the highest value in the set of alternatives is chosen as the ideal (e.g. tensile strength), whereas the lowest value is chosen where this is desirable (e.g. cost). If ideal values of material properties are known (e.g. a known CTE value to match an optical housing to a lens) then this value can also be used. The separation from the ideal and anti-ideal solutions is then calculated to give the relative closeness to the ideal solution and this enables a ranked list of alternatives to be determined (Opricovic and Tzeng, 2004).

2.3.2 VIKOR

Vise Kriterijumska Optimizacija Kompromisno Resenje (VIKOR), like TOPSIS, works by ranking and selecting from alternatives based on the criteria and uses the approach of closeness-to-ideal. This technique is very similar to TOPSIS however there are differences and these have been discussed by Opricovic and Tzeng (2004). One difference is how the methods use normalisation of the material property values. VIKOR uses a linear normalisation where the normalised value does not depend on the evaluation unit of the criterion, whereas TOPSIS uses vector normalisation where the normalised value can change for different evaluation units of a particular criterion. The aggregation function of each method is also different – VIKOR uses a function that factors in only the distance from the ideal value and TOPSIS uses the ideal and anti-ideal values. The material property being as far from the anti-ideal value may not be a goal and so using

VIKOR may be a more effective approach. Both of the techniques produce a ranked list of alternatives – the optimum alternative in VIKOR is the closest to the ideal, the optimum in TOPSIS has the best ranking index (calculated from the distance of both the ideal and anti-ideal values) (Opricovic and Tzeng, 2004).

VIKOR allows the analysis of the impact of modifying the importance weightings in the calculation (Opricovic and Tzeng, 2004). This allows some stability analysis in the results, reducing the possible bias in the chosen weighting values and being advantageous when the analyst is unsure of the weighting preference for each criterion.

2.3.3 ELECTRE

There are numerous forms of ELECTRE that exist, including ELECTRE I, II, III, IV and TRI, and these forms all use the same fundamental concepts but differ in operation and depending on the type of problem (Marzouk, 2011). According to Marzouk (2011), ELECTRE outperforms other MCDM methods due to its ability to use inaccurate and uncertain data – such as material properties or weightings. This is important in material selection as there are often uncertainties in the measurement of material properties (Shanian et al., 2008) and in the relative importance values of each property. ELECTRE is non-compensatory – meaning separate material properties cannot compensate for each other (Shanian et al., 2008). For example, a good Tensile Strength value does not compensate for a poor Young's Modulus. This is very different from the Weighted Property Method, for example, where the performance of a material is governed by the sum of the weighted material properties.

Shanian et al. (2008) suggest that the goal of MCDM in material selection should be to not only identify materials with high rankings, but to also ensure that the materials have the most stable ranks over several design scenarios. Sensitivity analysis in a revised Simos' importance weighting method (discussed later in this paper), combined with a post-operation group decision-making process using ELECTRE III is used by Shanian et al. (2008). Their findings showed that the approach allows the identification of materials with both high and stable ranks – two important requirements in the decision making process. They suggested that further study into applications of the proposed method would be worthwhile to further analyse it's effectiveness.

Shanian and Savadogo (2006) used ELECTRE IV for material selection of bipolar plates in a polymer electrolyte membrane fuel cell. ELECTRE IV was used due to the non-compensatory nature of the technique. Their findings suggest that ELECTRE IV is a worthwhile method for material selection and the results obtained agreed with available reported results for the component. Jahan et al. (2009) found that ELECTRE techniques have limitations of high amounts of calculations with increased number of alternatives, and ELECTRE does not give a comparable performance value for each alternative, it only gives the ranked shortlist.

2.3.4 AHP

Analytical Hierarchy Process (AHP) is a method that discriminates between alternatives where inter-related objectives should be met. It is based on straightforward maths formulae and is used in a range of fields (government, industry, education, etc.) for decision-making (Mayyas et al., 2011). AHP works by structuring the decision problem into a hierarchy of sub-problems which can be analysed. The decision-maker then compares the elements of the hierarchy against each other by pair-wise comparison. The alternative (material) with the highest importance is the optimum. As AHP uses pair-wise comparison, it is infeasible for use in a situation with a high number of alternatives and/or criteria, where other MCDM methods such as TOPSIS would be more suitable (Jahan et al., 2011).

AHP is an attractive technique for combining opinions from several groups of experts – either for obtaining criteria weights or for the final selection. According to Jahan et al. (2011) the stand-alone AHP technique has less attention than techniques integrating AHP with other methods, such as SMART (Edwards and Barron, 1994) which combines AHP with the simple additive weighting method. Mayyas et al. (2011) used AHP and Quality Function Deployment (QFD) in selecting a material for an auto-motive body-in-white. They found that QFD was the superior technique, but that AHP provides systematic selection and gives numerical priority vectors to the material candidates.

2.4 Preferential ranking methods

(Chatterjee and Chakraborty, 2012) state that although various MCDM methods have been successfully applied to material selection problems, there is still a requirement to search for other tools and techniques for accurate ranking of alternative materials in a given engineering application. Four alternative methods based on preference-ranking are proposed by Chatterjee and Chakraborty, (2012) for use in material selection (EXPROM2, COPRAS-G, ORESTE and OCRA), and applied to solve material selection for a gear. All of these proposed methods have the output of a list of best-to-worst suitable materials based on the criteria. The research from Chatterjee and Chakraborty (2012) shows that the four investigated methods have high potential in material selection problems. It was noted that the best and worst suited materials found by each of the trialled methods was the same, giving a good indication of consistency and showing that the preference ranking methods can be applied to any type of material selection problem. Further research into the applications of these four proposed methods would be valuable.

3. Material databases and data gathering

Any material selection method that is chosen requires data to give a property profile of the materials that are to be evaluated. Material data is available from several sources such as from material suppliers, manufacturing companies, consultants, internal sources (e.g. in-house testing) or from a database. Already-constructed databases provide a quick and efficient way of obtaining material data, however the data source should also be considered when assessing the accuracy. Material suppliers usually have their own database of data, however this will be limited to the materials that the company provides and so many suppliers will need to be researched in order to obtain the data required which is time consuming. In-house testing can be a lengthy and expensive process and the material samples need to be obtained first – for this reason it should not be used to fill an entire database for material selection but it could be used to further test promising materials for data that could not be obtained from other sources. Some material databases are reviewed here.

3.1 Granta CES selector software

Granta CES Selector combines a material selection utility involving charts and performance indices, with material databases such as Material Universe and CAMPUS plastics. The database has generic materials rather than tradename materials and the values are given in ranges rather than one specific value, to include all of the materials available of this type. Suppliers of each material are also listed to enable the user to efficiently purchase some material or contact the supplier for more information if required. The Material Universe covers a wide range of polymers, ceramics, metals and alloys, and composites. According to Ramalhete et al. (2010), there are over 3700 materials in the Selector Basic Edition database which includes most types except for Textiles, "Smart" materials, Aerogels and Nano-materials. There are more versions of the CES software such as the Polymer Selector, Aero Selector, Eco Selector and Medical Selector which offer more materials in the database. The database also includes information on fabrication and production processes such as Injection Moulding and Welding.

3.2 Matweb

At the time of writing, Matweb Online Materials Information Resource has data sheets for over 88,000 materials including metals, polymers, ceramics and composites. Ramalhete et al. (2010) carried out research on the digital tools available for material selection and found there were 74,000 materials available in Matweb – meaning that there has been an addition of 14,000 more materials in just two years. Matweb provides the highest number of different materials in the database, however there are other digital tools which are discussed by Ramalhete et al. (2010) such as IDES Prospector and Polymat.

The research from Ramalhete et al. (2010) gives substantial information on the digital tools and databases available and further investigation into more of the databases would be worthwhile. They classified the different databases into 'general', where more than one material family is included and 'spe-cific' which focuses on one class or subclass of material.

4. Property weighting methods

Determining the weights of criteria (material properties) is an important task in most material selection methods, especially in Decision–Making techniques. Weighting the

properties is subjective –it requires input of opinion from a decision maker which is then translated into quantitative data. The importance weightings of the material properties define the requirement profile of the product/ component. MCDM methods, such as TOPSIS, rely on the importance weightings in choosing an optimum material – this means that any change in the weightings will directly affect which material is output. In TOPSIS, the weightings are multiplied by the normalised property values and then summed to give a material property index – the value used as a comparison against other materials (B. Dehghan-Manshadi et al., 2007).

Weightings are subjective to the analyst that is applying them and this means that the designated decision-makers in the process should be chosen carefully. In the first instance, the material properties to be included in the weightings decision need to be chosen. The choice of important material properties to include depends upon the nature of the product or component or may depend upon whether the material property is intrinsic (such as Young's Modulus) or can be modified or designed in a way that counteracts the property (such as corrosion resistance and coefficient of thermal expansion). Bias can occur in material selection if specific group(s) of properties outweigh other included group(s), for example having 3 thermal properties (thermal conductivity, diffusivity and CTE), against 1 mechanical property (e.g. tensile strength). Even if the 3 thermal properties have a low weight, they may outweigh the 1 mechanical property and it must be ensured that this meets the functional requirements and objectives of the product. As many material selection techniques are very sensitive to weightings values, it is very important to obtain the most accurate values for weights. It is possible that an individual is designated to decide on the values, or a group of people, or separate individuals onto which some statistical calculations are carried out. Due to the wide-ranging implications of material change(s) in a business setting, it may be necessary to include analysts from several disciplines, for example Mechanical Design, Business Groups, Manufacturing Engineers and Material Scientists. For some analysts that do not have the suitable material knowledge to make a decision, information will need to be provided to them in order for them to make a decision on the weightings. Identifying these very important values is difficult but essential. Sensitivity analysis can also be carried out in some selection methods, such as in the case of research by Shanian et al. (2008). The weighting process can be done for an entire product, a component, or even parts of a component which can be split into a hybrid structure – this is more likely in a situation where materials in a current product are being re-evaluated for an identified benefit.

There have been a few proposed systematic methods of assigning weightings to criteria and these are reviewed in this section.

4.1 Simos' Card Play method

The card-play method proposed by Simos (1990) aims to obtain importance weightings for criteria using a hierarchal technique rather than assigning numbers from the outset, while also giving the decision-maker the informa-

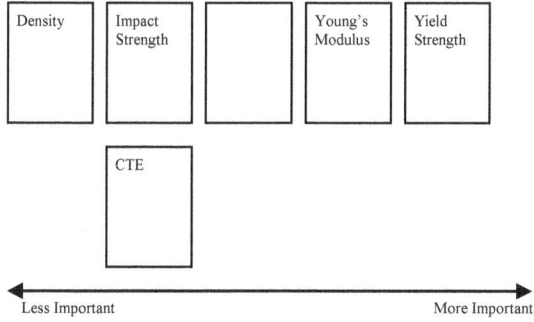

Figure 3. A schematic for Simos' Card-Play method

tion needed in order to decide on the weightings. The method Simos proposed is a simple and practical procedure that uses a set of pre-made cards to determine the numerical values of weightings indirectly and is a quick method of obtaining valuable information (Figuiera and Roy, 2002). The process works by firstly producing a set of cards with all of the criteria and any other necessary information for defining its importance. Designated analysts then rank the cards in order of importance, as shown in figure 3. Cards can be more important than others (on the right), or of the same importance (same horizontal level). Blank cards can then be placed between two successive cards to show even greater importance. For example, 1 blank card between 2 criteria cards means twice the difference between the criteria (Figuiera and Roy, 2002). After obtaining the ranked list of cards, a simple algorithm is used to assign numerical values to the criteria weightings.

Figueira and Roy (2002) found some problems with Simos' method and constructed an adapted procedure. One identified problem was that there is a piece of information lacking from Simos' procedure – the importance of the 'best' card compared to the 'worst' – i.e. how many times more important the most crucial criteria is compared to the least crucial. The modified technique from Figueira and Roy (2002) uses the same data collection method as Simos' original method; however the algorithm for calculating weights is modified to include a value 'z' – the ratio of importance of the highest ranked criterion to the lowest. The revised method also has some changes concerning rounding-off of figures in an optimal way and eliminating misprocessing of the blank card values. Figueira and Roy (2002) noted that their adapted technique has been applied successfully to real-life contexts, such as public transportation problems and environmental problems, and proved to be successful.

4.2 Digital Logic (Pair-wise comparisons)

The Digital Logic (DL) approach of weighting does not consider all criteria at the same time. Instead, the method uses comparisons between every pair of criteria, identifying which is most important in each case, to then find the overall most important and least important criteria for the requirements. For each pair to be evaluated, the maximum number of decisions is $N = n(n-1)/2$, where n is the number of criteria (properties) being considered (B. Dehghan-Manshadi et al., 2007). A matrix can be constructed using the number of decisions required to fully evaluate every criterion. If the property to be evaluated is

more important than the property it is being compared against, this column is assigned a '1', if it is less important, it is assigned '0'. The evaluation can be done by one individual, by a collaborative group, or by separate entities, on which some statistical analysis is carried out. To convert the DL matrix into weightings values for the Material Selection process, the values are scaled depending on whether it is beneficial to have a large or small value of each property.

A problem with the traditional DL method, found by Dehghan-Manshadi et al. (2007) is that if a property is found to be less important than every other, then it has acquired values of 0 in every comparison. This means that the overall weighting will then be 0 and it will not be of any importance in the material selection and is expelled. A modification by Dehghan-Manshadi et al. (2007) introduces relative values to DL – with a value of 1 assigned to a less important property and 3 to a more important one – ensuring that the least important remains in the selection list. Dehghan-Manshadi et al. (2007) used the modified version with the Weighted Property Method (WPM) and successfully applied it to material selection of a cryogenic tank and for a wing spar of a Human-Powered-Aircraft (HPA). The method provided more reasonable solutions for the wing spar than the existing WPM method.

5. Conclusions

There are numerous tools and techniques available to aid in material selection decisions. These include graphical techniques such as that proposed by Ashby, numerical techniques such as the MCDM methods, and also digital tools such as Matweb and Granta CES Selector. In addition to each individual technique, there are also several integrated or adapted methods that have been documented for material selection in order to improve the process. Despite the large amount of MCDM and other material selection methods available, no technique can be considered the most appropriate for any situation (Jahan et al. 2011) therefore it is necessary to understand the techniques in order to make a choice on which is most appropriate.

Several methods have been demonstrated to produce different outcomes in ranking a set of alternative materials/decisions. Jahan et al. (2011) propose that their aggregation method in MCDM has been developed to fill this gap, enhance the reliability of the chosen material and allow more robust decisions in material selection. There are also other integrated methods such as that proposed by Shanian et al. (2008) and Dehghan-Manshadi et al. (2007) as-well as others. There are some other documented methods that would be worthwhile to research further, these include: Z-transformation in statistics for normalisation of material properties (Fayazbaksh et al., 2009); Preference selection index method (Maniya and Bhatt, 2010); a novel method based on CES, adapted value engineering techniques, and TOPSIS (Thakker et al., 2008); Material filtration with multi-materials design (Giaccobi et al., 2010).

References

Ashby, M.F. (2011). *Materials Selection in Mechanical Design*. 4th ed, Elsevier.

Chatterjee, P. and Chakraborty, S. (2012). Material selection using pre-fer-ential ranking methods, *Materials and Design*, Vol. 35, pp. 384-393.

Cicek, K., Celik, M. and Ilker Topcu, Y. (2010). An integrated decision aid extension to material selection problem, *Materials and Design*, Vol 31, pp. 4398-4402.

Dehghan-Manshadi, B., Mahmudi, H., Abedian, A. and Mahmudi, R. (2007). A novel method for materials selection in mechanical design: Combination of non-linear normalization and a modified digital logic method, *Materials and Design*, Vol. 28, pp 8-15.

Edwards, W. and Barron, F.H. (1994). SMARTS and SMARTER: Improved simple methods for multiattribute utility measurement, *Organ Behav. Human Decis. Process*, Vol. 60, pp.306-325.

Fayazbaksh, K., Abedian, A., Dehghan-Manshadi, B. and Khabbaz, R.S. (2009). Introducing a novel method for materials selection in mechanical design using Z-transformation in statistics for normalization of material properties, *Materials and Design*, Vol. 30, pp.4396-4404.

Figuiera, J. and Roy, B. (2002). Determining the weights of criteria in the ELECTRE type methods with a revised Simos' procedure, *European Journal of Operational Research*, Vol. 139, pp. 317-326.

Findik, F. and Turan, K. (2012). Materials selection for lighter wagon design with a weighted property index method, *Materials and Design*, Vol. 37, pp.470-477.

Giaccobi, S., Kromm, F.X., Wargnier, H. and Danis, M. (2010). Filtration in materials selection and mulit-materials design, *Materials and Design*, Vol. 31, pp. 1842-1847.

Jahan, A., Mustapha, F., Ismail, M.Y., Sapuan, S.M. and Bahraminasab, M. (2010). A comprehensive VIKOR method for material selection, *Materials and Design*, Vol. 32, pp 1215-1221.

Jahan, A., Ismail, M.Y., Sapuan, S.M. and Mustapha, F. (2009). Material screening and choosing methods – A review, *Materials and Design*, Vol. 31, pp.696-705.

Jahan, A. Ismail, M.Y. Shuib, S., Norfazidah, D. and Edwards, K.L. (2011). An aggregation technique for optimal decision-making in materials selection, *Materials and Design*, Vol. 32, pp. 4918-4924.

Maniya, K. and Bhatt, M.G. (2010). A selection of material using a novel type decision-making method: Preference selection index method, *Materials and Design*, Vol. 31, pp. 1785-1789.

Matweb (n.d.) Online Materials Information Resource, [online] Available at www.matweb.com

Marzouk, M.M. (2011). ELECTRE III model for value engineering applications, *Automation in Construction*, Vol. 20, pp. 596-600.

Mayyas, A., Shen, Q., Mayyas, A., Abdelhamid, M., Shan, D., Qattawi, A. and Omar, M. (2011). Using quality function deployment and analytical hierarchy process for material selection of body-in-white, *Materials and Design*, Vol. 32, pp. 2771-2782.

Opricovic, S. and Tzeng, G. (2002). Compromise solution by MCDM methods: A comparative analysis of VIKOR and TOPSIS, *European Journal of Operational Research*, Vol. 156, pp.445-455.

Parate, O. and Gupta, N. (2011). Material selection for electrostatic microactuators using Ashby approach, *Materials and Design*, Vol. 32, pp. 1577-1581.

Ramalhete, P.S., Senos, A.M.R. and Aguiar, C. (2010). Digital tools for material selection in product design, *Materials and Design*, Vol. 31, pp.2275-2287.

Shanian, A., Milani, A.S., Carson, C. and Abeyaratne, R.C. (2008). A new application of ELECTRE III and revised Simos' procedure for group material selection under weighting uncertainty, *Knowledge-Based Systems*, Vol. 21, pp 709-720.

Shanian, A. and Savadogo, O. (2006). TOPSIS multiple-criteria decision support analysis for material selection of metallic bipolar plates for polymer electrolyte fuel cell, *Journal of Power Sources*, Vol. 159, pp.1095-1104.

Shanian, A. and Savadogo, O. (2006). A non-compensatory compromised solution for material selection of bipolar plates for polymer electro-lyte membrane fuel cell (PEMFC) using ELECTRE IV, *Electrochimica Acta*, Vol. 51, pp.5307-5315.

Simos, J. (1989). L'evaluation environnementale: Un processus cognitive negocie; These de doctorat. DGF-EPFL, Lausanne.

Thakker, A., Jarvis, J., Buggy, M. and Sahed, A. (2008). A novel approach to materials selection strategy case study: Wave energy extraction impulse turbine blade, *Materials and Design*, Vol. 29, pp.1973-1980.

The Polymeric Heterogeneous Matrix for Composites

Alexandr Muranov, Galina Malysheva, Anna Pilyugina

Rocket and Spacecraft Composite Structures Department, Faculty of Special Machinery, Bauman Moscow State Technical University, 5 2nd Baumanskaya Street, Moscow, 105005, Russia

Abstract: This paper presents the results of mechanical tests of the hybrid polymer matrix based on epoxy oligomer and thermoplastic polymers such as polysulfone and poly (arylene ether ketone). The optimum ratio of epoxy oligomer and thermoplastic polymer of the composite material, leading to the significant improvement of mechanical properties and without causing any decrease in thermal stability, was determined. Diaminodiphenylsulfone have been used as a hardener. Structures of hybrid polymer after curing were analysed using electron microscopy for various processing conditions (mixing parameters). The rates of failures and strength for the hybrid polymer were experimentally determined. Calculation of the failure durations (i.e. time to failure) were conducted using parametric models of reliability. The values of impact strength, flexural strength and elongation were used as output parameters.

Key Words: Polymer composite material, Epoxy oligomer, Polysulfone, Failure durations.

1. Introduction

The matrix of the polymer composite material (PCM) provides different functions. It provides rigidity to the material. It promotes even distribution of load between the reinforcing elements, which leads to inhibition of cracks growth, as well as transmission and redistribution of stresses. An ideal matrix should have high elasticity modulus, a relatively small elongation and high adhesion strength. One of the key requirements for matrix is the ductility of cured matrix and the corresponding deformation property of the filler. Failure strain of the polymer matrix should be somewhat higher than that of fibre.

When carbon and glass fibres of large diameter (15-20 microns) are used, the failure strain usually does not exceed 1.5-2%. Thin fibers (having diameter less than 10 microns) have significantly higher values of failure strain, i.e. 3-5%. Thus, the relative failure strain of the matrix must be between 1.5-5%.

Epoxy matrices are the most commonly used matrices for structural PCMs. A PCM based on epoxy is 15 times stronger than silicone, and several times stronger than one based on phenol-formaldehyde resins. They are only slightly inferior to matrices based on epoxyphenol resins in terms of thermal properties.

Some of the main advantages for using epoxy matrices are high adhesive strength, good processability, low shrinkage, swelling etc. However, standard epoxy are usually quite brittle and their failure strain, usually, does not exceed the value of 1%. Therefore, it is important to modify them in order to increase the deformation properties of PCMs.

Introduction of highly elastic caotchoucs allows to improve impact strength and fracture toughness of the PCM.

However this leads to the reduction of the elastic modulus and glass transition temperature.

At the same time, incorporation of thermoplastics leads to a substantial improvement in the deformation characteristics of epoxy matrix without causing any decrease in their performance. Thermoplastics are added to the epoxy oligomer at the stage of matrix preparation, often before adding the curing agent. (Gorbunova et al., 2009; Mikhailin et al., 2002).

2. Experimental methods

The goal of this study is to investigate the structure of a heterogeneous matrix and estimate time to failure for the epoxy products exposed to long term cyclic loading. The epoxy resin based on diglycidyl ether and hardener diaminodiphenylsulfone (30 parts by weight per 100 parts by weight of epoxy resin) were used. Polysulfone PSK-1 was selected as the thermoplastic polymer for this work. Polyetheretherketone (Table 1) was also used for this study.

Epoxy oligomer and polysulfone were mixed and then the hardener was added. Samples for bending tests σ_{bend} (GOST 626-90), elongation at stretching ε (GOST 11262-80) and impact toughness (GOST 14235-69) were produced. Adhesive strength was also determined using lap-shear test of the glued metal samples (GOST 14769-69). Curing was performed at 180°C for 3 hours. Mechanical properties of the matrix, depending on the quantity of the thermoplastics, are shown in Table 2.

3. Results and discussion

As it can be seen from Table 2, the introduction of thermoplastics in epoxy matrix leads to the increase of flex-

Table 1
Physical and mechanical properties of the thermoplastics

	Polysulfone (CPM-1)	Polyetheretherketone (PEEK)	Polyetherimide (PEI)
Operating temperature range, °C	−100 ... +175	−40 ... +230	−40 ... +170
Elastic modulus, MPa	2600	3000	3200
Coefficient of thermal expansion, $\times 10^{-5}$ 1/K	5.5	5	5
Water absorption, %	0.2	0.1	0.7
Solubility in ED-20	Complete	Partial	Partial

Table 2
Influence of the polysulfone content on the mechanical properties of
modified epoxy matrix

№	The content of CPM-1, %	t, MPa	e, %	σ_{bend}, MPa	A, KJ/m^2
1	0	24	0.2	25	4.8
2	5	24	1.2	32	10.4
3	10	25	1.5	42	19.5
4	20	26	1.9	49	28.2
5	30	22	2.2	54	37.4

Table 3
The calculated average of the no-failure time at different values
of the output parameter

The desired confidence level	No-failure time for different maximum permissible values of output parameters, h	
	$X_{lim} = 0.5X_0$	$X_{lim} = 0.6X_0$
Output parameter – toughness		
0.9	1808	1352
0.99	1662	1205
0.999	1480	1023
Output parameter – bending strength		
0.9	7382	5520
0.99	6789	4927
0.999	6048	4189
Output parameter – elongation		
0.9	988	729
0.99	894	635
0.999	776	517

ural strength and significant increase of impact toughness. Increase in the elongation, which is almost an order of magnitude, was also observed. This indicates a significant positive influence of thermoplastic polymer on the deformation properties of the matrix.

However, the adhesive strength, by contrast, is somewhat reduced when the amount of a thermoplastic polymer is increased. It is probably due to the increased viscosity of the system and resulting increase of the adhesive line thickness above its optimum value. At the same time, a slight reduction in adhesion strength is not critical here and does not lead to any reduction of the working properties of PCM made from this polymer hybrid matrix.

The glass transition temperature for the epoxy that does not contain polysulfone (Table 2 - for sample no. 1) is 180 °C. With the introduction of the polysulfone to the epoxy, in amounts of up to 10%, the glass transition temperature of the epoxy does not change. When the quantity of polysulphone is further increased up to 30%, the glass transition temperature continues to gradually fall to 165 °C (Table. 2 - for sample no. 5).

Broken samples of the hybrid matrix containing 30% of polysulphone were examined after bending tests using a Phenom electron microscope (Figure 1a and 1b). The brigher phase corresponds to epoxy oligomer, whereas the darker phase is polysulfone. As it can be seen from the SEM images, the degree of mixing is quite low. Therefore the development of new processing conditions that would lead to an even distribution of one polymer in another is required. However, even with such an inhomogeneous structure, deformation properties of the epoxy-polysulfone matrix were improved (see Table. 2).

Failure durations or time to failure of tested matrices were determined using the theory of reliability (Pronikov, 2002). The average failure durations T_{av} can be expressed in terms of the expected value of working time to failure.

$$T_{av} = \int_{-\infty}^{+\infty} P(t)dt \qquad (1)$$

where $P(t)$ is probability of failure-free operation during the time t.

In order to quantify this time, one has to determine the rate of change (destruction) γ, which in this case is taken as the value of the output parameter X.

$$\gamma = \frac{dX}{dt} \qquad (2)$$

Real products are normally characterised by a number of its strain-strength properties and in this case, the total rate of aging (destruction) would be defined as

$$X = f(\gamma_1, \gamma_2 ... \gamma_n) \qquad (3)$$

and the probability of failure of such product, respectively, can be represented by the following equation

$$P(t) = \prod_{i=1}^{n} P_i(t) \qquad (4)$$

where $P_i(t)$ is probability of failure of the i-th output parameter for the time t.

While the rate of failure can be found experimentally, the average value of the failure durations T_{av}, is given by

$$T_{av} = \frac{X_0 - \tau_p \sigma - X_{lim}}{\gamma} \qquad (5)$$

where X_0 is the initial value of the output parameter; X_{lim} is the maximum value of the same output parameter (usually taken to be 50% of the X_0); σ is the dispersion of the initial value of the output parameter (determined experimentally or given by the coefficient of variation); τ_p is quantiles, whose values vary with a given probability .

The results of calculating the average of the failure durations according to equation (5) for different values of the given confidence level are shown in Table 3. Mechanical properties from Table 2 have been used as the initial values for X_0. The dispersion values for each of the output parameters were determined by using the coefficient of variation of k, in accordance with Pronikov (2002), i.e. assumed to be equal $k = 0.08X_0$. For the calculation of alternating loads, we used the following values for the rate of failure for impact toughness $\gamma_1 = 0.0081$ kJ/m^2/h, for the elongation $\gamma_2 = 0.0085$ %/h and for the bending strength $\gamma_3 = 0.0029$ MPa/h.

It is known (Pronikov, 2002) that the rate of failures strongly depends on the aging conditions. Usually, in the experiments, the influence of each factor is tested separately (for example, the influence of high and low temperatures or cyclic loading), whereas in reality, these factors act simultaneously. Therefore, the values of the failure rates, which are determined experimentally for each of the working conditions, should be considered as conditional. These results provide the basis for a comparative

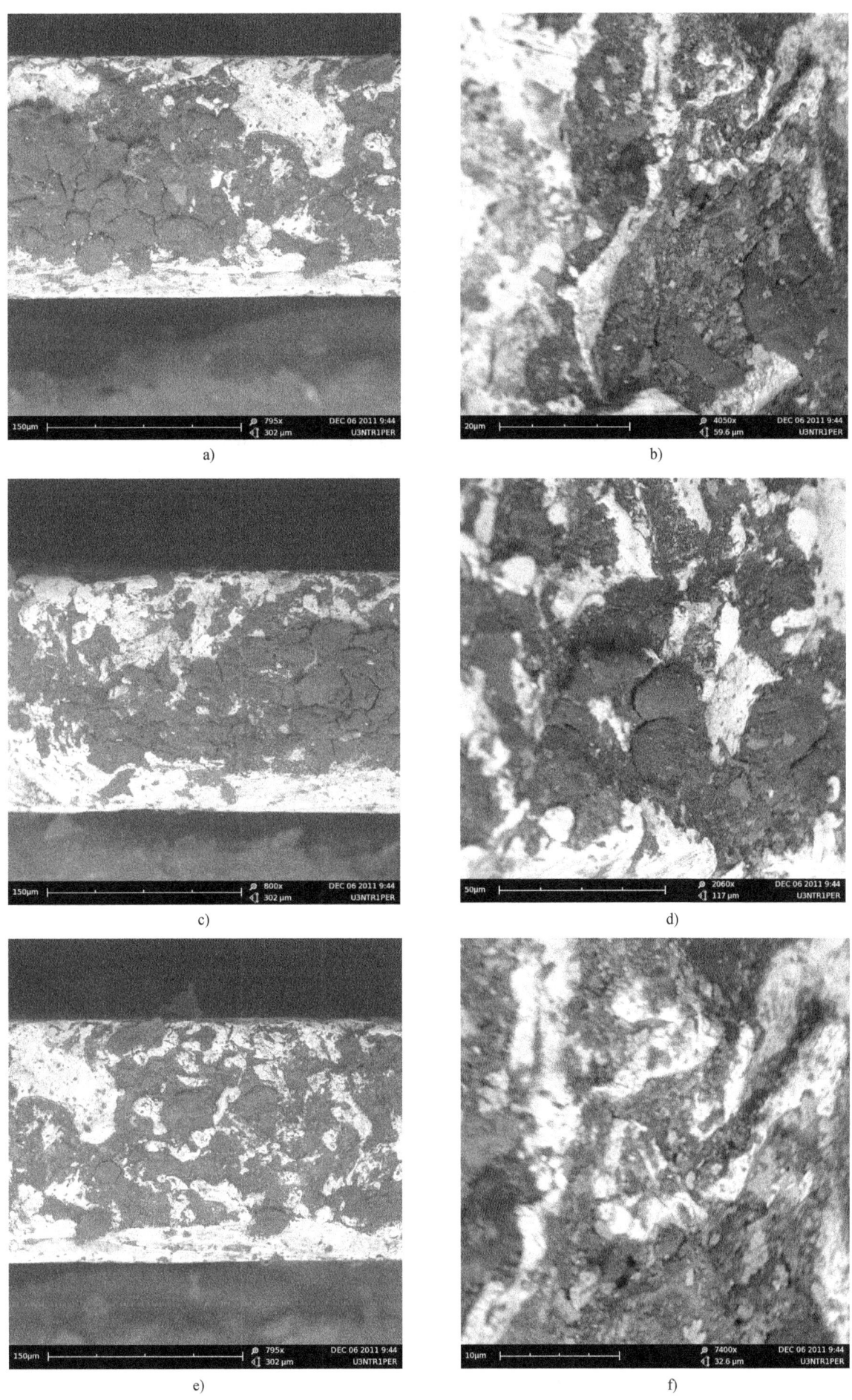

Figure 1. Fracture surfaces of epoxy matrix, content of 30% polysulfone for different microscope approach (a, c, e – x800) (b, d, f – x2060).

analysis of the effect for any of the output parameters on failure durations

4. Conclusion

1) The materials made from hybrid polymeric matrix based on epoxy oligomers and polysulfone have increased strain-strength properties. The most significant increase in the impact toughness was obtained for the matrix containing 30% polysulphone. It also significantly increased the elongation from 0.2% for the epoxy binder to 2.2% for the binder containing 30% polysulphone.

2) The microstructure of polymeric matrix based on epoxy oligomers and polysulfone is heterogeneous and requires the development of new mixing techniques.

3) Calculation of the failure durations (i.e. time to failure) at a given confidence level is performed by a parametric model of reliability. Impact toughness, flexural strength and elongation values were used as the output parameters. In order to improve the accuracy of calculated failure durations, it is required to determine the variation in each of the employed output parameters.

During this work, it was found that the hybrid polymer matrix formed by mixing thermoplastic and thermosetting materials, demonstrates increased deformation and strength properties. This also resulted in increasing the service life of products made of polymeric composite materials.

References

Gorbunova, I.Y., Kerber, M.L. and Kazakov, S.I. (2009). Effect of thermoplastics of different chemical structure to changes in viscosity during curing adhesives and binders based on epoksiamin, in *Proceeding of Tenth International Conference on Chemistry and Physical Chemistry of Oligomers*, Volgograd, pp. 3-4.

Mikhailin, Y.A., Kerber, M.L. and Gorbunova, I.Y. (2002). Binders for polymeric composite materials, *Plastics*, No 12, pp. 14-21.

Pronikov, A.S. (2002). *Reliability Theory of Machines*. Moscow: Bauman Moscow State Technical University.

A Parametric Study of the Low-Impulse Blast Behaviour of Fibre-Metal Laminates Based on Different Aluminium Alloys

Thuc Vo[1], Zhongwei Guan[2], Wesley Cantwell[2], Graham Schleyer[2]

[1] School of Mechanical, Aeronautical and Electrical Engineering, Glyndŵr University, Plas Coch, Mold Road, Wrexham, LL11 2AW, UK
[2] School of Engineering, University of Liverpool, Brownlow Street, Liverpool, L69 3GQ, UK

Abstract: A parametric study has been undertaken in order to investigate the influence of the properties of the aluminium alloy on the blast response of fibre-metal laminates (FMLs). The finite element (FE) models have been developed and validated using experimental data from tests on FMLs based on a 2024-O aluminium alloy and a woven glass-fibre/polypropylene composite (GFPP). A vectorized user material subroutine (VUMAT) was employed to define Hashin's 3D rate-dependant damage constitutive model of the GFPP. Using the validated models, a parametric study has been carried out to investigate the blast resistance of FML panels based on the four aluminium alloys, namely 2024-O, 2024-T3, 6061-T6 and 7075-T6. It has been shown that there is an approximation linear relationship between the dimensionless back face displacement and the dimensionless impulse for all aluminium alloys investigated here. It has also shown that the residual displacement of back surface of the FML panels and the internal debonding are dependent on the yield strength of the aluminium alloy.

Key Words: Fibre-metal laminates, Localised blast loading, Hashin's 3D failure criteria, Strain-rate effects, Finite element models.

1. Introduction

Fibre-metal laminates (FMLs) are hybridised metal and composite structural materials that have been attracting interest from a number of researchers due to their improved fatigue and impact resistance (Vlot, 1996; Krishnakumar, 1994; Vogelesang and Vlot, 2000; Compston et al., 2001; Reyes-Villanueva and Cantwell, 2004; Reyes and Cantwell, 2000). The most commonly used FML is GLARE, which comprises thin aluminium 2024-T3 sheets and a unidirectional or a biaxial glass-fibre-reinforced epoxy. The blast response of FMLs has received attention in a number of experimental studies. Fleischer (1996) presented data from blast test results on a lightweight luggage container based on GLARE and reported that it was capable of withstanding a bomb blast greater than that in the Lockerbie air disaster. Langdon et al. (2007a, 2007b) and Lemanski et al. (2007) carried out blast tests of FML panels based on a 2024-O aluminium alloy and a glass fibre reinforced polypropylene. Blast tests on FML panels based on other composites, such as a glass fibre polyamide matrix and GLARE were also been undertaken by Langdon et al. (2007c, 2009). Since experimental trials are usually very costly and time-consuming, it is evident that modelling the blast behaviour of FMLs using commercial finite element software would be great interest. Once these models are verified, they can be used to predict the response of FMLs based on different configurations, lay-ups, loading and boundary conditions without the need to undertake a large number of experimental tests. However, in spite of the fact that there have been a number of experimental studies on the blast behaviour of FMLs, relatively little work has been conducted to model their response. Kotzakolios et al. (2011a) used LS-DYNA to investigate the blast response of GLARE laminates-comparison against experimental results. Later, Kotzakolios et al. (2011b) extended their research to investigate the damage induced in a typical commercial fuselage based on aluminium and GLARE, when subjected to an explosive charge. Soutis et al. (2011) investigated the

structural response of fully clamped GLARE panels to blast loads using LS-DYNA. Karagiozova et al. (2010) modelled the blast response of FML panels based on various stacking configurations using ABAQUS/Explicit in order to predict the influence of the loading parameters and structural characteristics on their overall behaviour. However, damage occurring inside the composite material was not considered in the analysis.

In this paper, rate-dependent failure criteria for a unidirectional composite are developed by modifying Hashin's 3D failure criteria (Hashin, 1980). The constitutive model and failure criteria are then implemented in ABAQUS/Explicit using the VUMAT subroutine. The FE models were developed and validated using experimental data from tests on FMLs based on a 2024-O aluminium alloy and a woven glass-fibre/polypropylene composite. Using the validated models, a parametric study was carried out to investigate the influence of the properties of the aluminium alloy on the blast resistance of FMLs. Particular attention is given to predicting the front and back displacements and the energies dissipated during the blast process. In total, thirty-six cases are studied.

2. Geometric and blast loadings of FML panels

For verification purposes, the FML panels previously subjected to localised blast loading in the experimental study by Langdon et al. ([8]-[10]) are used to validate the current FE models. These 400×400 mm panels (300×300 mm exposed area), were manufactured from sheets of 0.025 in. (approximately 0.6 mm) thick 2024-O aluminium alloy and a woven glass-fibre/polypropylene composite. The FML panels are identified using the notation, AXTYZ-#, as described in (Langdon, 2007a), where A = aluminium, X = number of aluminium layers, T = GFPP, Y = number of blocks of GFPP, Z = number of plies of GFPP per block and # indicates the panel number.

In order to investigate the influence of the properties of the aluminium alloy on the low-impulse blast behavior, FML panels based on the four aluminium alloys, namely

69

Table 1
Details of the lay-ups and impulses for verification.

Lay-ups	No. of layers	Thickness (mm)	Impulse (Ns)
A2T18-4	10	5.60	7.94
A3T24-8	11	6.06	7.85
A3T26-3	15	8.10	9.54
A3T28-5	19	9.82	10.34
A4T32-4	10	5.85	7.23
A4T34-5	16	8.73	7.01
A4T36-2	22	11.48	11.61
A4T38-2	28	13.90	11.13
A5T42-4	13	7.46	8.87

2024-O, 2024-T3, 6061-T6 and 7075-T6, subjected to an impulse I = 8 Ns were considered. All four alloys are widely used in the aerospace industry. Details of the lay-ups and impulses investigated in this study are listed in Table 1.

Blast pressure load, calculated from the measured impulse, is a function of both time and distance from the plate centre. The pressure–time history is idealized as a uniform function over a small central region and follows an exponentially decaying function as:

$$P(r,t) = p_1(r) p_2(t) \qquad (1)$$

where:

$$p_1(r) = \begin{cases} P_0 & r \le r_0 \\ P_0 e^{-k(r-r_0)} & r_0 < r < r_b \\ 0 & r > r_b \end{cases} \qquad (2)$$

$$p_2(t) = e^{-2t/t_0}$$

where $r_0 = 15$ mm is the radius of the explosive disc used in the experiments, $r_b < L/2$, L is the length of the panel and $t_0 = 0.008$ ms is the characteristic decay time for the pulse and k is an exponential decay parameter. The decay parameter is not constant, but a function of the total impulse (Karagiozova et al., 2010). The total impulse is defined as:

Table 2
Johnson-Cook constants for aluminium alloys

Johnson-Cook constants	A (MPa)	B (MPa)	n	C
Al 2024-O (Karagiozova et al., 2010)	85	325	0.40	0.0083
Al 2024-T3 (Lesuer, 2000)	369	684	0.73	0.0083
Al 6061-T6 (Corbett, 2006)	324	114	0.42	0.0020
Al 7075-T6 (Brar et al., 2009)	546	678	0.71	0.0240
Damage constants	D_1	D_2	D_3	D_4
Al 2024-O (Karagiozova et al., 2010)	0.130*	0.130*	-1.500*	0.011*
Al 2024-T3 (Lesuer, 2000)	0.130	0.130	-1.500	0.011
Al 6061-T6 (Corbett, 2006)	-0.770	1.450	-0.470	0.000
Al 7075-T6 (Brar et al., 2009)	-0.068	0.451	-0.952	0.036

* Damage constants for Al 2024-T3 were used due to the lack of available data.

$$I = 2\pi \int_0^\infty \int_0^{r_b} P(r,t)\, dr\, dt \qquad (3)$$

A user subroutine VDLOAD was used to model the pressure distribution over the exposed area of the plate.

3. Material modelling

3.1 Aluminium layers

The aluminium alloy was modelled as an elasto-plastic material, exhibiting rate-dependent behaviour. Temperature effects in the aluminium alloy were not taken into account. The Johnson-Cook material model was used in the form:

$$\sigma = \left[A + B(\overline{\varepsilon}_{pl})^n \right] \left[1 + C \ln\left(\frac{\dot{\overline{\varepsilon}}_{pl}}{\dot{\varepsilon}_0} \right) \right] \qquad (4)$$

where $\overline{\varepsilon}_{pl}$ is the equivalent plastic strain; $\dot{\overline{\varepsilon}}_{pl}$ and $\dot{\varepsilon}_0$ are the equivalent plastic and reference strain rate and A, B, C and n are material parameters.

Damage in the Johnson-Cook material model is predicted using the following cumulative damage law:

$$D = \sum \left(\frac{\Delta \overline{\varepsilon}_{pl}}{\overline{\varepsilon}_f^{pl}} \right) \qquad (5)$$

in which:

$$\overline{\varepsilon}_f^{pl} = \left[D_1 + D_2 \exp\left(D_3 \sigma^* \right) \right] \left[1 + D_4 \ln\left(\frac{\dot{\overline{\varepsilon}}_{pl}}{\dot{\varepsilon}_0} \right) \right] \qquad (6)$$

where $\Delta \overline{\varepsilon}_{pl}$ is the increment of equivalent plastic strain during an increment in loading and σ^* is the mean stress normalised by the equivalent stress. The parameters D_1, D_2, D_3, and D_4 are constants. Failure is assumed to occur when $D = 1$. Hence the current failure strain, $\overline{\varepsilon}_f^{pl}$, and thus the accumulation of damage, D, is a function of the mean stress and the strain rate. The constants in the Johnson-Cook model for the four alluminium alloys used in this study are given in Table 2. The Young's modulus, Poisson's ratio and density of the various aluminium alloys were taken as $E = 73.1$ GPa, $\nu = 0.3$ and $\rho = 2690$ kg/m^3, respectively.

3.2 Glass fibre reinforced composite layers

3.2.1 The 3D damage model for the composite material

Given that a woven glass-fibre/polypropylene composite layer is produced by placing fibres in a $[0^0/90^0]$ pattern, the material behaviour within the plane of the laminate is similar in those two directions. There is therefore no need to separate the fibre and resin in order to simulate the overall response of the composite ply. Besides, the material tests carried out in this paper were based on the composite laminates, i.e. no individual tests to address fiber and resin separately. Therefore, Hashin's 3D failure criteria (Hashin, 1080) are sufficient to simulate woven glass-fibre/polypropylene composite layer. The failure functions may be expressed as follows:

Fibre tension $(\sigma_{11} \geq 0)$:

$$F_f^t = \left(\frac{\sigma_{11}}{X_{1t}}\right)^2 + \left(\frac{\sigma_{12}}{S_{12}}\right)^2 + \left(\frac{\sigma_{13}}{S_{13}}\right)^2, \ d_{ft} = 1$$

Fibre compression $(\sigma_{11} < 0)$:

$$F_f^c = \frac{|\sigma_{11}|}{X_{1c}}, \ d_{fc} = 1$$

Matrix tension $(\sigma_{22} + \sigma_{33} \geq 0)$:

$$F_m^t = \frac{(\sigma_{22} + \sigma_{33})^2}{X_{2t}^2} + \frac{\sigma_{23}^2 - \sigma_{22}\sigma_{33}}{S_{23}^2} + \frac{\sigma_{12}^2 + \sigma_{13}^2}{S_{12}^2}, \ d_{mt} = 1 \quad (7)$$

Matrix compression $(\sigma_{22} + \sigma_{33} < 0)$:

$$F_m^c = \left[\left(\frac{X_{2c}}{2S_{23}}\right)^2 - 1\right]\frac{(\sigma_{22} + \sigma_{33})}{X_{2c}}$$

$$+ \frac{(\sigma_{22} + \sigma_{33})^2}{4S_{23}^2} + \frac{(\sigma_{23}^2 - \sigma_{22}\sigma_{33})}{S_{23}^2} + \frac{\sigma_{12}^2 + \sigma_{13}^2}{S_{12}^2}, \ d_{mc} = 1$$

where X_{1t}, X_{1c}, X_{2t}, X_{2c}, S_{12}, S_{13} and S_{23} are the various strength components (Hashin, 1980) and d_{ft}, d_{fc}, d_{mt} and d_{mc} are the damage variables associated with the four failure modes.

The response of the material after damage initiation (which describes the rate of degradation of the material stiffness once the initiation is satisfied) is defined by the following equation:

$$\sigma = C(d) \cdot \varepsilon \quad (8)$$

where $C(d)$ is a 6×6 symmetric damaged matrix, whose non-zero terms can be written as:

$$C_{11} = (1 - d_f)E_1(1 - \nu_{23}\nu_{32})\Gamma$$
$$C_{22} = (1 - d_f)(1 - d_m)E_2(1 - \nu_{13}\nu_{31})\Gamma$$
$$C_{33} = (1 - d_f)(1 - d_m)E_3(1 - \nu_{12}\nu_{21})\Gamma$$
$$C_{12} = (1 - d_f)(1 - d_m)E_1(\nu_{21} + \nu_{31}\nu_{23})\Gamma$$
$$C_{23} = (1 - d_f)(1 - d_m)E_2(\nu_{32} + \nu_{12}\nu_{31})\Gamma \quad (9)$$
$$C_{13} = (1 - d_f)(1 - d_m)E_1(\nu_{31} + \nu_{21}\nu_{32})\Gamma$$
$$C_{44} = (1 - d_f)(1 - s_{mt}d_{mt})(1 - s_{mc}d_{mc})G_{12}$$
$$C_{55} = (1 - d_f)(1 - s_{mt}d_{mt})(1 - s_{mc}d_{mc})G_{23}$$
$$C_{66} = (1 - d_f)(1 - s_{mt}d_{mt})(1 - s_{mc}d_{mc})G_{13}$$

where the global fibre and matrix damage variables as well as the constant Γ are also defined as:

$$d_f = 1 - (1 - d_{ft})(1 - d_{fc})$$
$$d_m = 1 - (1 - d_{mt})(1 - d_{mc})$$
$$\Gamma = \frac{1}{1 - \nu_{12}\nu_{21} - \nu_{23}\nu_{32} - \nu_{13}\nu_{31} - 2\nu_{21}\nu_{32}\nu_{13}} \quad (10)$$

where E_i is the Young's modulus in the i direction, G_{ij} is the shear modulus in the i-j plane and ν_{ij} is the Poisson's ratio for transverse strain in the j-direction, when the stress is applied in the i-direction. The Young's moduli, shear's moduli, Poisson's ratios and strengths of the GFPP are given in Table 3.

The factors s_{mt} and s_{mc} in the definitions of the shear moduli are introduced to control the reduction in shear stiffness caused by tensile and compressive failure in the matrix respectively. The following values are recommended in ABAQUS (2009): $s_{mt} = 0.9$ and $s_{mc} = 0.5$.

3.2.2 Strain-rate effects in the mechanical properties

The effects of strain-rate on the mechanical properties of a composite material are typically modelled using strain-rate dependent functions for both the elastic modulus and the strength. Yen (2002) developed logarithmic functions to account for strain-rate effects in a composite material as follows:

$$\{S_{RT}\} = \{S_0\}\left(1 + C_1 \ln\frac{\dot{\bar{\varepsilon}}}{\dot{\varepsilon}_0}\right)$$
$$\{E_{RT}\} = \{E_0\}\left(1 + C_2 \ln\frac{\dot{\bar{\varepsilon}}}{\dot{\varepsilon}_0}\right) \quad (11)$$

where:

$$\{\dot{\bar{\varepsilon}}\} = \{|\dot{\varepsilon}_1| \ \ |\dot{\varepsilon}_2| \ \ |\dot{\varepsilon}_1| \ \ |\dot{\varepsilon}_2| \ \ |\dot{\varepsilon}_{12}| \ \ |\dot{\varepsilon}_{13}| \ \ |\dot{\varepsilon}_{23}|\}^T$$
$$\{S_{RT}\} = \{X_{1t} \ \ X_{2t} \ \ X_{1c} \ \ X_{2c} \ \ S_{12} \ \ S_{13} \ \ S_{23}\}^T \quad (12)$$
$$\{E_{RT}\} = \{E_1 \ \ E_2 \ \ E_3 \ \ G_{12} \ \ G_{13} \ \ G_{23}\}^T$$

and the subscript 'RT' refers to the rate-adjusted values, the subscript '0' refers to the static value, $\dot{\varepsilon}_0 = 1 \ \text{s}^{-1}$ is the reference strain-rate, $\dot{\bar{\varepsilon}}$ is the effective strain-rate, C_1 and C_2 are the strain-rate constants, respectively.

3.2.3 Implementation of the material model in ABAQUS/ Explicit

The material model and failure criteria described in the previous sections were implemented in ABAQUS/ Explicit using the VUMAT subroutine. This subroutine is compiled and enables ABAQUS/Explicit to obtain the required information regarding the state of the material and the material mechanical response during each time step, at each integration point of each element. The stresses are computed within the VUMAT subroutine using the given strains and the material stiffness coefficients. Based on these stresses, Hashin's 3D failure criteria outlined in Eq.(7) are calculated, and the elastic

Table 3
Properties of the GFPP layers

Elastic properties	Values	Progressive failure	Values
ρ (kg/m³)	1800	X_{1t} (MPa)	300
E_1 (GPa)	13.0	X_{1c} (MPa)	200
E_2 (GPa)	13.0	X_{2t} (MPa)	300
E_3 (GPa)	2.40	X_{2c} (MPa)	200
G_{12} (GPa)	1.72	S_{12} (MPa)	140
G_{13} (GPa)	1.72	S_{13} (MPa)	140
G_{23} (GPa)	1.69	S_{23} (MPa)	140
ν_{12}	0.1		
ν_{13}	0.3		
ν_{23}	0.3		

modulus and strength values are adjusted for strain-rate effects using Eq.. When an element fails, as determined by the failure criteria, the element status is then changed from 1 to 0. At this point, the stresses at that material point are reduced to zero and it no longer contributes to the model stiffness. When all of the material status points of an element have been reduced to zero, the element is removed from the mesh.

3.3 Cohesive elements and material properties

Debonding at the interface between the composite and aluminium layers was modelled using cohesive elements available in ABAQUS (2009). The elastic response was defined in terms of a traction-separation law with uncoupled behaviour between the normal and shear components. The default choice of the constitutive thickness for modelling the response, in terms of traction versus separation, is 1.0, regardless of the actual thickness of the cohesive layer. Thus, the diagonal terms in the elasticity matrix and density should be calculated using the true thickness of the cohesive layer as follows:

$$K_{nn} = \frac{E_n}{t_c}$$

$$K_{ss} = \frac{E_s}{t_c}$$

$$K_{tt} = \frac{E_t}{t_c}$$

$$\rho = \rho_c t_c$$

(13)

The quadratic nominal stress and energy criterion were used to model damage initiation and damage evolution, respectively. Damage initiated when a quadratic interaction function, involving the nominal stress ratios, reached unity. Damage evolution was defined based on the energy conjunction with a linear softening law. The mechanical properties of the cohesive elements were obtained from Karagiozova et al. (2010) and are given in Table 4.

4. Finite element modelling

The 3D FML panel consisted of the aluminium alloy, the composite and the cohesive layers as three separate parts.

Table 4
Properties of the cohesive layers with thickness 1mm.

Elastic properties			
ρ_c (kg/m³)	E_n (GPa)	E_s (GPa)	E_t (GPa)
920	2.05	0.72	0.72
Damage initiation			
t_n^0 (MPa)	t_s^0 (MPa)	t_t^0 (MPa)	
140	300	300	
Damage evolution			
G_n^c (J/m²)	G_s^c (J/m²)	G_t^c (J/m²)	
2000	3000	3000	

G_n^c, G_s^c and G_t^c are the critical fracture energies in the normal, the first, and the second shear directions.
t_n^0, t_s^0 and t_t^0 are the critical nominal normal stress, the first and the second shear stresses.

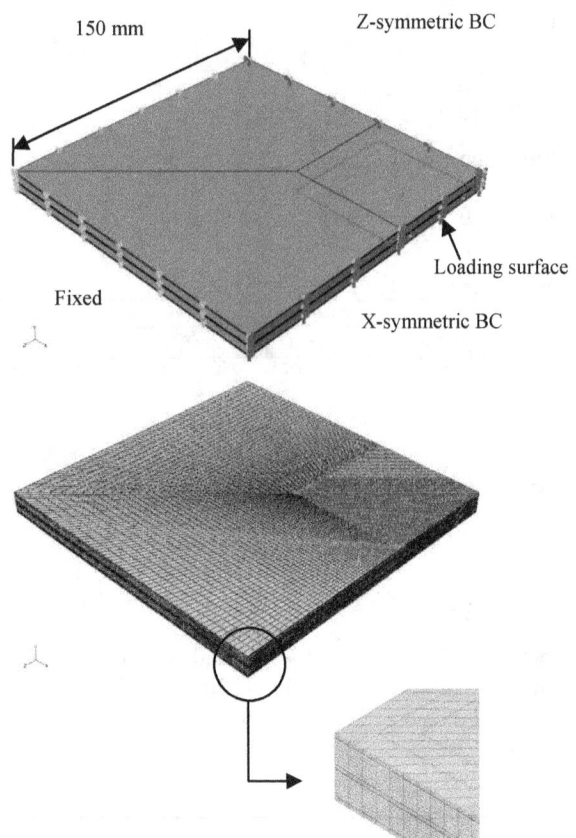

Figure 1. Dimensions, loading, boundary conditions and mesh generation for typical 3/2 FML panel.

The aluminium and composite layers were meshed using C3D8R elements, which are eight-noded, linear hexahedral elements with reduced integration and hourglass control. The individual aluminium and composite plies were discretized with two elements through the thickness. The interfaces between the aluminium and the composite layers were created using eight-node 3D cohesive elements (COH3D8). As the structure has symmetry in both the directions, only a quarter of each FML panel was modeled with the appropriate boundary conditions applied along the planes of symmetry, as shown in Fig. 1. A mesh size of 1×1 mm for a central area of 60×60 mm (Fig. 1) was found to be the most appropriate for these FML panels. Symmetric boundary conditions were applied to the nodes lying on the XY and YZ planes, while the other two edges were fully fixed. The general contact algorithm was used

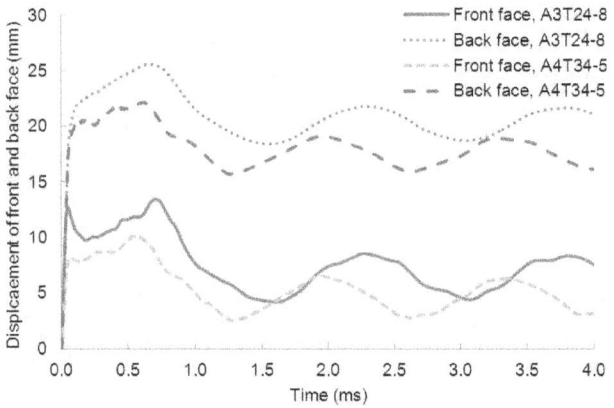

Figure 2. Back and front face displacements versus time for panels A3T24-8 and A4T34-5.

Table 5
Comparison of experimental data from Lemanski et al. (2007) and numerical simulation results of displacements of front and back faces for verification (mm).

Panel	Impulse (Ns)	Experiment		ABAQUS	
		Front	Back	Front	Back
A2T18-4	7.94	4.95	22.19	9.10	16.30
A3T24-8	7.85	6.06	20.34	10.10	19.30
A3T26-3	9.54	4.32	20.35	7.20	23.70
A3T28-5	10.34	3.23	22.59	2.20	20.90
A4T32-4	7.23	9.72	21.73	13.80	17.20
A4T34-5	7.01	4.57	17.48	7.60	14.00
A4T36-2	11.61	1.55	27.21	5.40	22.60
A4T38-2	11.13	1.29	24.59	4.30	20.70
A5T42-4	8.87	7.95	25.31	11.50	19.90

for the definition of contact between the two neighbouring layers of the aluminium and the composite. Surface-based tie constraints were imposed between either the aluminium or the composite layer and the cohesive layer to model adhesion between the adjacent layers.

5. Results and Discussion

Since there are no experimental data available in the literature to describe strain-rate effects in the woven glass-fibre/polypropylene composite, rate-dependent material models, with different values of the strain-rate constant, were investigated in this study. A material model incorporating strain-rate effects in the strength, shear and the through-thickness modulus values was chosen. Strain-rate constant values that agreed well with the experimental results were $C_1 = C_2 = 0.35$. This material model is consistent with results of McCarthy et al. (2004) and Gama and Gillespie (2011). Initially, two FML panels, A3T24-8 and A4T34-5, were studied to investigate their transient and residual displacements. After conducting a number of convergence studies, numerical simulations were carried

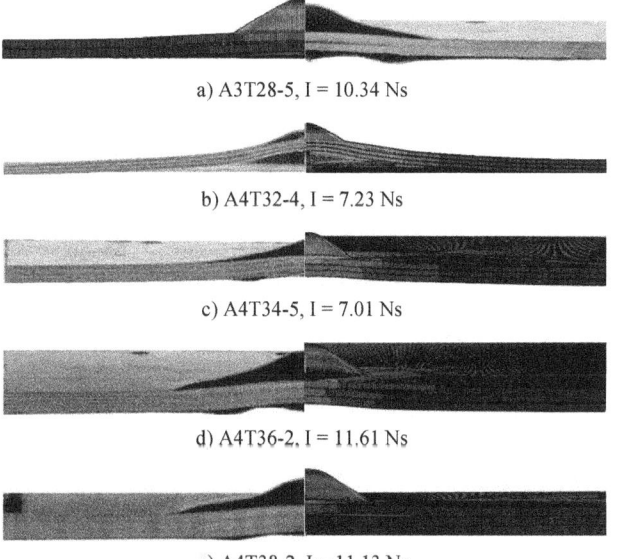

a) A3T28-5, I = 10.34 Ns

b) A4T32-4, I = 7.23 Ns

c) A4T34-5, I = 7.01 Ns

d) A4T36-2, I = 11.61 Ns

e) A4T38-2, I = 11.13 Ns

Figure 3. Comparison between the experiments and numerical simulations for five FML panels.

Table 6
Summary of permanent front and back face displacements of FMLs based on the four aluminium alloys.

Panel	Aluminium types	Dimensional parameters		
		Impulse (Ns)	Displacement (mm)	
			Front	Back
A2T18	Al 2024-O	8.00	5.53	22.92
	Al 2024-T3	8.00	5.17	18.19
	Al 6061-T6	8.00	5.71	20.84
	Al 7075-T6	8.00	4.50	15.20
A3T24	Al 2024-O	8.00	6.63	18.77
	Al 2024-T3	8.00	7.10	16.29
	Al 6061-T6	8.00	7.20	18.10
	Al 7075-T6	8.00	5.56	12.07
A3T26	Al 2024-O	8.00	4.58	16.21
	Al 2024-T3	8.00	4.42	13.15
	Al 6061-T6	8.00	4.21	15.13
	Al 7075-T6	8.00	2.91	9.30
A3T28	Al 2024-O	8.00	3.28	17.99
	Al 2024-T3	8.00	2.41	12.09
	Al 6061-T6	8.00	2.81	14.80
	Al 7075-T6	8.00	1.20	9.74
A4T32	Al 2024-O	8.00	9.48	24.29
	Al 2024-T3	8.00	8.52	16.42
	Al 6061-T6	8.00	8.04	19.12
	Al 7075-T6	8.00	4.91	12.57
A4T34	Al 2024-O	8.00	4.95	20.42
	Al 2024-T3	8.00	3.02	13.75
	Al 6061-T6	8.00	3.43	15.87
	Al 7075-T6	8.00	1.88	10.82
A4T36	Al 2024-O	8.00	2.36	18.41
	Al 2024-T3	8.00	1.21	13.49
	Al 6061-T6	8.00	1.45	15.64
	Al 7075-T6	8.00	0.47	10.44
A4T38	Al 2024-O	8.00	1.18	17.88
	Al 2024-T3	8.00	0.68	12.92
	Al 6061-T6	8.00	0.71	14.12
	Al 7075-T6	8.00	0.28	10.30

out over a time period of 4 ms. The transient displacement relates to the first peak in the displacement time trace and the residual displacement is taken as the average after more than three cycles following unloading. The variation of the front and back displacements with time are shown in Fig. 2. As expected, the deflections of the thinner A3T24 panel are greater than those of its stiffer A4T34 counterpart. It is worth noting that the difference between the front and back surface displacements is greater than the initial thickness of the panel. This increase in the effective thickness of the FML is associated with the opening up of planes of delamination within the volume of the laminate. FE models of other types of FML panels subjected to a low impulse were also developed to broaden the validation. The experimental and numerical results are presented in Table 5. Reasonable agreement between the predicted and experimental mid-point displacements is observed. Comparing the experimental and numerical failure modes of five typical panels, as shown in Fig. 3, the simulations accurately capture the primary failure mechanisms in the FMLs, which include large out-of-plane plastic displacements, debonding of the back face and local buckling of the internal aluminium layer.

The numerical results corresponding to FML panels based on the four aluminium alloys are presented in Table 6. It can be seen that the front and back displacements of those panels based on the aluminium 7075-T6 are the smallest, whereas those based on the aluminium 2024-O are the largest. This suggests that the properties of the aluminium alloy, most particularly its yield stress, greatly influence the blast response of these hybrid materials. It can be seen that the permanent displacements tend to decrease with increasing yield strength of the aluminium alloy.

6. Conclusions

A parametric study of the low-impulse blast behaviour of FMLs based on different aluminium alloys is presented. Here, three dimensional finite element models of FML panels based on a 2024-O aluminium alloy and a woven glass-fibre/polypropylene composite subjected to low-impulse localised blast loading are developed and validated against previously-published experimental data. Hashin's 3D failure criteria, incorporating strain-rate effects in the GFPP is implemented into ABAQUS/Explicit. Using the validated models, a parametric study is used to investigate the influence of the properties of the aluminium alloy on the blast resistance of FMLs based on the four aluminium alloys, namely 2024-O, 2024-T3, 6061-T6 and 7075-T6. The residual back displacement of the FML panels decreases with the increasing yield strength of the aluminium alloy.

References

ABAQUS (2009). *Theory Manual*, Version 6.9, Hibbitt, Karlsson and Sorensen.

Brar, N.S., Joshi, V.S. and Harris, B.W. (2009). Constitutive model constants for Al7075-T651 and Al7075-T6. In: *Proceedings of AIP Conference*, Vol. 1195, No 1, pp. 945-948.

Compston, P., Cantwell, W.J., Jones, C. and Jones, N. (2001). Impact perforation resistance and fracture mechanisms of a thermoplastic based fiber-metal laminate. *Journal of Materials Science Letters*, Vol. 20, pp. 597-599.

Corbett, B. (2006). Numerical simulations of target hole diameters for hypervelocity impacts into elevated and room temperature bumpers. *International Journal of Impact Engineering*, Vol. 33, No 1-12, pp. 431-440.

Fleisher, H.J. (1996). Design and explosive testing of a blast resistant luggage container. In: *International Conference on Structures under Shock and Impact*, pp. 51-60.

Gama, B.A. and Gillespie, J.W., Jr. (2011). Finite element modelling of impact, damage evolution and penetration of thick-section composites. *International Journal of Impact Engineering*, Vol. 38, No 4, pp. 181-197.

Hashin, Z. (1980). Failure criteria for unidirectional fiber composites, *Journal of Applied Mechanics*, Vol. 47, pp. 329-334.

Karagiozova, D., Nurick G.N. and Langdon, G.S. (2009). Behaviour of sandwich panels subject to intense air blasts – Part 2: Numerical simulation. *Composite Structures*, Vol. 91, No 4, pp. 442-450.

Karagiozova, D., Langdon, G.S., Nurick, G.N. and Yuen, S.C.K. (2010). Simulation of the response of fibre-metal laminates to localised blast loading. *International Journal of Impact Engineering*, Vol. 37, No 6, pp. 766-782.

Kingery, C.N. and Bulmash, G. (1984). *Air Blast Parameters from TNT Spherical Air Burst and Hemispherical Surface Burst*, ARBRL-TR-02555. Aberdeen, ML: Ballistic Research Laboratories.

Kotzakolios, T., Vlachos, D. and Kostopoulos, V. (2011a). Investigation of blast response of GLARE laminates: comparison against experimental results. *Plastics, Rubber and Composites*, Vol. 40, No 6-7, pp. 349-355.

Kotzakolios, T., Vlachos, D. and Kostopoulos, V. (2011b). Blast response of metal composite laminate fuselage structures using finite element modelling. *Composite Structures*, Vol. 93, No 2, pp. 665-681.

Krishnakumar, S. (1994). Fibre metal laminates-the synthesis of metals and composites. *Materials and Manufacturing Processes*, Vol. 9, No 2, pp. 295-354.

Langdon, G.S., Lemanski, S.L., Nurick, G.N., Simmons, M.C., Cantwell, W.J. and Schleyer, G.K. (2007a). Behaviour of fibre-metal laminates subjected to localised blast loading: Part I–Experimental observations. *International Journal of Impact Engineering*, Vol. 34, No 7, pp. 1202-1222.

Langdon, G.S., Nurick, G.N., Lemanski, S.L., Simmons, M.C., Cantwell, W.J. and Schleyer,G.K. (2007b). Failure characterisation of blast-loaded fibre-metal laminate panels based on aluminium and glass-fibre reinforced polypropylene. *Composites Science and Technology*, Vol. 67, No 7-8, pp. 1385-1405.

Langdon, G.S., Cantwell, W.J. and Nurick, G.N. (2007c). Localised blast loading of fibre-metal laminates with a polyamide matrix. *Composites Part B: Engineering*, Vol. 38, No 7-8, pp. 902-913.

Langdon, G.S., Chi, Y., Nurick, G.N. and Haupt, P. (2009). Response of GLARE© panels to blast loading, *Engineering Structures*, Vol. 31, No 12, pp. 3116-3120.

Langdon, G.S., Cantwell, W.J. and Nurick, G.N. (2005a). The blast response of novel thermoplastic-based fibre-metal laminates – some preliminary results and observations. Composites Science and Technology, Vol. 65, No 6, pp. 861-872.

Langdon, G.S., Yuen, S.C.K. and Nurick, G.N. (2005b). Experimental and numerical studies on the response of quadrangular stiffened plates. Part II: localised blast loading. International Journal of Impact Engineering, Vol. 31, No 1, pp. 85-111.

Lemanski, S.L., Nurick, G.N., Langdon, G.S., Simmons, M.C., Cantwell, W.J. and Schleyer, G.K. (2007). Behaviour of fibre metal laminates subjected to localised blast loading–Part II: Quantitative analysis. *International Journal of Impact Engineering*, Vol. 34, No 7, pp. 1223-1245.

Lesuer, D.R. (2000). *Experimental Investigations of Material Models for Ti-6Al-4V Titanium and 2024-T3 Aluminum*, Livermore, CA: Lawrence Livermore National Laboratory.

McCarthy, M.A., Xiao, J.R., Petrinic, N., Kamoulakos, A. and Melito, V. (2004). Modelling of bird strike on an aircraft wing leading edge made from fibre metal laminates–Part 1: Material modelling. *Applied Composite Materials*, Vol. 11, pp. 295-315.

Reyes-Villanueva, G. and Cantwell, W.J. (2004). The high velocity impact response of composite and FML-reinforced sandwich structures. *Composites Science and Technology*, Vol. 64, No 1, pp. 35-54.

Reyes, G. and Cantwell, W.J. (2000). The mechanical properties of fibre-metal laminates based on glass fibre reinforced polypropylene. *Composites Science and Technology*, Vol. 60, No 7, pp. 1085-1094.

Soutis, C., Mohamed, G. and Hodzic, A. (2011). Modelling the structural response of glare panels to blast load. *Composite Structures*, Vol. 94, No 1, pp. 267-276.

Vlot, A. (1996). Impact loading on fibre metal laminates. *International Journal of Impact Engineering*, Vol. 18, No 3, pp. 291-307.

Vogelesang, L.B. and Vlot, A. (2000). Development of fibre metal laminates for advanced aerospace structures. *Journal of Materials Processing Technology*, Vol. 103, No 1, pp. 1-5.

Yen, C.F. (2002). Ballistic impact modeling of composite materials. In: *Proceedings of the 7th International LS-DYNA Users Conference*, Vol. 6, pp. 15-23.

Time-Dependent Behaviour of Carbon Fibre Reinforced Laminates

Alexander Dumansky[1], Lyudmila Tairova[2]

[1] Institute of Machines Science of the Russian Academy of Sciences, 4 Maly Kharitonievsky Per., Moscow, 101990, Russia
[2] Rocket and Spacecraft Composite Structures Department, Faculty of Special Machinery, Bauman Moscow State Technical University, 5 2nd Baumanskaya Street, Moscow, 105005, Russia

Abstract: A method of construction of constitutive equations was developed based on classical laminate theory and algebra of resolvent operators. Angle ply specimens were tested to determine the elastic and viscoelastic properties of the layer. It was assumed that the time-dependent properties of angle ply and multidirectional laminates are defined by shear in the plane of the layer. Rheological properties of the layer were derived from relatively short-time creep tests of $[\pm40]_{2s}$ lay-up specimens. The time-dependent properties of the layer under in-plane shear can be described by Abel's operator with the parameter of the kernel singularity close to -0.9. The rheological properties of the carbon fibre reinforced laminates were described based on Abel's and Rabotnov's fraction-exponential operators. The expressions of interrelation between the material functions were presented. The convenience of algebra of resolvent operators for strain analysis under variable loading was shown. The stress-strain curve of $[\pm40]_{2s}$ lay-up specimen under three cycles of trapezoidal loading was constructed. Satisfactory agreement of predicted and experimental data was obtained. Using this approach it is possible to take into account nonlinear behaviour of carbon fibre reinforced plastics.

Key Words: Constitutive equation, Resolvent operator, Abel's operator, Rabotnov's fraction-exponential function, Creep, Relaxation.

1. Introduction

Structural carbon fibre reinforced composites are widely used in aerospace applications and are of great importance in setting up structural-phenomenological models describing time-dependent behaviour of laminates. It is obvious that time-dependent properties of fibrous laminated composites, to a large extent, are defined by the properties of polymer matrix and its adhesion to the fibres, these properties clearly appear under off-axis loading. Testing of angle ply and off-axis specimens is typical to determine rheological and nonlinear proper-ties of carbon fibre reinforced laminates. Analysis of creep strain of angle ply and off-axis specimens has been performed by a number of workers (Charentenay and Zaidi, 1982; Deng et al., 2003; Ma et al., 1997; Potter, 1974). The most pronounced creep effect was observed for $[\pm45]$ lay-up. In particular, an analysis of the experimental data of tensile loading of $[\pm45]_{2s}$ carbon fibre reinforced specimens show that rheological and nonlinear properties are determined by shear properties in the plane of the layer (Charentenay and Zaidi, 1982; Potter, 1974) and creep strain can be approximated by the Findley power approximation (Findley et al., 1976). The power approximation was used (Charentenay and Zaidi, 1982; Deng et al., 2003; Guedes et al., 1998; Ma et al., 1997) to describe creep of carbon-epoxy $[\pm45]_{2s}$. In order to satisfy some physical restrictions the time exponent parameter of the equation must lie between 0 and 1. Moreover, in case of carbon fibre reinforced composite laminates, experiments indicate that the exponent parameter is close to 0.1. Carbon fibre reinforced laminates are sensitive to strain rate (Guedes et al., 1998; Hsiao and Daniel, 1998; Potter, 1974). Constitutive equations of hereditary mechanics of solids in operator form are most convenient to describe mechanical behaviour under variable loading with time (Dumansky and Strekalov, 1999; Oza et al., 2003). The first stage of viscoelastic behaviour investigation is the estimation of elastic properties of the material. In an elastic body strain instantly follows stress and their time dependencies are similar. In viscoelastic bodies, the effect of retardation takes place and is of great importance in dividing elastic and time-dependent strain. In measuring elastic properties the effect of time-dependent strain is decreased by shortening the time of loading. Potter (1974) observed that to determine elastic characteristics it is necessary to load the specimen to failure within a few seconds. It should be noted that the elastic characteristics do not coincide with the characteristics defined under quasi-static loading.

When predicting mechanical properties of carbon-fibre reinforced laminates, the properties of the layer are used as a baseline. The majority of the papers of World-Wide Failure Exercise (Hinton et al., 2004) devoted to mechanical behaviour of polymer composites under quasi-static loading are based on elastic properties of the layer and classical lamination theory relations. The inverse problem (Zinoviev and Tairova, 1995) is to define elastic properties of the layer from experimental data of the multilayered laminates. The application of classical lamination theory to describe time-dependent behaviour of carbon fibre reinforced laminates was considered in (Guedes et al., 1998; Korontzis and Vellios, 2000; Dumansky and Tairova, 2007, 2008). Finally, it is our opinion, that use of operator representation in the constitutive equations is convenient for computing algorithm. The algebra of resolvent operators developed by (Rabotnov, 1979) allows one to simplify the functions of the operators for practical use.

2. Elastic properties of the layer

The tension test of angle ply carbon reinforced plastics based on viscoplastic resin was the subject of investigation. Flat specimens of $[0]_4$, $[\pm20]_{2s}$, $[\pm40]_{2s}$, $[\pm50]_{2s}$, $[\pm70]_{2s}$, $[90]_4$ lay-ups were tested under some triangular cycles of quasi-static tensile strain. Stress-strain diagrams under quasi-static uniaxial tension with unloading up to

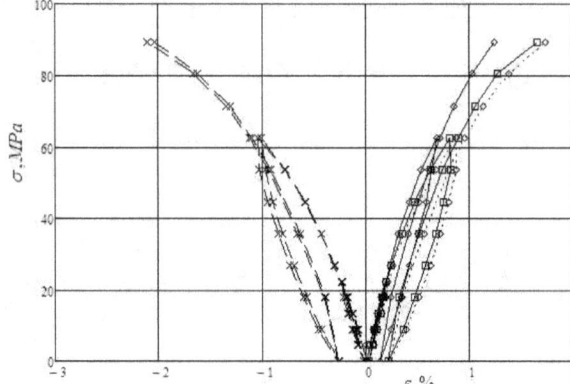

Figure 1. Stress-strain diagrams for *[±40]₂ₛ* lay-up.

0.3-0.7 of the failure stress in longitudinal and transverse directions were obtained. These diagrams are shown in Fig. 1. Elastic moduli and Poisson's ratio were determined within the linear range of stress-strain using the method proposed by Zinoviev and Tairova (1995).

All the data apart from that for the *[0]₂ₛ, [90]₂ₛ* lay-ups revealed nonlinear viscoelastic properties and this was especially significant for *[±40]₂ₛ, [±50]₂ₛ*, lay-ups.

To determine the technical characteristics of the layer the minimization of the following residual function was performed:

$$\sum_k \left(\varepsilon_k^{exp} - \varepsilon_k^{calc} \left(E_1, E_2, G_{12}, \nu_{12} \right) \right)^2 \rightarrow \min \quad (1)$$

The values (Dumansky and Tairova, 2007) appeared to be equal: $E_1 = 150$ MPa, $E_2 = 3.95$ MPa, $G_{12} = 2.39$ MPa, $\nu_{12} = 0.315$.

3. Algebra of resolvent operators

The operator form of the constitutive hereditary equation under uniaxial loading can be written in the form

$$E\varepsilon = \sigma + K^*\sigma = \left(1 + K^*\right)\sigma \quad (2)$$

where $K^*\sigma = \int_0^t K\left(1 - \tau\right)\sigma\left(\tau\right)d\tau$ is a kernel of the operator.

The constitutive equation (2) (Rabotnov, 1979) can be inversed and specified by

$$\sigma = E\left(\varepsilon - R^*\varepsilon\right) = E\left(1 - R^*\right)\varepsilon \quad (3)$$

where operator R^* is resolvent in relation to the operator K^*.

Substituting the expression for σ from (3) into (2) yields the equation for the interrelation between the initial operator and its resolvent

$$\left(1 + K^*\right)^{-1} = 1 - R^* \quad (4)$$

Solving equation (4) the common expression for the resolvent operator can be represented by

$$R^* = \frac{K^*}{1 + K^*} = K^* - K^{*2} + K^{*3} \quad (5)$$

The series in relation (5) is Neumann series for the resolvent (Rabotnov, 1979). Rabotnov's fraction-

exponential function was obtained as the resolvent of Abel's operator using the expression for kernel of the operators' multiplication

$$N\left(t - \tau\right) = \int_\tau^t M\left(t - x\right)L\left(x - \tau\right)dx \quad (6)$$

For Abel's kernel $K\left(t\right) = I_\alpha\left(t\right) = \dfrac{t^\alpha}{\Gamma\left(1 + \alpha\right)}, -1 < \alpha < 0$, the expression for Abel's operator is given by

$I_\alpha^* \cdot 1 = I_\alpha^* = \dfrac{t^{1+\alpha}}{\Gamma\left(2 + \alpha\right)}$, 1 is a unity function.

With the aid of formula (6) the power of Abel's operator can be written as

$$I_\alpha^{*m} = I_{m-1+m\alpha}^* \quad (7)$$

Substituting in place of the operator K^* in relation of Neuman series fraction-exponential function takes the following form

$$Z_\alpha^*\left(-\beta\right) = -\beta I_\alpha^* + \beta^2 I_\alpha^{*2} - \beta^3 I_\alpha^{*3} + \ldots \quad (8)$$

Then, using the explicit form for power of Abel's operator (7), Rabotnov's fraction-exponential function can be written as

$$Z_\alpha^*\left(-\beta\right) \cdot 1 = t^{1+\alpha} \sum_{n=0}^{\infty} \frac{\left(-\beta t^{1+\alpha}\right)^n}{\Gamma\left[1 + \left(1 + \alpha\right)\left(1 + n\right)\right]} \quad (9)$$

where $\Gamma()$ is gamma-function. Series (9) converges at $\beta > 0$.

These are two main relations of the algebra of fraction-exponential functions (Rabotnov, 1979)

$$\left(1 - xZ_\alpha^*\left(y\right)\right)^{-1} = 1 + xZ_\alpha^*\left(y + x\right);$$
$$Z_\alpha^*\left(x\right)Z_\alpha^*\left(y\right) = \frac{1}{x - y}\left[Z_\alpha^*\left(x\right) - Z_\alpha^*\left(y\right)\right]. \quad (10)$$

The particular cases of Rabotnov's fraction-exponential function are the following: $Z_\alpha^*\left(0\right) = I_\alpha^*$ is Abel's operator, $Z_0^*\left(-\beta\right)$ is exponential operator, and $Z_0^*\left(0\right)$ is integration operator. It should be noted that kernel as a sum of the exponential functions (Prony series) also allows a resolvent operator.

4. Constitutive equations

Stress and strain in a viscoelastic material can be connected with the aid of Stieltjs convolutions

$$\varepsilon\left(t\right) = J * d\sigma = \int_0^t J\left(t - \xi\right)d\sigma;$$
$$\sigma\left(t\right) = G * d\varepsilon = \int_0^t G\left(t - \xi\right)d\varepsilon \quad (11)$$

Then under creep and relaxation relations (11) can be rewritten as follows

$$\varepsilon\left(t\right) = J\left(t\right)\sigma;$$
$$\sigma\left(t\right) = G\left(t\right)\varepsilon, \quad (12)$$

where $J(t)$, $G(t)$ are creep and relaxation functions.

Similar to (12) under stress and strain rate loading constitutive equations take the following form

$$\varepsilon(t) = \eta(t)\dot{\sigma};$$
$$\sigma(t) = \varsigma(t)\dot{\varepsilon}. \qquad (13)$$

Using relations (12) and (13) we can describe strain and stress under stepwise loading. For loading which can be defined with the following forms

$$\sigma(t) = \sum_{k=1}^{n} H(t - t_k)\Delta\sigma_k;$$
$$\varepsilon(t) = \sum_{k=1}^{n} H(t - t_k)\Delta\varepsilon_k, \qquad (14)$$

where $H(\ )$ is the Heaviside unit function, $\Delta\sigma_k$, $\Delta\varepsilon_k$ are step-wise changes of stress and strain. The corresponding change of strain and stress in consequence with (12) (Bugakov, 1973) can be represented as

$$\varepsilon(t) = \sum_{k=1}^{n} J(t - t_k)\Delta\sigma_k;$$
$$\sigma(t) = \sum_{k=1}^{n} G(t - t_k)\Delta\varepsilon_k. \qquad (15)$$

Obviously similar relations can be obtained for relations (13)

$$\varepsilon(t) = \sum_{k=1}^{n} \eta(t - t_k)\Delta\dot{\sigma}_k;$$
$$\sigma(t) = \sum_{k=1}^{n} \varsigma(t - t_k)\Delta\dot{\varepsilon}_k. \qquad (16)$$

There are relations for connection between the material functions

$$\eta(t) = \int_0^t J(\xi)d\xi;$$
$$\varsigma(t) = \int_0^t G(\xi)d\xi. \qquad (17)$$

These are in turn connected with creep and relaxation kernels by the following relations

$$J(t) = \frac{1}{E}\left(1 + \int_0^t K(\xi)d\xi\right);$$
$$G(t) = E\left(1 - \int_0^t R(\xi)d\xi\right). \qquad (18)$$

5. Creep and relaxation $[\pm 40]_{2s}$ lay-up

Creep and relaxation of plane $[\pm 40]_{2s}$ lay-up specimens were made on an Instron 8800 servo hydraulic testing machine. The process of testing longitudinal and transverse strain was made using 5 mm extensions at 5 minute intervals. Clearly pronounced creep and relaxation of the CFRP were observed. The level of load under creep was close to limit of elastic strain and equal to 65 MPa. The initial level strain of the relaxation was equal to 0.5%.

Experimental data were analyzed and treated to describe the creep curve of the $[\pm 40]_{2s}$ lay-up. Similarly to the work of Charentenay and Zaidi (1981) a power law representing Abel's kernel of the constitutive equation was taken. The linear constitutive hereditary equation with Abels's kernel is given by

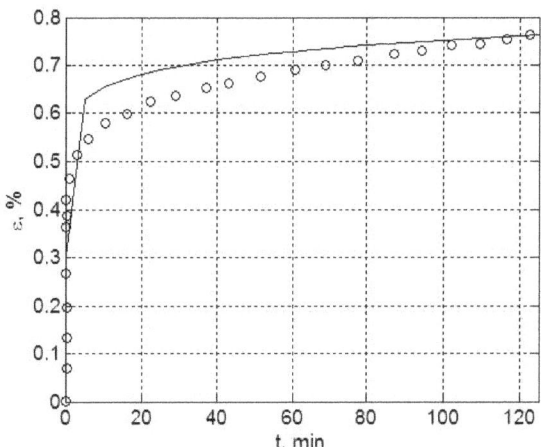

Figure 2.

$$\varepsilon(t) = \frac{1}{E_0}\left(1 + kI_\alpha^*\right)\sigma \qquad (19)$$

where E_0 is instant modulus, k and α are parameters of the equation. In the case of creep constitutive equation (19) can be rewritten as

$$\varepsilon(t) = \varepsilon_0\left(1 + kI_\alpha^*\right) = \varepsilon_0\left(1 + \frac{k}{\Gamma(2+\alpha)}t^{1+\alpha}\right) \qquad (20)$$

where ε_0 is instant elastic strain. The parameters of the constitutive equation were calculated by minimizing of the following expression

$$\sum_k \left(\varepsilon_k^{\exp} - \varepsilon_k^{calc}(E_0, k, \alpha)\right)^2 \rightarrow \min \qquad (21)$$

The parameters are as follows: $E_0 = 19900$ MPa, $\alpha = -0.894$, $k = 0.873$ min$^{-(1+\alpha)}$. The prediction of creep strain shown in Fig. 2 is in satisfactory agreement with the experimental data.

Using the above calculated parameter values the relaxation curve was obtained. The experimental values and predicted curve are shown in Fig. 3.

The linear constitutive equation is obtained by use of relation (10) and can be written as

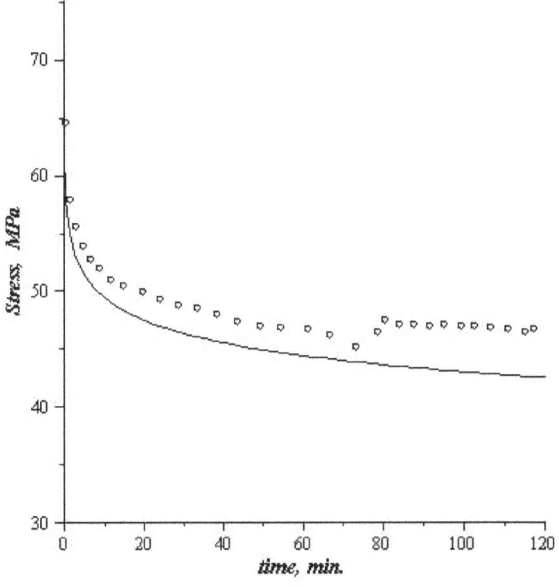

Figure 3.

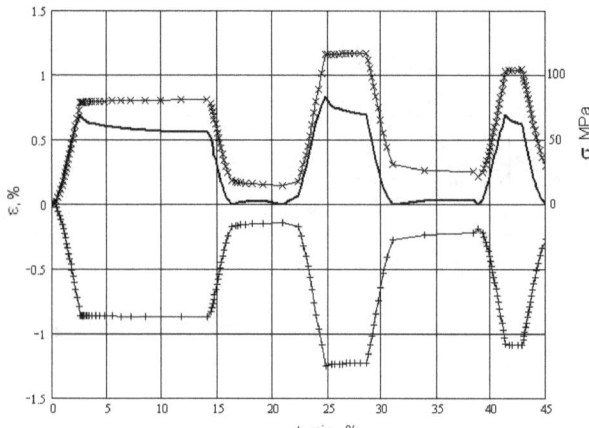

Figure 4. Stress response of $[\pm40]_{2s}$ lay-up to strain input.

$$\sigma = E_0\left(1 - kZ_\alpha^*\left(-k\right)\right)\varepsilon \tag{22}$$

Using fraction-exponential function representation (9) the constitutive equation for relaxation is as follows

$$\sigma = E_0\varepsilon_0\left(1 - kt^{1+\alpha}\sum_{n=0}^{\infty}\frac{\left(-\beta t^{1+\alpha}\right)^n}{\Gamma\left[1+\left(1+\alpha\right)\left(1+n\right)\right]}\right) \tag{23}$$

6. Time-dependent behaviour under variable loading

Experimental studies under variable loading were performed for $[\pm40]_{2s}$ specimens. There were three trapezoidal cycles of strain and the corresponding strain in longitudinal and transverse directions were measured. The result of the testing is shown in Fig. 4.

Within the permanent strain ranges there is relaxation of stress. Isochronic curves obtained from the $[\pm40]_{2s}$ lay-up are shown in Fig 5. In Fig. 5 one can see that hysteresis effects are determined partly by rheological properties. These are reflected by vertical parts of the curves in Fig. 5. The residual strain determining the shift of the diagrams, to some extent, depends on the horizontal shift dependent on the time of recovery.

To describe the stress response under trapezoidal strain input it is convenient to represent strain in strain rate-time coordinates.

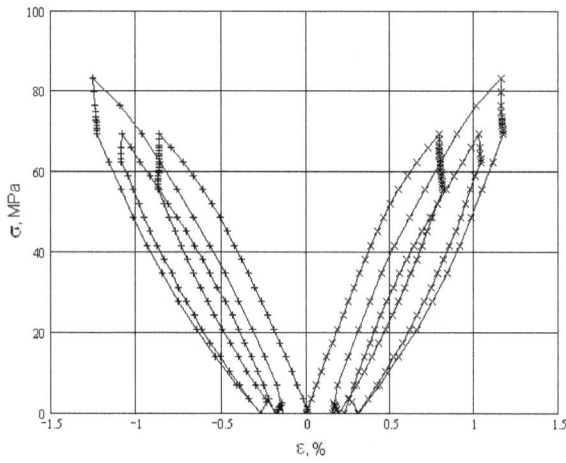

Figure 5. Isochronic curves of carbon fibre reinforced composite laminates with $[\pm40]_{2s}$ lay-up.

From all the array of the strain values only eleven points were chosen in which the strain values have derivative discontinuity. In this case the strain rate is given by the stepwise function:

$$\dot{\varepsilon} = \sum_{k=1}^{n} H\left(t-t_k\right)\Delta\dot{\varepsilon}_k \tag{24}$$

And using the second relation from (16) the stress dependence can be described. In the case of Abel's creep kernel, the material function in the constitutive equation in (16) is taken as follows

$$\varsigma\left(t\right) = E\left(t - Z_\alpha^*\left(-\beta\right)\cdot t\right) \tag{25}$$

By integration of power functions in series (9) it is not hard to obtain the explicit form of material function $\varsigma\left(t\right)$. The form of the operator yields

$$Z_\alpha^*\left(-\beta\right)\cdot t = t^{2+\alpha}\sum_{n=0}^{\infty}\frac{\left(-\beta t^{1+\alpha}\right)^n}{\Gamma\left[2+\left(1+\alpha\right)\left(1+n\right)\right]} \tag{26}$$

Using constitutive equation (26) and the second equation in (16) it is possible to describe the stress-strain diagram under loading shown in Fig. 4. The comparison of experimental data and calculated values is presented in Fig. 6. The strain rate increment is as follows:

$$\Delta\dot{\varepsilon}_k = \frac{\Delta\varepsilon_{k+1}}{\Delta t_{k+1}} - \frac{\Delta\varepsilon_k}{\Delta t_k} \tag{27}$$

The difference in Fig. 5 can be explained by the presence of nonlinear effects especially in the second and third cycles of loading in which the level of stresses increase above the linear limit.

7. Nonlinear behaviour

Nonlinear effect can be taken into account by change of left part of constitutive equation (2) (Rabotnov, 1979) and rewritten as

$$\varphi\left(\varepsilon\right) = \left(1+K^*\right)\sigma \tag{28}$$

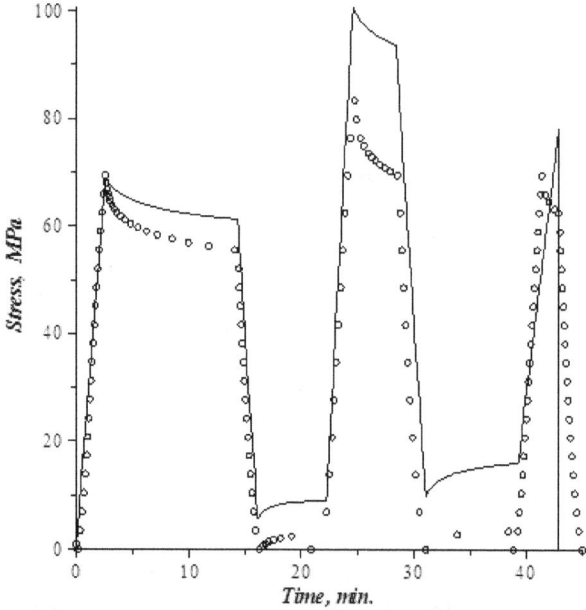

Figure 6.

where $\varphi(\varepsilon)$ is instant strain curve which is defined by data treatment. A possible way of its representation is as follows

$$\varphi(\varepsilon) = E_0\varepsilon - \Delta E_0 \cdot (\varepsilon - \varepsilon_1) \cdot H(\varepsilon - \varepsilon_1) \qquad (29)$$

where ΔE_0 is change of instant modulus E_0 at strain ε_1.

Substituting (29) into (28) and by use of (3) the constitutive equation for stress specified by

$$\sigma = E_0(1 - R^*)\varepsilon - \Delta E_0 R^*(\varepsilon - \varepsilon_1) \qquad (30)$$

An application of such approach to carbon fibre reinforced composite laminates under quasi-static loading was considered in (Dumansky et al., 2011).

8. Results and discussion

The possibility of application of mechanics of hereditary solids in describing of time-dependent behaviour of carbon fibre reinforced composite laminates has been shown. The use of algebra of resolvent operators and theory of generalized functions allow significant simplifying the construction the constitutive equations. Fraction-exponential function combining properties of power and exponential functions can be successfully used to characterize rheological properties of carbon fibre reinforced composites. A preliminary estimation of the singularity parameter α which is connected with the time exponent in the Findley's law is equal to –0.9. The possibility of model generalization to take into account the nonlinear behaviour of carbon fibre reinforced composite laminates is shown.

References

Bugakov, I.I. (1973). *Creep of Polymer Materials*. Moscow: Nauka (in Russian)

Charentenay, F.X. and Zaidi, M.A. (1982). Creep behaviour of carbon-epoxy (±45°)$_{2s}$ laminates, in *Progress in Science and Engineering of Composites ICCM-IV*, Tokyo, pp. 787-793.

Deng, S., Li, X. and Weitsman, Y.J (2003). Time-dependent deformation of stitched T300 Mat/Urethane 420 IMR cross-ply composite laminates, *Mechanics of Time-Dependent Materials*, Vol. 7: pp. 41-69.

Dumansky, A.M. and Tairova, L.P. (2007). The prediction of viscoelastic properties of layered composites on example of cross ply carbon reinforced plastic, in *World Congress on Engineering*, July 2-4, London, pp. 1346-1351.

Dumansky, A.M. and Tairova, L.P. (2008). Construction of hereditary constitutive equations of composite laminates, in *Proceedings of the Second International Conference on Advances in Heterogeneous Materials Mechanics*, June 3-8, Huangshan, China, pp. 934-937.

Dumansky, A.M., Tairova, L.P., Gorlach, I. and Alimov, M.A. (2011). A design-experiment study of nonlinear properties of coal-plastic, *Journal of Machinery Manufacture and Reliability*, Vol. 40, No 5. pp. 483-488

Dumansky, A.V. and Strekalov, V.B. (1999). Creep and relaxation of heritable orthotropic continua under the plane stress state, *Journal of Strain Analysis for Engineering Design*, Vol. 34, No 5, pp. 361-367.

Findley, W.N., Lai, J.S. and Onaran, K. (1976). *Creep and Relaxation of Nonlinear Viscoelastic Materials*. New York: Dower Publications.

Guedes, M.R. and Marques, A.T. (1998). Analytical and experimental evaluation of nonlinear viscoelastic-viscoplastic composite laminates under creep, creep-recovery, relaxation and ramp loading, *Mechanics of Time-Dependent Materials*, No 2, pp. 113-128.

Hinton, M.J., Kaddour, A.S. and Soden, P.D. (Eds) (2004). *Failure Criteria in Fibre Reinforced Polymer Composites: The World Wide Failure Exercise*. Amsterdam: Elsevier.

Hsiao, H.M. and Daniel, I.M., (1998). Strain rate behaviour of composite materials, *Composites: Part B*, Vol. 29, pp. 521-533.

Korontzis, D.Th., Vellios, L. and Kostopoulos, (2000). On the viscoelastic response composite laminates, *Mechanics of Time-Dependent Materials*, No 4, pp. 381-405.

Ma, C.C.M., Tai, N.H., Wu, S.H., Lin, S.H., Wu, J.F. and Lin, J.M., (1997). Creep behaviour of carbon-fiber reinforced polyetheretherketone (PEEK) *[±45]$_{4s}$* laminated composites, *Composites: Part B*, Vol. 28, pp. 407-417.

Oza, A., Vanderby, R. and Lakes, R.S. (2003). Interrelation of creep and relaxation for nonlinearly viscoelastic materials: application to ligament and metal, *Rheol. Acta*, Vol. 42, pp. 557-568.

Potter, R.T. (1974). Repeated loading and creep effects in shear property measurements on unidirectional cfrp, *Composites*, November, pp. 261-265.

Rabotnov, Yu.N. (1979). *Elements of Hereditary Solid Mechanics*, Moscow: Mir Publishers.

Zinoviev, P. and Tairova, L. (1995). Identifying the properties of individual plies constituting hybrid composites, *Inverse Problems in Engineering*, Vol. 2, pp. 141-154.

Carbon Nanotubes for Epoxy Nanocomposites: A Review on Recent Developments

Nataliia Luhyna, Fawad Inam

Advanced Composite Training and Development Centre and School of Mechanical and Aeronautical Engineering, Glyndŵr University, Plas Coch, Mold Road, Wrexham, LL11 2AW, UK

Abstract: Carbon nanotubes (CNTs) are one of the strongest and stiffest engineering fibres. Due to their unique combination of chemical and physical properties at an incredibly small size, they possess great potential to be used as nanofillers for many structural and functional materials, particularly in aerospace sector. Depending on the type, geometrical parameters, concentration, dispersion and many other factors, CNTs can significantly modify the mechanical, electrical and thermal properties of epoxy based materials. This review paper, covering methods of synthesis, composite processing techniques and properties, presents an overview of developments in the field of CNT/ epoxy nanocomposites in recent years.

Key Words: Carbon nanotubes (CNTs), CNT/ epoxy nanocomposites, Properties.

1. Introduction and historical perspective

The relatively recent discovery (Iijima, 1991) of carbon nanotubes (CNTs) has opened new possibilities for the production of advanced novel materials. Due to their unique mechanical, electrical and the thermal properties, CNTs is subject of intense research, which is evident from a growing number of publications in this area. CNT based materials possess potential to be used in almost any of the leading industries such as aeronautics, electronics, optics, medicine, architecture/ construction, automotive, mechatronics and biotechnology to name a few (Endo et al., 2004).

CNTs were discovered accidently in 1991 (Iijima, 1991). CNTs were formed at the cathode during sputtering of graphite by electronic arc. Derived nanotubes were multi-walled nanotubes (MWNTs) with an inner cylindrical diameter about 4 nm (Iijima, 1991). After two years, Iijima and his colleagues synthesised single-walled carbon nanotubes (SWNTs). SWNTs were prepared using the method originally researched by Iijima (1991), but with the addition of metal particles on the carbon electrodes (Bethune et al., 1993).

The discovery of CNTs belongs to the most remarkable achievements of modern science and technology. This form of carbon structure is intermediate between graphite and fullerene. However, many properties of CNTs have nothing in common either with graphite or fullerene, which explains uniqueness of CNTs in many ways.

This review paper has two parts. In the first part, it elucidates the properties, synthesis and uniqueness of CNTs. The second part provides a summary of recent development in the field of CNT/ epoxy nanocomposites.

2. Classifications and types of CNTs

CNTs have chicken-wire structures with an incredibly small size (diametre: 0.1 nm to 100 nm). CNTs are extended cylindrical structures composed of collapsed sheets of graphene (Geim and Novoselov, 2007). The main unit of graphene sheets is hexagon with carbon atoms arranged at the corners (Fig. 1).

Chirality or helicity is an important property, indicating the incompatibility of the object with its mirror image. It is characterised by two integers (n, m), which indicate the location of the hexagonal grid (Belin and Epron, 2005). Chirality of the nanotube can also be uniquely determined by the angle α, formed by the folding direction of the nanotube. The value of (n, m) determines the chirality of CNT, which affects the optical, mechanical and electronic properties (Fig. 2).

Subject to the chirality, SWNTs possess electrical properties of semiconductors as well as that of metals (Belin

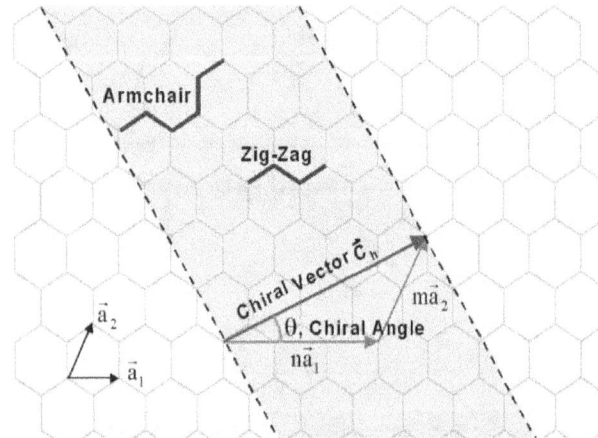

Figure 1. Schematic diagram showing how a hexagonal sheet of graphite is 'rolled' to form a carbon nanotube (Thostenson et al., 2001).

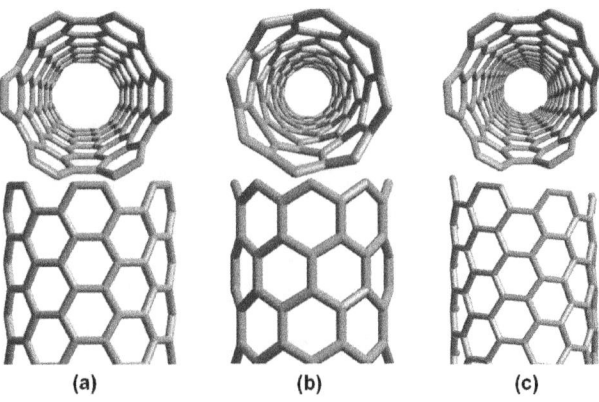

Figure 2. Different types of CNTs based on their chirality: a) armchair (metallic); b) zigzag (metallic or semiconducting); and c) chiral (metallic or semiconducting).

and Epron, 2005). As evident from Fig. 2, the parameters (n, m) can change the type of CNTs, i.e.

1) Direct (achiral) nanotubes:
  a) "Armchair" with values n = m = 0 and the angle α = 0;
  b) "Zigzag" with m = 0 or n = 0 and the angle α = 30.
2) Helical (chiral) nanotube with an angle α of 0 to 30.

Other than chirality, CNTs can be synthesised in many different physical forms, i.e. short, long, thick, thin, single wall, multi wall, functionalised, open, capped, stacked, containing different structural defects and rolling structures. Depending of number of rolled graphene shells, CNTs can be classified as single walled nanotubes (SWNTs), double walled nanotubes (DWNTs) and multi walled nanotubes (MWNTs). They can also be further classified depending on the structure - straight, branched, curled, cup-stacked and herringbone; and variety of the crystalline structure - well aligned and distorted (Hayashi and Endo, 2011). Each type has its own advantages and disadvantages and possesses potential to be used for different types of applications.

Fig. 3 shows two of the most common type of CNTs. SWNT is a rolled up sheet of graphene, which is having half hemisphere fullerene molecule (Fig. 3a). The ends of the CNT are not always closed as shown in Fig. 3b. MWNT (Fig. 3b) possess cylinders, inserted one into each other of SWNTs. The number of cylinders can be from 2 to 30 and the outer diameters are around 3-100 nm (Fig. 3b).

3. Properties of CNTs

A perfectly crystalline CNT possesses excellent electrical conduction similar to that of copper. CNTs also have tensile strengths 100 times greater than steel having nearly 1/6 the density of steel (Zolotuchin, 1999). Apart from this, CNTs were also found to have thermal conductivity higher than the purest diamond (Holister et al., 2003). These unique properties of nanotubes and small sizes make them indispensable materials for modern nanotechnology (Breeuer and Sundararaj, 2004, Holister et al., 2003). Tab. 1 shows a basic comparison of CNTs with graphite, where CNTs are distinguished by their excellent mechanical, thermal and electrical properties.

As compared to SWNTs, DWNTs are distinguished by their heat-stable properties. Upon heating to temperatures of 2000 °C (in vacuum), they maintain their crystallinity and cylindrical structure. When heated above 2000 °C the outer layers of CNTs start degrading. The mechanical properties of the DWNT are superior as compared to

SWNT and MWNTs, which makes them an excellent filler/reinforcement material for nanocomposites (Hayashi and Endo, 2011).

Depending on the type, CNTs could be both conductors and/ or semiconductors. It is related to the topological defects in the structure (Xie et al., 2005). CNTs with metallic properties conduct electricity at absolute zero temperature, whereas the conductivity of semiconducting nanotubes is zero at absolute zero and increases with increasing temperature. Different types of metallic and semiconducting CNTs are shown in Fig. 2.

Diamond is another allotropic form of carbon. It has lower modulus (20 GPa) than graphite (1000 GPa). However, compressive strength of diamond is 14 GPa and of graphite is 105 MPa (Smith, 1987). As compared to diamond and graphite, the higher values of the elastic modulus and compressive strengths of CNTs allows possibility of creating a composite material with very different set of mechanical properties. Such combination of superior properties cannot be observed for any other allotropic form on carbon (Xie et al., 2005). Tensile strength of CNTs is significantly higher (i.e. 11 to 63 GPa) than any of the known engineering materials (Yu et al., 2000). This is explained by the lack of micro-defects in the CNT crystal structure and high degree of interaction forces between the atoms in its crystal lattice.

Graphite is thermodynamically more stable than diamond at room temperature, and diamond is thermodynamically more stable at high pressure. In terms of thermodynamic stability, CNTs stand in between. Because of large specific surface area (up to 2600 m^2g^{-1}) of CNTs (Eletskii, 2007), they possess unique absorption capabilities for different types of gases.

The rigidity of CNTs under radial compression is much lower than the longitudinal compression or tension (Hertel et al., 1998). Therefore, the tubular structure may

Table 1
Properties of carbon nanotubes (Xie et al., 2005, Sengupta et al., 2011, Li et al., 2008)

Property	CNTs	Graphite
Specific gravity	0.8 g/cm^3 for SWCNT; 1.8 g/cm^3 for MWCNT (theoretical)	2.26 g/cm^3
Elastic modulus	1 TPa for SWCNT; 0.3–1 TPa for MWCNT	1 TPa (in-plane)
Tensile strength	50–500 GPa for SWCNT; 10–60 GPa for MWCNT	130 GPa
Resistivity	5–50 μΩcm	50 μΩcm (in-plane)
Thermal conductivity	3000 W/ mK (theoretical)	3000 W/ mK (in-plane), 6 W/mK (c-axis)
Magnetic susceptibility	22×10^6 EMU/g (perpendicular with plane), 0.5×10^6 EMU/g (parallel with plane)	
Thermal expansion	Theoretically negligible	-1×10^6 1/K (in-plane), 29×10^{-6} 1/K (c-axis)
Thermal stability	>700 °C (in air); 2800 °C (in vacuum)	450–650 °C (in air)
Specific surface area	10–20 m^2/g	7 m^2/g

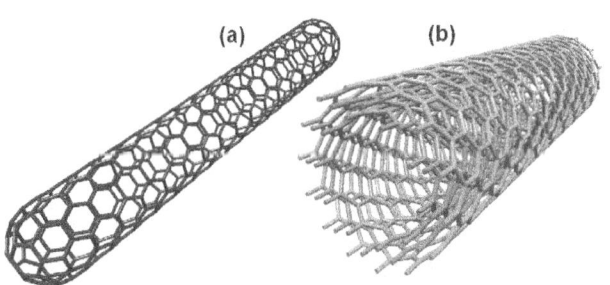

Figure 3. Different types of CNTs based on the number of graphene layers/ cylinders: (a) capped SWNT; and (b) open MWNT.

be distorted under the influence of van der Waals forces with the substrate or as a result of exposure of nanotubes to each other. Such radial defects increase with the increase in the diametres of CNTs. The deformation of MWNTs decreases with increasing the inner layers of the nanotube and vice versa.

4. Synthesis of CNTs

The most common methods for producing CNTs are the arc discharge method, the laser ablation, the gas phase catalytic growth, and chemical vapor deposition (CVD). During synthesis, impurities from the catalyst particles, amorphous carbon, and tubular fullerenes are produced. Gas phase methods are more suitable for large scale production of CNTs since it is necessary to receive a good amount of reinforcement materials for composite materials (Thostenson et al., 2001).

For producing CNTs, arc discharge technique involves thermal decomposition of graphite electrodes in arc discharge plasma. It occurs in the helium atmosphere under a very high voltage. It is important to use anode and cathode of high-purity graphite rods. The final dimensions of CNTs depend on various process parameters. During combustion plasma, there is an intensive thermal evaporation of the anode, resulting in a build-up formed at the cathode which generates CNTs (Journet et al., 1997). The production yield is highest when the plasma current is minimal, and its density is around 100 A/cm^2 (Zolotuchin, 1999).

In laser ablation method, a laser is directed to the target for producing CNTs. It is usually performed in a horizontal tube furnace in a stream of inert gas (with pressure) at a temperature of around 1200 °C. The target is usually of the carbon origin and CNTs are deposited on one of the ends. Like other synthesizing methods of CNTs, the quality and type of CNTs depend on the type of employed catalysts (Rinzler et al., 1998).

Both techniques, arc discharge and laser ablation are restricted in terms of the production quantity. When using the arc discharge method, the limitation is the size of the anode and in the laser ablation it is the graphite target. Also, there are issues with the purification of CNTs, which need to be integrated with the primary synthesis methods (Thostenson et al., 2001). Elimination of such drawbacks is possible by employing CVD method, in which nanotubes are formed by decomposition of carbon containing gases like ferrocene (Singh et al., 2002). The method is based on the deposition of CNTs on a catalytic substrate with a constant controlled flow of hydrocarbon gas. CNTs are obtained by the decomposition of carbon containing gas at 700-900 °C and nanotubes are subsequently grown on the metallic catalyst (Khare and Bose, 2005). This method also offers the advantage of controlling the length and diametre of CNTs with no side additives. Therefore, CVD method is an ideal way for mass production of CNTs (Ren et al., 1998).

5. CNTs vs. carbon fibres

Carbon fibers used to reinforce materials such as polymers, carbon-carbon composites and carbon fiber reinforced materials (Smith, 1987). These micro-sized fibers have high specific strength and specific modulus which make them ideal for many applications. Carbon fibers are also significant for their excellent thermal and electrical conductive properties, and low coefficient of thermal expansion (Donnet et al., 1990). Materials based on carbon fibers are widely used in aeronautical applications, because of its high strength, stiffness, durability, low weight and most important reliability (Chand, 2000). All this make carbon fibres one of the light-weight reinforcements for epoxy matrix, especially for new aerospace initiatives like Boeing 787 and Airbus A350 programmes.

As compared to carbon fibres, CNTs are unique for having smaller dimensions which makes them supplementary for filling carbon fibre reinforced epoxy composites (Inam et al., 2010). It is not possible to substitute CNTs with well-matured carbon fibre for aerospace technology at this stage. This is because of the issues with the mass production of long and perfectly crystalline CNTs. Currently, it is not possible to produce defect-free long CNTs. Some attempts have been made for synthesizing CNT ropes (Kis et al., 2004, Koziol et al., 2007, Liu et al., 2010, Zhang et al., 2002), but it is still subject of intense research. Therefore, CNTs cannot be substituted against well matured micro-carbon fibre technology. However, CNTs may offer some added advantages to carbon fibre reinforced composites like enhanced adhesion between matrix and carbon fibre and improved matrix-dominated properties (Qian et al., 2010, Hayashi and Endo, 2011, Kim et al., 2011). For example, 0.2 wt.% CNTs were added to woven carbon fibre reinforced epoxy and significant improvements in strengths and modulus were reported (Kim et al., 2011). A comprehensive study was conducted by Inam et al. (2010), where the concept of multi-scale hybrid CNT/ carbon fibre reinforced epoxy micro-nanocomposite was thoroughly discussed.

6. CNT filled epoxy nanocomposites

CNTs are currently widely used as reinforcements for various matrices such as ceramics, polymers and metals. However, significant interest can be observed in the field CNT filled epoxy matrices (Kim et al., 2008). The main objectives of these researches have been to improve the manufacturing CNT reinforced epoxy composites, de-agglomeration of CNT bundles, homogenous CNT dispersion, CNT alignment and interfacial bonding between the nanotubes and the matrix (Breeuer and Sundararaj, 2004, Quin et al., 2010).

It is now well established that CNTs influence the curing of thermoset polymers (Hussain et al., 2006). Vega et al. (2009) studied the influence of CNTs on the curing of epoxy nanocomposites. During curing, it was found that the time to gelation (t_{gel}) and time to vetrification (t_{vit}) is slightly higher for neat epoxy samples as compared to CNT/ epoxy nanocomposites (Tab. 2). The addition of 0.2 wt.% CNTs with the matrix does not significantly affect the gelation and vitrification times, however it was pointed that further increase in CNT concentration could affect these parameters significantly. There was not much difference in the glass-transition temperature. Furthermore, it was shown that thermal compressive strain developed in the epoxy above the glass transition temperature (T_g) was due to the presence of CNTs.

Table 2
Curing parameters of the CNT/ epoxy nanocomposites
(Vega et al., 2009)

System	CNT (wt.%)	t_{gel} (min)	t_{vit} (min)	T_g (°C)
Epoxy system 1	0	36.3 ± 0.8	69.4 ± 1.1	130 ± 0
	0.2	36.1 ± 0.7	68.5 ± 2	128.5 ± 1.3
Epoxy system 2	0	59.5 ± 4.9	106 ± 0	177.0 ± 0.1
	0.2	59.0 ± 5.6	95.5 ± 0.7	176.9 ± 0.9
Epoxy system 3	0	~880	~980	26
	0.2	~850	~970	26

6.1. Mechanical properties of CNT/ epoxy nanocomposites

It is well known that it is very difficult to achieve a good homogenisation of CNTs in epoxy matrix, therefore the properties of composites are often lower than expected. Significant research has been carried out on the manufacturing techniques to produce good composite materials. Resin transfer moulding (RTM) is one of the most common ways of producing CNT/ epoxy nanocomposites. RTM consists of pouring mould with liquid resin under high pressure and then it is lead to subsequent curing. The method allows manufacturing complex shape and large size in shorter durations. Cheng et al. (2009) evaluated the mechanical and physical properties of the CNT composites with a high concentration of CNTs (up to 16.5 wt.%) which was prepared by RTM method. MWNT/ epoxy materials were recently prepared using hot melt prepreg method (Ogasawara et al., 2011). High tensile strength and Young's modulus were achieved for samples prepared with hot melt prepped method as compared to composites prepared by conventional methods. For hot melt prepreg technique, a very good penetration (wet-out) between the matrix and CNTs was observed. This method of producing composite materials showed excellent output results and it can be applied for mass production as well. Experiments have confirmed that mechanical and electrical properties of nanocomposite were improved with the increase in CNT dispersion temperature (Martone et al., 2010). Glaskova et al. (2012) researched the effects of parameters like ultrasonication duration, temperature and power level and reported significant differences in the final properties of the nanocomposites. Among these processing variables, temperature and duration of the ul-

Table 3
Effect of CNT concentration on the mechanical properties of epoxy based materials (Montazeri and Montazeri, 2011)

Material	Elastic module (MPa)	Ultimate tensile strength (MPa)	Ultimate tensile strain (%)
Epoxy	3430	64	6.1
0.1 wt. % MWNT	3458	67	5
0.5 wt. % MWNT	3705	69	4.45
1 wt. % MWNT	3951	71	5
2 wt. % MWNT	4225	75	7.5

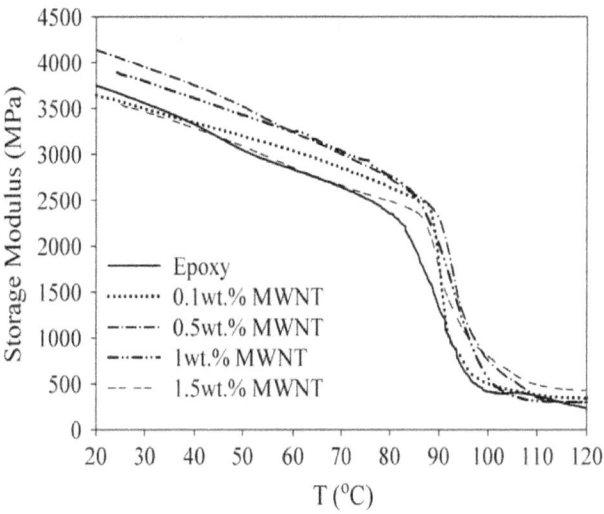

Figure 4. Storage modulus for epoxy and nanocomposites with different weight percent at 10 Hz frequency (Montazeri et al., 2012).

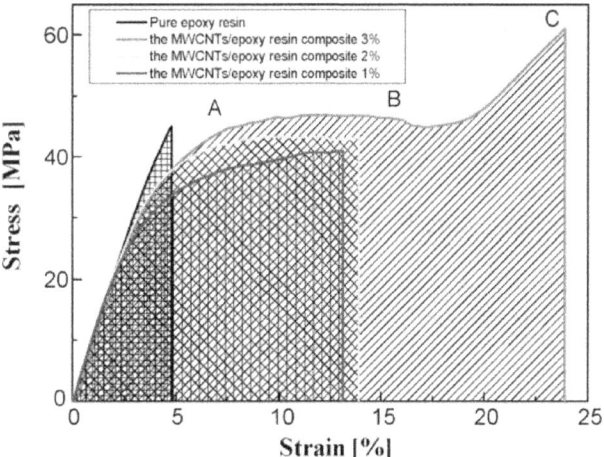

Figure 5. Stress–strain curves of pure epoxy resin and MWNT/ epoxy composites (Guo et al., 2009).

trasonication were found to be the major positive contributors.

Recently, Montazeri and Montazeri (2011) conducted the investigation on viscoelastic and tensile properties of MWNT/ epoxy composites. The study showed a major influence of CNT concentration (0.1, 0.5, 1 and 2 wt.%) on the mechanical properties of the nanocomposites. The tensile properties of composites with different amount of CNTs are presented in Tab. 3. It was also reported that with addition of CNTs, epoxy composites became more brittle. Recently, Montazeri et al. (2012) also reported the effect of CNTs on the storage modulus of the composite (Fig. 4).

Guo et al. (2009) investigated the interphase between CNTs and epoxy and reported increase in the strength and toughness (Fig. 5). Fracture elongation and ultimate tensile strength (UTS) were increased with the increase in MWNT concentration (i.e. 1 wt.%, 2 wt.%, 3 wt.% of CNTs). Prolongo et al. (2011) reported the presence of good adhesion between CNTs and matrix were the main attributes for these results. To form a strong interface, amino functionalised MWNTs were dispersed in epoxy matrix. It was evident that the functionalisation treatment increased thermo-mechanical and flexural properties of the nanocomposite (Prolongo et al., 2011). The thermally pre-cured samples (with 0.25 wt.% functionalised CNTs) showed 58% improvement in the flexural strength, whereas samples having non-functionalised CNTs showed only 45% improvement as compared to the pure resin (Prolongo et al., 2011). Therefore, functionalisation of CNTs has significant positive effect on the mechanical properties of the epoxy nanocomposites.

Epoxy nanocomposites processed with acid treated CNTs showed have higher tensile strength and fracture strain (Montazeri et al., 2010). Moreover, untreated MWNT/ epoxy were found to be brittle. Montazeri et al. (2010) investigated the properties of MWNT/ epoxy composites reinforced with CNTs treated with nitric and sulfur acids. Young's modulus is higher in the untreated samples, possibly due to the aggregation of carbon nanotubes. The purified or acid treated MWNTs have modified interface as compared to the original nanotubes. After modification and acid treatment of CNTs, they have excellent structure and found to be adhering well with the matrix. It was shown by Montazeri et al. (2010) that the composites with purified CNTs have capability of absorbing greater stresses (i.e. higher fracture elongation and ultimate tensile strength), which is due to good interfacial adhesion between CNTs and the matrix. Modifying surfaces of CNTs based pre-pregs also contribute in achieving superior mechanical properties. Chen et al. (2010) showed that the modified CNT-added nano-prepreg sheets have better properties due to superior dispersion and the absence of CNT clusters. Tensile strength, flexural strength and impact strength were improved with the increase in CNT content. Moreover, the modified composite material exhibits high electrical conductivity compared to the unmodified, and it can be further increased with the addition of CNTs.

Currently, it is difficult to align CNTs in epoxy matrices because of the dimensions of CNTs. Uncontrolled orientation is responsible for low degree of dispersion. Thus, the resulting composite indexes of properties were found to be much lower due to these deficiencies (Cheng et al., 2009). However, for some properties, like hardness random oriented CNTs contribute positively as reported by Felisberto et al. (2011) in Fig. 6.

6.2. Electrical and thermal properties of CNT/epoxy nanocomposites

To achieve good dispersion and electrical/ thermal properties, optimal resin viscosity is essential. Recently, Pereira et al. (2010) reported significant increase in the electrical and thermal conductivity by incorporating MWNTs in low-viscosity epoxy composite. Such improvements are only possible when CNTs are thoroughly de-bundled and homogenously dispersed as demonstrated by Pereira et al. (2010). Just like hardness (Fig. 6), there is a strong dependence of composite's electrical conductivity on the alignment of nanofillers, as seen in Fig. 7. Recently, Inam et al. (2011) reported the dependence of CNT aspect ratio on the percolation threshold and electrical conductivity of the CNT/ epoxy nanocomposites. It was found that CNTs having higher aspect ratios have lower percolation threshold for electrical conductivity and vice versa (Inam et al., 2011).

CNT is an interesting material to research. Recently, Felisberto et al. (2011) analysed the electrical conductivity of CNT nanocomposites during curing. The results showed that CNTs were mobile at ~60 °C because of the decrease in viscosity. This increased the number of pathways and subsequently increased electrical conductivity of the composite. After 50 mins of processing, the resin viscosity raised sharply which stopped CNT mobility and stabilised electrical conductivity. In this way, CNT mobility can be monitored during different stages of curing CNT/ thermoset nanocomposites. This approach can also be used to characterise curing cycles of new thermoset formulations.

Recently, Chang et al. (2012) produced CNT nanocomposites by using microwaves for curing. The group also reported improved thermal and mechanical properties of microwaved composites as compared to the samples prepared by conventional oven heating. Microwave curing imparted very high dielectric constant and low dielectric loss to the samples. Using microwaves for curing CNT/ epoxy nanocomposites nanotubes is also advantageous because of shorter curing durations. This would help in reducing re-agglomeration during curing as reported by Inam and Peijs (2006). As a result, Chang et al. (2012) reported superior dispersion of carbon nanotubes in epoxy matrix, which contributed towards significantly increasing the electrical conductivity of the nanocomposite. Therefore, good dispersion and consistency in the direction of the nanotubes provides high dielectric properties to the

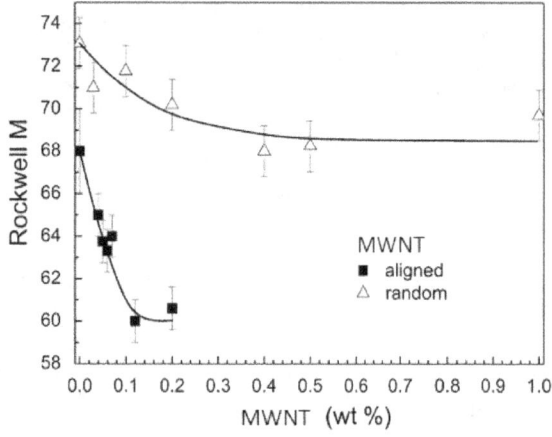

Figure 6. Rockwell hardness of aligned and randomly oriented CNT/ epoxy nanocomposites (Felisberto et al., 2011).

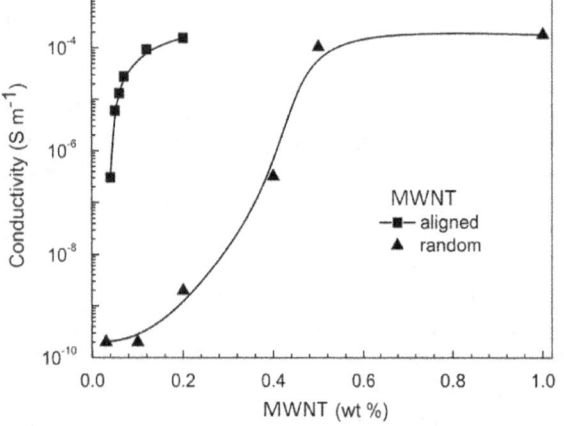

Figure 7. Electrical conductivity of aligned and randomly oriented CNT/ epoxy nanocomposites (Felisberto et al., 2011).

Table 4
Room temperature two-probe electrical and thermal conductivity of 60±5 wt.% CNT/ epoxy composites at its highest value (Park et al., 2012).

CNT sheet for epoxy composite	Electrical conductivity (S/cm)	Thermal conductivity (W/mK)
Random short-MWNT	100	6
Random long-MWNT	640	55
25% stretched long-MWNT	1300	83
40% stretched long-MWNT	800	103
Functionalised long-MWNT	160	22

Table 5
CTEs of the neat epoxy and the nanocomposite samples (Abdalla et al., 2010).

Sample	CTE (ppm/°C), n = 5
Neat epoxy system	74.04 ± 3
1% MWNT/ epoxy (random)	65.43 ± 2
1% MWNT/ epoxy (perpendicular magnetic field)	66.57 ± 2
1% MWNT/ epoxy (parallel magnetic field)	64.71

nanocomposite. A comprehensive review of dispersion of CNT in polymers was presented by Xie et al. (2005).

Park et al. (2012) dispersed different types of CNTs into epoxy to study electrical and thermal conductivities. In order to improve these properties, CNT/ epoxy nanocomposite sheets were mechanically stretched and aligned. It was shown that the thermal conductivity increased with the increase in temperature, and the highest electrical and thermal conductivities were observed for 40% stretched samples (Tab. 4).

Abdalla et al. (2010) cured CNT/ epoxy nanocomposites in the presence of magnetic field and reported thermal properties. CNTs were aligned perpendicularly and parallel to the magnetic field during curing. For these anisotropic nanocomposites, mechanical properties were also found to vary significantly in different directions. In the parallel direction, modulus increased to 72% compared with a net resin, in the perpendicular - up to 24%, and in a random direction, it rose to 32%. It was also found that thermal diffusivity and conductivity are strongly dependent on the alignment of CNTs. CNTs were also responsible for decreasing co-efficient of thermal expansion for epoxy matrices as shown in Tab. 5.

7. Conclusions

This paper provides detailed introduction of CNTs and a review of the recent developments in the field of CNT/ epoxy nanocomposites. Because of their unique properties, CNTs are indispensable nanofillers for novel and advanced epoxy nanocomposites. Significant developments have been made recently in the field of CNT/ epoxy composites. CNT/ epoxy materials have remarkably

high thermal and electrical conductivities. Such composites also have improved strength, ultimate tensile strength, viscoelastic properties and thermo mechanical characteristics. Because of their low densities, CNTs reinforced epoxy nanocomposites possess great potential to be used for many aerospace applications. Carbon fibre is a well-mature technology and replacing them with CNTs is not feasible at this stage. More development is required in the field of CNTs before they can be substituted against carbon fibre. However, CNTs certainly offer many supplementary benefits to carbon fibre reinforced epoxy composites. Recent research reported that CNTs' types, properties and dispersion profile have a strong effect on the final mechanical, electrical and thermal properties of the CNT/ epoxy composites.

References

Abdalla, M., Dean, D., Theodore, M., Fielding, J., Nyairo, E. and Price, G. (2010). Magnetically processed carbon nanotube/ epoxy nanocomposites: Morphology, thermal, and mechanical properties. *Polymer*, Vol. 51, No 7, pp. 1614–1620.

Belin, T. and Epron, T. (2005). Characterization methods of carbon nanotubes: a review, *Materials Science and Engineering B*, Vol. 119, No 2, pp. 105–118.

Bethune, D. S., Klang, C. H., de Vries, M. S., Gorman, G., Savoy, R., Vazquez, J. and Beyers, R. (1993). Cobalt-catalysed growth of carbon nanotubes with single-atomic-layer walls, *Nature*, Vol. 363, No 6430, pp. 605 – 607.

Breeuer, O. and Sundararaj, U. (2004). Big returns from small fibers: A review of polymer/carbon nanotube composites, *Polymer Composites*, Vol. 25, No 6, pp. 630–645.

Chand, S. (2000). Review carbon fibers for composites, *Journal of Materials Science*, Vol. 35, No 6, pp. 1303 – 1313.

Chang, J., Liang, G., Gu, A., Cai, S. and Yuan, L. (2012). The production of carbon nanotube/epoxy composites with a very high dielectric constant and low dielectric loss by microwave curing. *Carbon*, Vol. 50, No 2, pp. 689-698.

Chen, W.-J., Li, Y.-L., Chiang, C.-L., Kuan, C.-F., Kuan, H.-C., Lin, T.-T. and Yip, M.-C. (2010). Preparation and characterization of carbon nanotubes/epoxy resin nano-prepreg for nanocomposites. *Journal of Physics and Chemistry of Solids*, Vol. 71, No 4, pp. 431–435.

Cheng, Q. F., Wang, J.P., Wen, J.J., Liu, C.H., Jiang, K.L., Li, Q.Q. and Fan, S.S. (2009). Carbon nanotube/epoxy composites fabricated by resin transfer molding. *Carbon*, Vol. 48, No 1, pp. 260–266.

Donnet, J., Bansal, R. and Stoeckli, F. (1990). *Carbon Fibers*, Marcel Dekker, Inc., New York.

Eletskii, A. (2007). Mechanical properties of carbon nanostructures and related materials. *Uspekhi Fizicheskikh Nauk*, Vol. 50, No 3, pp. 233-274.

Endo, M., Hayashi, T., Kim, Y.A., Terrones, M. and Dresselhaus, M.S. (2004). Applications of carbon nanotubes in the twenty-first century. *Philosophical Transactions of the Royal Society A*, Vol. 362, No 1832, pp. 2223-2238.

Felisberto, M., Arias-Duran, A., Ramos, J.A., Mondragon, I., Candal, R., Goyanes, S. and Rubiolo, G.H. (2011). Influence of filler alignment in the mechanical and electrical properties of carbon nanotubes/ epoxy nanocomposites. *Physica B: Condensed Matter*. In Press, Corrected Proof, Available online 16 December 2011.

Geim A. K. and Novoselov K. S. (2007). The rise of grapheme. *Nature Materials*, Vol. 6, No 3, pp. 183-191.

Glaskova, T., Zarrelli, M., Aniskevich, A., Giordano, M., Berzina, B. and Trinkler, L. (2012). Quantitative optical analysis of filler dispersion degree in MWCNT-epoxy nanocomposite. *Composites Science and Technology*, Vol. 72, No 4, pp. 477–481.

Guo, P., Song, H. and Chen, X. (2009). Interfacial properties and microstructure of multiwalled carbon nanotubes/epoxy composites. *Materials Science and Engineering*, Vol. 517, No 1–2, pp. 17–23.

Hayashi, T. and Endo, M. (2011). Carbon nanotubes as structural material and their application in composites. *Composites Part B: Engineering*, Vol. 42, No 8, pp. 2151-2157.

Hertel, T., Walkup, R. E. and Avorius, P. (1998). Deformation of carbon nanotubes by surface van der Waals forces. *Physical Review B*, Vol. 58, No 20, pp. 13870-13873.

Holister, P., Harper, T. and Román Vas, C. (2003), Nanotubes. White Paper. [e-journal] Las Rozas: CMP Científica. Available through: website < http://cientifica.eu/blog/ >, Accessed 10 March 2012.

Hussain, F., Hojjati, M., Okamoto, M. and Gorga, R.E. (2006). Review article: polymer-matrix nanocomposites, processing, manufacturing, and application: an overview. *Journal of Composite Materials*, Vol. 40, No 17, pp. 3107-3123.

Iijima, S. (1991). Helical microtubules of graphitic carbon. *Nature*, Vol. 354, No 6348, pp. 56-58.

Inam, F. and Peijs, T. (2006). Re-aggregation of carbon nanotubes in two-component epoxy system. *Journal of Nanostructured Polymers and Nanocomposites*, Vol. 2, No 3, pp. 87-95.

Inam, F., Wong, D.W.Y., Kuwata, M. and Peijs, T. (2010). Multi-scale hybrid micro-nanocomposites based on carbon nanotubes and carbon fibers. *Journal of Nanomaterials*, Published online, doi:10.1155/2010/453420.

Inam, F., Reece, M.J. and Peijs, T. (2011). Shortened carbon nanotubes and their influence on the electrical proeprties of polymer nanocomposites. *Journal of Composite Materials*, Published online, doi: 10.1177/0021998311418139.

Journet, C., Maser, W. K., Bernier, P., Loiseau, A., Lamy de la Chapelle, M., Lefrant, S., Deniard, P., Lee, R. and Fischer, J. E. (1997). Large-scale production of single-walled carbon nanotubes by the electric-arc technique. *Nature*. Vol. 388, No 6644, pp. 756-758.

Khare, R. and Bose, S. (2005). Carbon nanotube based composites - A review. *Journal of Minerals & Materials Characterization & Engineering*, Vol. 4, No 1, pp. 31-46.

Kim, B., Park, S. and Lee, D. (2008). Fracture toughness of the nano-particle reinforced epoxy composite. *Composite Structures*, Vol. 86, No 1–3, pp. 69–77.

Kim, M. T., Rhee, K. Y., Lee, J. H., Hui, D. and Lau, A. K. (2011). Property enhancement of a carbon fiber/epoxy composite by using carbon nanotubes. *Composites: Part B*, Vol. 42, No 5, pp. 1257–1261.

Kis, A., Csanyi, G., Salvetat, J.-P., Lee, T.-N., Couteau, E., Kulik, A. J., Benoit, W., Brugger J. and Forro, L. (2004). Reinforcement of single-walled carbon nanotube bundles by intertube bridging. *Nature Materials*, Vol. 3, No 3, pp. 153 – 157.

Koziol, K., Vilatela, J., Moisala, A., Motta, M., Cunniff, P., Sennett, M. and Windle, A. (2007). High-performance carbon nanotube fiber. *Science*, Vol. 318, No 5858, pp. 1892-1895.

Li, H. Q., Wang, Y. G., Wang, C. X. and Xia, Y. Y. (2008). A competitive candidate material for aqueous super capacitors: High surface-area graphite. *Journal of Power Sources*, Vol. 185, No 2, pp. 1557-1562.

Liu B., Liu Q., Ren W., Li F., Liu C. and Cheng H.-M. (2010). Synthesis of single-walled carbon nanotubes, their ropes and books. *Comptes Rendus Physique*, Vol. 11, No 5-6, pp. 349-354.

Martone, A., Formicola, C., Giordano, M., and Zarrelli, M. (2010). Reinforcement efficiency of multi-walled carbon nanotube/epoxy nano composites. *Composites Science and Technology*, Vol. 70, No 7, pp. 1154–1160.

Montazeri, A. and Montazeri, N. (2011). Viscoelastic and mechanical properties of multi walled carbon nanotube/epoxy composites with different nanotube content. *Materials & Design*, Vol. 32, No 4, pp. 2301–2307.

Montazeri, A., Pourshamsian, K. and Riazian, M. (2012). Viscoelastic properties and determination of free volume fraction of multi-walled carbon nanotube/epoxy composite using dynamic mechanical thermal analysis, *Materials and Design*, Vol. 36, pp. 408–414.

Ogasawara, T., Moona, S., Inoue, Y., and Shimamura, Y. (2011). Mechanical properties of aligned multi-walled carbon nanotube/epoxy composites processed using a hot-melt prepreg method. *Composites Science and Technology*, Vol. 71, No 16, pp. 1826–1833.

Park, J.G., Cheng, Q., Lu, J., Bao, J., Li, S., Tian, Y., Liang, Z., Zhang, C. and Wang, B. (2012). Thermal conductivity of MWCNT/epoxy composites: The effects of length, alignment and functionalization. *Carbon*, Vol. 50, No 6, pp. 2083-2090.

Pereira, C.M., Novoa, P., Martins, M., Forero, S., Hepp, F. and Pamba-guian, L. (2010). Characterization of carbon nanotube 3D-structures infused with low viscosity epoxy resin system. *Composite Structures*, Vol. 92, No 9, pp. 2252–2257.

Prolongo, S., Gude, M. and Urena, A. (2011). Improving the flexural and thermomechanical properties of amino-functionalized cabon nanotube/epoxy composites by using a pre-curing treatment. *Composites Science and Technology*, Vol. 71, No 5, pp. 765–771.

Qian, H., Shaffer, M.S.P., Bismarck, A. and Greenhalgh, E.S. (2010). Carbon nanotube-based hierarchical composites: A review. *Journal of Materials Chemistry*, Vol. 20, No 23, pp. 4751-4762.

Quin, H., Greenhalgh, E.S., Shaffer, M.S.P. and Bismarck, A. (2010). Carbon nanotube-based hierarchical composites: a review. *Journal of Materials Chemistry*, Vol. 20, No 23, pp. 4751-4762.

Ren, Z.F., Huang, Z.P., Xu, J.W., Wang, J.H., Bush, P., Siegal, M.P. and Provencio, P.N. (1998). Synthesis of large arrays of well-aligned carbon nanotubes on glass. *Science*, Vol. 282, No 5391, pp. 1105-1107.

Rinzler, A., Liu, J., Dai, H., Nikolaev, P., Huffman, C.B., Rodrıguez-Macıas, F.J., Boul, P.J., Lu, A.H., Heymann, D., Colbert, D.T., Lee, R.S., Fischer, J.E., Rao, A.M., Eklund, P.C. and Smalley, R.E. (1998). Large-scale purification of single-wall carbon nanotubes: Process, product and characterization. *Applied Physics*, Materials Science and Processing Vol. 67, No 1, pp. 29–37.

Sengupta, R., Bhattacharya, M., Bandyopadhyay, S. and Bhowmick, A. K. (2011). A review on the mechanical and electrical properties of graphite and modified graphite reinforced polymer composites. *Progress in Polymer Science*, Vol. 36, No 5, pp. 638-670.

Singh, C., Shaffer, M., Kinloch, I. and Windle, A. (2002). Production of aligned carbon nanotubes by the CVD injection method. *Physica B*, Vol. 323, No 1-4, pp. 339-340.

Smith, W. (1987). *Engineered Materials Handbook*: Composites. Ohio: ASM International, Vol. 1, p. 49.

Thostenson, E.T., Renb, Z. and Choua, T.-W. (2001). Advances in the science and technology of carbon nanotubes and their composites: a review, *Composites Science and Technology*, Vol. 61, No 13, pp. 1899–1912.

Vega, A., Kovacs, J. Z., Bauhofer, W. and Schulte, K. (2008). Combined Raman and dielectric spectroscopy on the curing behaviour and stress buildup of carbon nanotube-epoxy composites, *Composites Science and Technology*, Vol. 69, No 10, pp. 1540–1546.

Xie, X., Mai, Y. and Zhou, X. (2005). Dispersion and alignment of carbon nanotubes in polymer matrix: A review. *Materials Science and Engineering*, Vol. 49, No 4, pp. 89–112.

Yu, M.-F., Lourie, O., Dyer, M., Moloni, K., Kelly, T. F. and Ruoff, R. S. (2000). Strength and breaking mechanism of multiwalled carbon nanotubes under tensile load. *Science*, Vol. 287, No. 5453, pp. 637-640.

Zhang, X., Cao, A., Li, Y., Xu, C., Liang, J., Wu, D. and Wei, B. (2002). Self-organized arrays of carbon nanotube ropes. *Chemical Physics Letters*, Vol. 351, No 3-4, pp. 183-188.

Zolotuchin, I. (1999). Carbon nanotubes, *Soros Educational Journal*: Voronezh State University, Vol.3, No 3, pp. 111-115.

Mathematical Model of the Heat in a Dielectric Composite Cylinder during Microwave Treatment

Sergey Rumyantsev, Sergey Reznik, Tatyana Guzeva

Rocket and Spacecraft Composite Structures Department, Faculty of Special Machinery, Bauman Moscow State Technical University, 5 2nd Baumanskaya Street, Moscow, 105005, Russia

Abstract: The traditional curing technology of epoxy based composites requires a long thermal process. Microwave heating could be used to improve curing technology by reducing treatment time with no detriment in the material quality. In this paper, mathematical model of temperature distribution inside the epoxy composite part during microwave treatment is investigated and theoretical, experimental and simulation results are discussed.

Key Words: Heat model, Composite materials, Microwave heating.

1. Introduction

Increased global demands for polymer matrix composite in aerospace and aviation are drawing attention towards rapid production of composite. Recently it has become more important to upgrade and optimize the technological processes for curing composite parts in shorter time without compromising product quality. One of the prospective technologies is utilization of microwave irradiation for accelerating production rate for the polymerization process of fibre reinforced epoxy composites.

In general, conventional thermal heating technology depends on the thermal conductivity of the materials. Heat transfer from surface to the bulk of materials leads to large process times and high energy requirements because the thermal conductivity of polymeric composites is low. So the rate of change in the temperature has to be slow and local overheating may cause the surface to scorch and internal stresses to appear. These factors could cause a decrease in material quality.

Theoretically, microwave curing of fibre reinforced epoxy composites has a few significant advantages versus traditional heating technology. Firstly, due to volumetric origin of heat generation, thermal gradients in material reduce to minimum. This fact allows to reduce the curing time up to 4 times or more compared to traditional curing technology (Yusoff et al., 2007). Secondly, microwave energy is mainly absorbed by the dielectric material and not spent on heating the chamber and other tooling; therefore, it is possible to increase process control and reduce overall energy consumption (Papagyris et al., 2008). And

thirdly, by the means of significant thermal gradients reduction and more uniform heating, a microstructure of better quality can be obtained in the cured material (Nightingale and Day, 2002).

In general, microwave heating is the absorption of electromagnetic field energy by the whole volume of a dielectric material. The electromagnetic wave penetrates into material and interacts with charged particles. A sequence of microscopic processes leads to consumption of electromagnetic field energy in the material.

The molecules of dielectric materials could be polar or non-polar depending on the charge location. In some molecules, charge locations are so symmetric, that they have zero dipole moment in the absence of an external electric field. Polar molecules have some dipole moment without an external electric field applied. Non-polar molecules become polarized and acquire the dipole moment after electric field imposition. Polar molecules do not only change their electric moment value in presence of electric fields, but they also align with the direction of the field. That way, heat generation is implemented even without a conductive current.

Fig. 1 shows the character of microwave penetration in different types of material.

High efficiency of microwave heating in epoxy-based composites is caused by high values of dielectric proper-

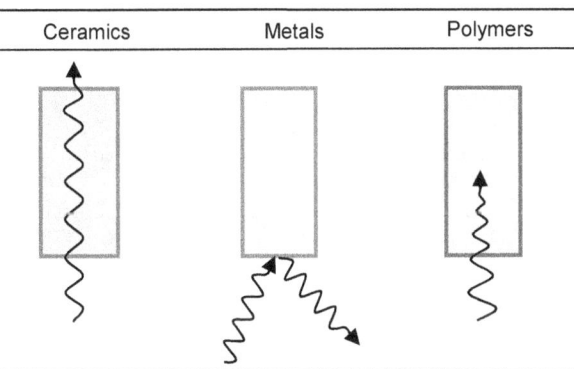

Figure 1. Patterns of microwave penetration.

$$-O-CH_2-CH-CH_2$$

$$-O-CH_2-CH-CH_2-O-$$
$$\qquad\qquad OH$$

$$-O-CH_2-CH-CH_2-N-$$
$$\qquad\qquad OH \qquad\qquad H$$

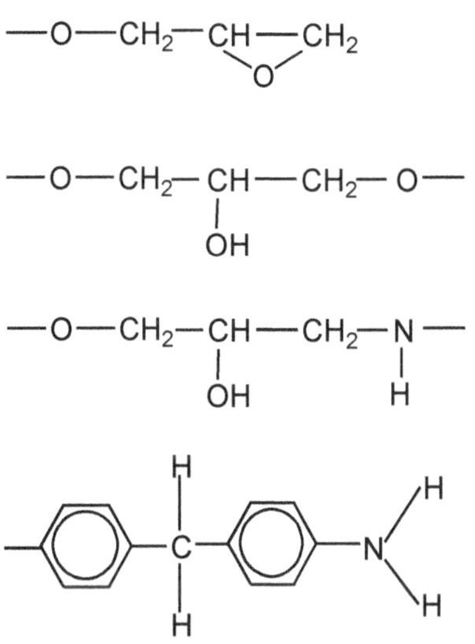

Figure 2. Polar groups in epoxy chemical chains.

ties. The dielectric characteristics have a chemical origin and strictly depend on the polar group's content. Epoxy resins usually have the following sets of polar groups in its composition: epoxy ($-OCH_2CH-$), hydroxyl ($-OH$), imine ($-NH$) and amino ($-NH_2$).

The amount of energy generated in the heated material depends on a number of factors and one of the main is frequency. It is well known that microwave frequencies range from 300 MHz to 300 GHz. For polymer curing, the frequency of 2.45 GHz is mostly used, the same as already used in all domestic microwave ovens. There are also some laboratory settings with Variable Frequency Microwave (VFM) generators, in which working frequency changes from 1 MHz to 20 GHz. Papargyris et al. (2008) and Qiu et al. (2001) used VFMs as microwave generators in their experiments.

2. Literature review

Methods of mathematical modeling form the basis of the modern design of complex technical systems, machinery and technologies. A lot of information about applying mathematical models to solve specific technological problems, including the appropriate equipment design and rational choice of technological modes during microwave treatment of dielectric materials has appeared in the literature during the last few years.

The problems considered were related to various fields of application, e.g.: thermoset and thermoplastic resin polymerization in glass- and carbon fibre composites (Morozov et al., 2011), multi-core composite cables drying, polymeric electrical insulation drying (Shvorobey, 1992), wood products drying (Afanasiev and Sipliviy, 2010), extraction of substances from solid materials (Beloborodov, 1999), separation of oil-water emulsion, oil well pipe cleaning (Morozov et al., 2010), snow melting (Afinagentov and Tahauv, 2011). The usage of microwaves in pultruded lines (Lin and Hawley, 1993) has also been proven to be useable for improved curing.

For example, Shvorobey et al. (1992) considered a mathematical model of a polymeric transformer electrical insulation in a microwave field, with the following assumptions: an infinitely extended flat wall heating model was used; the process of heat transfer was considered as one-dimensional and transient; the electromagnetic field was considered uniform; the Dirichlet boundary condition

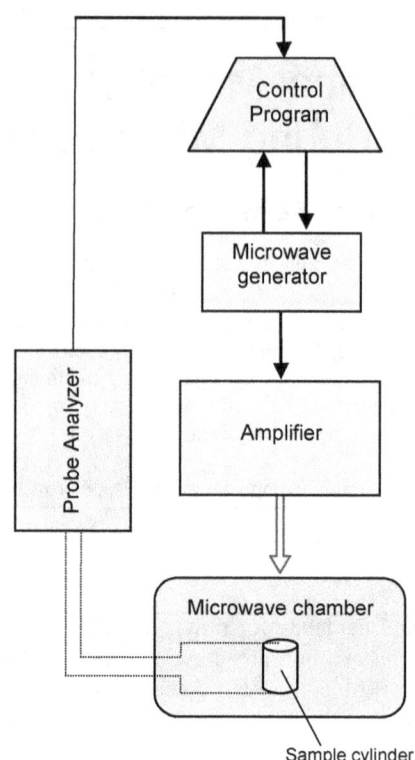

Figure 4. Experiment scheme.

on the outer surface was given as a temperature ramp; thermal radiation from the surface was not taken into account due to a relatively low temperature level; the boundary condition on the inner surface wasn't given in an explicit form; thermo physical material properties were taken constant. The Laplace integral transform method was used to solve this model. As is often accepted in analytical solutions, only the first term of the series was taken into account.

The analytical solution can give just a preliminary model of the technological process because thermo physical properties change with the temperature. New opportunities of accounting for all physical features of technological process were opened with the development of numerical methods, like the Finite Element Method. In many cases the most productive way of mathematical modeling of nonlinear and transient processes is accomplished with the use of commercial software systems such as CTS Microwave Studio, AWR Microwave office; multipurpose simulation software: Comsol Multiphysics, MSC Patran, Ansys and others.

A mathematical model of this process could help to predict temperature levels during microwave curing and exercise control for the polymerization reaction.

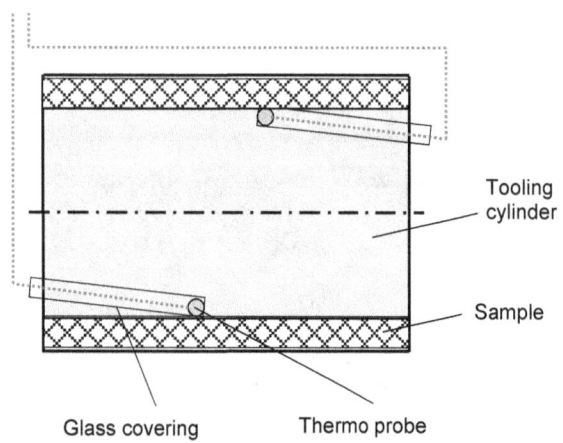

Figure 3. Temperature probe location.

Table 1
Heat capacity and thermal conductivity vs temperature

T,°C	20	30	40	50	60
C, J/g·K	0.77	0.85	0.92	0.98	1.04
λ, W/m·K	0.23	0.25	0.26	0.28	0.29
T,°C	80	100	120	140	150
C, J/g·K	1.14	1.20	1.24	1.26	1.26
λ, W/m·K	0.31	0.31	0.31	0.31	0.30

3. Experimental

To obtain the temperature distribution in the composite part over the period of polymerization, a series of experiments have been done.

The sample has a cylindrical shape. Geometry dimensions of the sample are: $l = 55$ mm, outer radius = 18 mm, inner radius = 15 mm. The sample is unidirectional carbon fibre/epoxy-prepreg Hexcel M21. The mandrel was made from solid tooling plate – a combination of epoxy resins, glass micro balloons and hardeners, cured together. The prepreg was wound by hand on the cylindrical mandrel and fixed by shrinking tape. The number of layers in the sample was 34. The mandrel was not pulled off from the sample cylinder until the end of the experiment.

Close attention should be paid to the choice of devices for temperature control during the process. It is difficult to use any electrical conductors inside the microwave chamber to measure the temperature. Remote pyrometers are not satisfactory either because they only scan the surface and are not able to show internal temperature level. This is why fibre optic (fluoroptic) sensors are widely used for temperature measurement during microwave treatment: they have a dielectric nature and also do not absorb microwave energy.

Two thermal fluoroptic sensors were provided to monitor the temperature distribution a cross the sample during the test. The sensitive probe tips have to be positioned as close to material area as possible. The probe tips were located on the inner surface of the sample on the opposite sides – as it shown in Fig. 3. For this purpose notches were made in the mandrel. Two tiny glass tubes covered the tips to keep sensors isolated from uncured resin.

The experimental layout (shown in the Fig. 4) consisted with microwave generator, microwave amplifier, brass chamber and temperature analyzer for fluoroptic probes. All parts of the system were connected to and controlled by a PC program.

The purpose of the experiment was to detect the heating rate and, as a consequence, the curing kinetics of the carbon fibre cylinder under microwave treatment. In the first step a few optimal frequencies which returns less reflective power were chosen – that means that most microwave energy absorbed by the sample. Variable frequency heating was used in order to more uniformly heat

the sample – in this experiment three frequencies were used: 2.38, 2.403, 2.407 GHz.

Additional experiments were made to obtain the thermal characteristics of the material: heat capacity and conductivity. Differential scanning calorimetry studies were carried out on a Netzsch STA 449 F1; laser flash analysis was carried out on a Netzsch LFA 457. Small size square samples (3×3 mm) were cut from cured sample and heated from 20°C up to 150°C.

4. Mathematical model

For the mathematical model the equal cylindrical shape of the sample was chosen. To describe the model let us consider that the sample is located in the center of the chamber and is subjected to a uniform energy flux from the microwave generator.

The sample model is a hollow cylinder with outer radius R_1 and inner radius R_2. The cylinder is subject to asymmetrical heating from the exposure to the uniform outer electromagnetic field. Let us consider that microwave energy is released within the entire volume of the sample with two-dimensional, transient heating process. Physical and mechanical properties of the material are simulated as orthotropic. Outward heat transfer is by surface radiation.

This physical model corresponds to the heat equation in cylindrical coordinate system:

$$C_i(T)\frac{\partial T_i}{\partial \tau} = \frac{\partial}{\partial z}\left(\lambda_{z,i}(T)\frac{\partial T_i}{\partial z}\right) +$$
$$+\frac{1}{r}\frac{\partial}{\partial r}\left(r \cdot \lambda_{r,i}(T)\frac{\partial T_i}{\partial r}\right) + q_{mw}(T) + q_{ch}(\eta), \quad (1)$$
$$i = \overline{1,2}$$

where C – volumetric heat capacity; T – temperature; τ – time; λ – thermal conductivity; q_{mw} – energy generated by microwaves; q_{ch} – energy generated by exothermic reaction of polymerization; η – degree of polymerization; i – index (1 – composite part, 2 – mandrel cylinder).

Initial conditions are:
$$\tau = 0,\ T = T_0 \quad (2)$$

Boundary conditions are as follows:

$$r = 0,\ \frac{\partial T_2}{\partial r} = 0 \quad (3)$$

$$r = R_2,$$
$$\lambda_{r,1}(T)\frac{\partial T_1}{\partial r} = \lambda_{r,2}(T)\frac{\partial T_2}{\partial r}; \quad (4)$$
$$T_{w1,2} = T_{w2,1}$$

$$r = R_1,$$
$$\lambda_{r,1}(T)\frac{\partial T_1}{\partial r} = -\alpha_f\left(T_{w1,1} - T_f\right) - \varepsilon_{ef}\sigma_0\left(T_{w1,1}^4 - T_f^4\right) \quad (5)$$

$$z = 0,\ \frac{\partial T_1}{\partial z} = \frac{\partial T_2}{\partial z} = 0 \quad (6)$$

$$z = \pm\frac{1}{2},$$
$$\lambda_{z,1}(T)\frac{\partial T_1}{\partial z} = -\alpha_f\left(T_b - T_f\right) - \varepsilon_{ef}\sigma_0\left(T_b^4 - T_f^4\right) \quad (7)$$

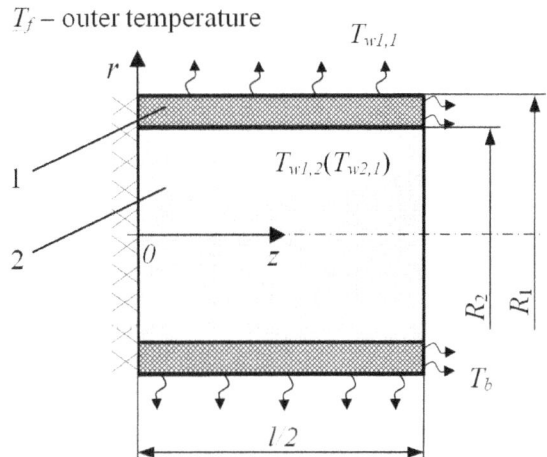

Figure 5. Composite cylinder model
(1 – composite part, 2 – tooling cylinder).

The amount of microwave energy generated in the material is:

$$q_{mw}(T) = \varepsilon_0 \varepsilon(T) \tan \delta(T) f E^2 \qquad (8)$$

where ε_0 – electric constant; ε – dielectric permittivity; $\tan\delta$ – dielectric loss tangent; f – frequency; E – electric field intensity.

Energy which is generated in the material by exothermic reaction during the curing process is quite difficult to calculate. This parameter is need to be get from experimental DSC thermal analysis data for similar epoxy resin and heating rate.

The electric field intensity goes down while wave penetrates into material and gives energy for heating. Penetration depth of the electromagnetic wave is a distance in which electric field intensity is reduced to half the original value. The penetration depth into the dielectric depends on dielectric and thermal parameters of material.

$$v(T) = \frac{\lambda_r(T)}{\pi \sqrt{2\varepsilon(T)\left(\sqrt{1 + \tan\delta(T)^2} - 1\right)}} \qquad (9)$$

This effect could be neglected, however, since the dimensions of cylinder are more than 50 times less than penetration depth in epoxy resin.

5. FEM analysis

The above model was taken as a basis for Finite Element Model of cylinder under microwave treatment. Simulation of temperature distribution was made using AnsysMultiphysics v.14 and the temperature change rate during the process in MSC Patran 2010. The geometry used was exactly the same as the experimental sample; thermal and physical characteristics of material were put in simulation program with respect to an orthotropic composite structure. Thermal conductivity and heat capacity of the epoxy resin were taken from experimental data as a function of temperature.

The impact of electromagnetic radiation was modelled as an internal heat generation of variable magnitude through all of the material volume. Heat radiation from outer surface to ambient – with coefficients of emissivity and absorptivity 0.89 and 0.9 respectively – was consid-

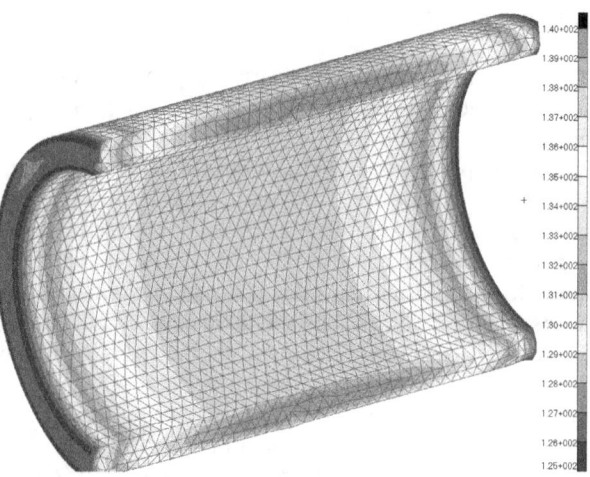

Figure 7. Temperature distribution (cross-section showed).

ered with subjection to reciprocal irradiation inside cylinder walls.

Two simulation models were made: the first mode does not consider energy, generated by exothermic reaction; the second mode applies this energy.

6. Results

Fig. 6 shows the temperature growth on the inner surface of sample during microwave treatment which was extracted from experiment. Also in the Fig. 6 is shown two theoretic graphs from FEM analysis in two modes. Mode 1 – with consideration of heat generated chemically $q_{ch}(\eta)$, mode 2 – without $q_{ch}(\eta)$. The discrepancy between graphs is about 15-20°C, but the rate of the curve is similar.

Following FEM analysis, the temperature gradient inside the sample (except edge of the end) does not exceed the value of 5-8°C – hence validating the volumetric nature of the microwave heating.

7. Conclusions

The mathematical model of a fibre reinforced composite cylinder has been developed to examine the effect of microwave heating. A set of experiments was made and expected temperature rate was obtained. Qualitative agreement of this model with experiment suggested that more precise thermal and dielectric characteristic of material should be obtained. Refined values of temperature dependent parameters subjected to intermediate curing phases should be calculated.

8. Acknowledgements

The authors would like to thank Professor Richard Day (Glyndwr University) for support in microwave experiments and Dr. M.O. Zabezjailov (ONPP Technologia) for help with DSC and LFA analysis.

References

Afanasev, A.M. and Sipliviy, B.N. (2010). Electromagnetic drying edge effects of extended samples with rectangular cross-section, *Physics of Wave Process and Radio Systems*, Vol. 13, No 1, pp. 90-94.

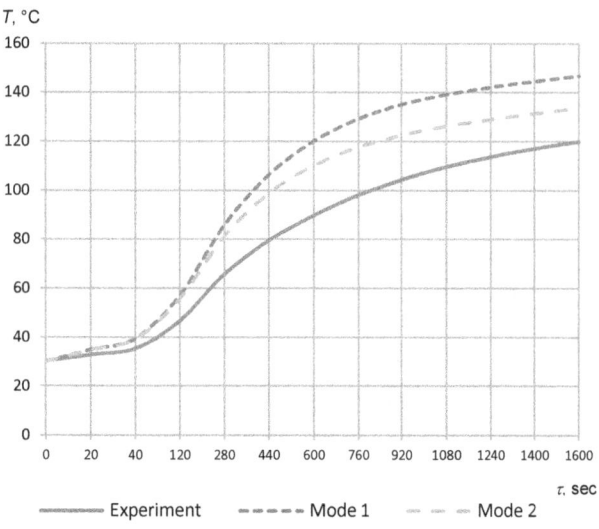

Figure 6. Temperature vs time

Afinogentov, V.I. and Tahauv, A.A. (2011). The phase interface controlling under SHF snow heating, *Physics of Wave Process and Radio Systems*, Vol. 14, No 1, pp. 66-70.

Bai S.L., Djafari V., Andreani, M. and Francois, D. (1995). A comparative study of the mechanical behavior of an epoxy resin cured by microwaves with one cured thermally, *European Polymer Journal*, Vol. 31, No. 9, pp. 875-88.

Balzer, B.B. and McNabb, J. (2008). Significant effect of microwave curing on tensile strength of carbon fiber composites, *Journal of Industrial Technology*, Vol. 24, No 3, pp. 1-9.

Beloborodov, V.V. (1999). Extracting from solid materials inside SHF electromagnetic field, *Engineering Physics Journal*, Vol. 72, No 1.

Berlov, A.V. (2010). Determination of temperature field in moist composite materials during Super High Frequency heating, *System Technologies*, Vol. 2, No 67, pp. 3-9.

Hubbard, R.L., Ahmad, I. and Toleno, B. (2006). Low temperature curing of epoxies with microwaves. *IMAPS International Conference,* March 20-23, Scottsdale, AZ.

Morozov, G.A., Morozov, O.G., Nasibullin, A.R. and Samigullin, R.R. (2011). Microwave treatment of thermoset and thermoplast polymers, *Physics of Wave Process and Radio Systems*, Vol. 14, No 3, pp. 114-121.

Morozov, G.A., Orlov, I.G. and Shakirov, A.S. (2010). The models of cleaning pipe deposits with microwave technology, *Physics of Wave Process and Radio Systems*, Vol. 13, No 3, pp. 125-130.

Nightingale, C. and Day, R.J. (2002). Flexural and interlaminar shear strength properties of carbon fibre/epoxy composites cured thermally and with microwave radiation, *Composites: Part A: Applied Science and Manufacturing*, No 33, pp. 1021-1030.

Okress, E. (1968). *Microwave Power Engineering.* New York and London: Academic Press.

Papargyris, D.A., Day, R.J., Nesbitt, A. and Bakavos, D. (2008). Comparison of the mechanical and physical properties of a carbon fibre epoxy composite manufactured by resin transfer moulding using conventional and microwave heating, *Composites Science and Technology*, No 68, pp. 1854-1861.

Qiu, Y. and Hawley, M. (2001). Uniform processing of V-shaped and Tri-planar composite parts using microwaves, *Journal of Composite Materials*, No 35, pp. 1062-1078.

Shvorobey, U. (1992). *The Basis of Microwave Heating Technology of Polymer and Composite Materials.* Moscow: TSNII NTIKPK.

Trigorliy, S.V. (2000). Numerical modeling and process optimization of dielectric Super High Frequency heat treatment, *Applied Mechanic and Technical Physics*, Vol. 41, No 1, pp. 112-119.

Wallace, M., Attwood, D., Day, R.J. and Heatley, F. (2006). Investigation of the microwave curing of the PR500 epoxy resin, *Journal of Material Science*, No 41, pp. 5862-5869.

Yusoff, R., Aroua, M.K., Nesbitt, A. and Day, R.J. (2007). Curing of polymeric composites using microwave RTM, *Journal of Engineering Science and Technology*, Vol. 2, No 2, pp. 151-163.

Zong, L., Kempel, L.C. and Hawley, M.C. (2005). Dielectric studies of three epoxy resin systems during microwave cure, *Polymer*, No 46, pp. 2638-2645.

Direct Computational Method for Composites Based-on Coupled Plasticity and Continuous Damage Mechanics

Dmytro Vasiukov[1], Sephane Panier[1], Abdelkader Hachemi[2]

[1] Polymers and Composites Technology and Mechanical Engineering Department, Ecole des Mines de Douai, 941 rue Charles Bourseul, Douai, 59508, France
[2] Institute of General Mechanics - RWTH Aachen University, 55 Templergraben, Aachen, 52056, Germany

Abstract: Fatigue damage of fibre-reinforced polymers is of interest to researchers and designers. In this paper new direct computational method was developed in order to reduce computational costs for life prediction of composites. Constitutive model is based-on plasticity and damage mechanics. Macro-micro transition stress concentration tensor has been used in order to take into account influence of the micro-structure. Plasticity yield surface is based-on the modified Raghava criterion. Damage of matrix and transverse cracking are governed by quadratic damage surface. Method was examined on a fibre-reinforced polymer composite plate with unidirectional plies.

Key Words: Direct computation, Fatigue damage, Fibre-reinforced polymers, Life prediction.

1. Introduction

Nowadays composite materials have many applications in most sectors of industry. For the last decades the fatigue behaviour of fibre-reinforced polymers (FRP) has been extensively investigated due to intensive implementation of composite structures in aeronautical applications. FRP components demonstrate damageability during service loading conditions so fatigue damage analysis of FRP is important and complex task. The complexity of the task is increased by the fact that properties depend on loading condition and current level of degradation. A number of various numerical methods has been developed for analysis of composite materials which include simulation and prediction of non-linear behaviour. In terms of condition these methods could be grouped into: micromechanical, continuous and multi-scale.

Micromechanical approaches consider mechanisms that occur at the scale of fibre and matrix. Macroscopic behaviour is evaluated from micro-model (unit cell). It is efficient in predicting material properties but limited in structure modelling. On another hand, the majority of commercial and scientific software are considering material as continuous. The knowledge about micro-structure is underlying in the stiffness/compliance tensor as was described by Lubineau and Ladevèze (2008). The third type is multi-scale methods that are kind of uniting methods and consist of both previous approaches (Raghavan and Ghosh, 2004). Described above methods require significant computational resources.

However non-linearity of composite can be described by plasticity or continuous damage theory or combination of plasticity with damage. According to continuous damage mechanics (CDM) framework a new formulation has been discussed in Maire and Chaboche (1997) where authors considered the bases that should be involved into model development.

Abdelal et al. (2002) have developed a model for fatigue damage evolution of FRP. Orthotropic damage model has been used for evaluation damage within cycle. For damage evolution author used thermodynamic force potential which depends on material's orientation and described by Tsai-Wu criterion.

Supporting the concept of FRP non-linearity due to damage Hassan and Batra (2008) considered following main modes of composites failure: fibre breakage, matrix cracking, fibre/matrix debonding, and delamination/sliding. Critical values of thermodynamic forces were evaluated using the test data available in the literature.

The work by Vyas et al. (2011) gives a typical example of constitutive modelling of composites based on the plasticity only. Authors used modified Raghava criterion as yield function. Using experimental data available in the literature, the importance of hydrostatic sensitivity was argued. Translation of the yield surface was modelled using non-linear hardening law.

Abu Al-Rub and Voyiadjis (2003) have developed constitutive model based on strong coupling between damage and plasticity where authors decomposed total strain on plastic, damage and elastic components. Plastic dissipation potential expressed in terms of effective stresses. Damage dissipation potential has a general form as plastic one. Translation of both damage and plasticity yield surfaces have been modelled with exponential laws. Authors revealed that in case of multiaxial loading the residual strain is decomposed on two terms related to plastic and damage.

In the present paper the direct numerical simulation method based on the micro-mechanical damage model of FRP and plasticity coupled with continuous damage mechanics is developed. Coupling plasticity and damage are based on Abu Al-Rub and Voyiadjis (2003) model, whereas plasticity model has used modified Raghava criterion. Finally the life prediction FRP can be performed using implemented constitutive behaviour and direct simulation method

2. Constitutive model

Non-linear behaviour of composites materials occurs due to plasticity of the matrix and damage (degradation of the matrix, fibre/matrix deboding, and transverse cracking). To model plasticity and damage mechanisms two potentials were predefined. Then evolution laws of residual strains are governed by plastic and damage potentials. Evolution of damage depends on both potentials that imply coupled effect. Residual strains are separated in two parts: one due to plasticity and another due to damage. Then total strain written as follow:

$$\varepsilon = \varepsilon^e + \varepsilon^p + \varepsilon^d \tag{1}$$

2.1 Plasticity

To model the plasticity in FRP the modified Raghava yield function has been involved. Ability to apply this criterion to FRP has been discussed by Vyas et al., (2011). Here the criterion has been extended to case of coupled plasticity with CDM. The plastic yield function is expressed in effective stress space at the micro level as:

$$f = \sqrt{\frac{1}{6}\left(\xi_{22} - \xi_{33}\right)^2 + \xi_{12}^2 + N\xi_{23}^2 + \xi_{31}^2 + s} - \sigma_0 \tag{2}$$

where $\xi = \sigma - \alpha$; σ – stress tensor; α – back stress tensor; N – parameter which characterised difference in shear loading, σ_0 – material parameters, s – part that depends on material parameter μ which is varying form yield stresses.

$$s = \frac{\mu}{2}\left(\tilde{\sigma}_{22} + \tilde{\sigma}_{33}\right) \tag{3}$$

Based-on yield function the plastic potential could be defined as:

$$\dot{p} = \dot{\lambda}^p \frac{\partial F}{\partial \sigma} \tag{4}$$

where $p = (2/3\ \varepsilon^p : \varepsilon^p)^{1/2}$ – equivalent plastic deformations, F – plastic potential. If $F = f$ the flow rule is associated else non-associated.

Macroscopic stress tensor decomposed into microscopic and back-stress tensor is defined as follow:

$$\Sigma = A : \sigma + \alpha \tag{5}$$

Macro-micro concentration tensor A is a four order tensor which varying for different cases of the microstructure. The numerical method of definition of the concentration tensor and application to two types of micromechanical models (square and hexagonal) has been discussed in Sun et al., (2011).

Back stress tensor is self-equilibrated and following condition should be satisfied:

$$\nabla \cdot \alpha = 0 \tag{6}$$

In order to satisfy experimentally observed behaviour kinematic hardening is correlated to micro plastic strain through exponential law:

$$\alpha = \frac{c}{\gamma}\left(1 - e^{-\gamma p}\right) \tag{7}$$

where c and γ are material parameters defined from fitting experimental data.

2.2 Damage

Continuous damage mechanics has been applied in order to describe degradation of the material. At first effective stress tensor is defined as:

$$\tilde{\sigma} = M : \sigma \tag{8}$$

where M – four order damage effect tensor.

Different types of the damage effect tensor and symmetrisation problem have been discussed in work by Voy-

iadjis and Park (1997). Regarding to Chow and Chen (1992) thermodynamic forces driven by damage are defined as second-order tensor as follows:

$$y = \sigma : \left[\tilde{C}^{-1} : M : \frac{\partial M}{\partial d}\right]^s : \sigma \tag{9}$$

where $\tilde{C}$ – effective stiffness tensor, d – second order damage tensor.

In this work the effective compliance tensor defined by Lubineau and Ladevèze (2008) has been used in order to evaluate effective stiffness tensor. In Voight notation effective compliance tensor can be written as follows:

$$\begin{bmatrix} S_{11}f_1 & S_{12}f_1 & S_{12}f_1 & & & \\ S_{12}f_1 & S_{22}f_2 & S_{23}f_3 & & 0 & \\ S_{12}f_1 & S_{23}f_3 & S_{33}f_4 & & & \\ & & & S_{44}f_5 & & \\ & 0 & & & S_{55}f_6 & \\ & & & & & S_{66}f_7 \end{bmatrix} \tag{10}$$

where S_{ij} – compliance tensor components, f_i – damage effect functions:

$$\begin{aligned} f_1 &= \left(1 - d_f\right)^{-1} \\ f_2 &= \left(1 - \langle\sigma_{22}\rangle d_t\right)^{-1}\left(1 - \langle\sigma_{22}\rangle d_m\right)^{-1} \\ f_3 &= \left(1 - \langle\sigma_{22}\rangle d_t\right)^{-1} \\ f_4 &= \left(1 - \langle\sigma_{33}\rangle d_m\right)^{-1} \\ f_5 &= \left(1 - \overline{d}_t\right)^{-1}\left(1 - \tilde{d}_m\right)^{-1} \\ f_6 &= \left(1 - \tilde{d}_m\right)^{-1} \\ f_7 &= \left(1 - \overline{d}_t\right)^{-1}\left(1 - d_m\right)^{-1} \end{aligned} \tag{11}$$

where d_f – fibre damage, d_m – matrix damage, d_t – transverse cracking.

As shown in Lubineau and Ladevèze (2008) transverse damage and density of cracks have linear dependence while critical value of the transverse cracks not reached. Above that the threshold value dependence is non-linear.

Flow rule for damage variables is written as:

$$\dot{r} = \dot{\lambda}^d \frac{\partial G}{\partial \sigma} \tag{12}$$

where $r = (\varepsilon^d : \varepsilon^d)^{1/2}$ – equivalent damage deformations, G – damage potential which is defined by quadratic surface in following form:

$$g = \sqrt{\left(y - h\right)\left(y - h\right)} - \gamma_0 - k_d \tag{13}$$

where h – back thermodynamic forces tensor as analogue to back-stress tensor for plasticity, k_d – isotropic hardening, γ_0 – damage threshold.

Flow rule for damage can be written based-on two potentials:

$$\dot{d} = \dot{\lambda}^p \frac{\partial F}{\partial y} + \dot{\lambda}^d \frac{\partial G}{\partial y} \tag{14}$$

Equation (14) defines evolution damage law and represents main idea of Abu Al-Rub and Voyiadjis (2003) that damage could occur in three cases: plastic flow and damage flow are non zero; no plastic flow; no damage flow.

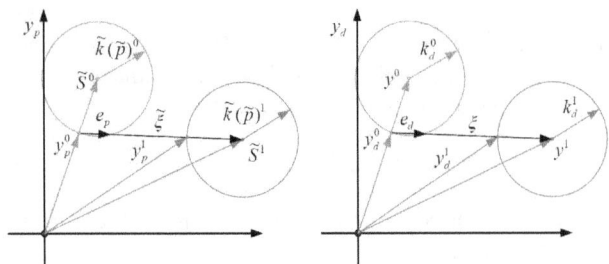

Figure 1. Numerical calculation scheme: a) plasticity; b) damage.

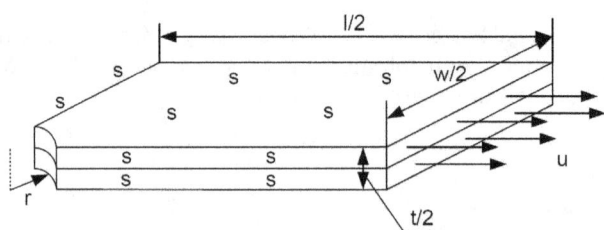

Figure 3. Geometry of model and boundary condition.

3. Implementation

In this paper the high cycle fatigue (HCF) life of the composites is considered within 10^5-10^7 cycles. In order to adopt direct computational method to HCF composites it has to be imposed that fatigue limits could be defined. This condition was dictated by the fact that during gigacycle regimes the composites continue to degradate and fatigue limits are lower than in HCF regime (Michel et al 2006).

Method developed has been named as simplified approach and was characterized as particular case of direct methods in Staat and Heitzer (2002). It has obvious advantages in comparison with step-by-step method when the shakedown analysis is performed.

Numerical procedure of the method has been described by Jabbado and Maitournam (2007). Authors applied direct calculation to predict life of metals. In this paper the method has been extended to case of FRP. The scheme of the numerical calculation procedure is shown in Fig. 1. In accordance of the main idea of the method the new variable named as y_p and y_d have been implemented. Those variables represent residual components of stresses and residual forces conjugated with damage for plastic and damage potentials accordingly. In Fig. 1 e_p and e_d are normal to plastic and damage surfaces. Upper indexes show step by time during one cycle. In Fig. 1a S is stress deviator; k is elastic limit as a function of hydrostatic stress. In Fig. 1b y is forces conjugated with damage, k_d is damage threshold function, that can represent an isotropic hardening due to damage.

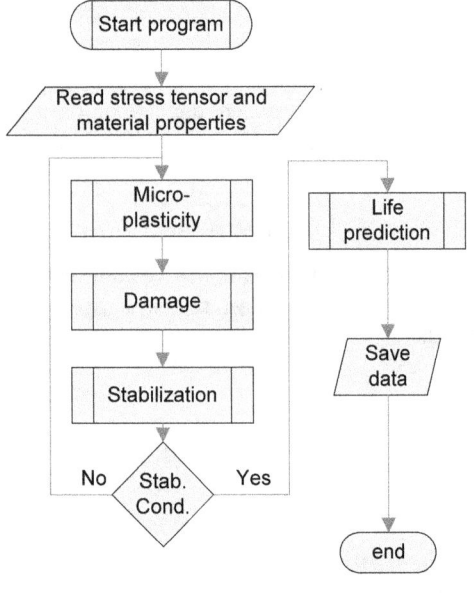

Figure 2. Algorithm of program.

For developed procedure it is enough to perform one elastic finite element analysis. The program is post-treatment software dealing here with data given after performed calculation in ABAQUS. Input data are stress tensor components and information about integration points. Output variables of program are residual strain, damage tensor and number cycles to failure. A significant reduction in computational costs can be achieved as a result. Efficiency increases when FRP is considered subjected to multiaxial non-proportional loading condition and level of stresses is close to elastic limit.

Algorithm of program is shown in Fig.2. At the beginning the stress tensor is read from input file for one integration point within cycle. Next step is evaluation of the residual strains and damage. During stabilisation procedure program checks if stabilised cycle is reached. Simple condition of stability is equality of the residual strains of current cycle with previous one. In case of the proportional loading two iterations are sufficient to obtain solution.

4. Simulation

Before execution of the method an elastic analysis has performed in ABAQUS and stress tensors components were obtained at every integration points at each increment. An example with the symmetric holed plate $[0/90]_s$ with continuous fibre plies has been considered for verification. The sketch of the model is shown in Fig.3. Dimensions of the plate in (mm): width (w) – 40, length (l) – 80, thickness (t) – 2.5, diameter of hole (2r) – 5 mm. Material parameters are presented Tab.1. Model has three planes of symmetry (symbols s in Fig.3) therefore one fourth part has been modelled.

Part of the results is presented on the Fig. 4. In left pictures present damage distribution for 0° - ply and in right ones for 90° - ply. Fig. 4a shows fibre damage. At maximum level of load the fibre breakage observed on high point of the hole. Fig. 4b and 4c show significant degradation of material and initiation of the splitting tangent to the hole.

Table 1
Material properties

Elastic		Damage		Plastic	
E_1 (GPa)	148	Y_c (MPa)	15	γ	140
E_2 (GPa)	9.57	Y_0 (MPa)	0.02	η_γ	0.007
v_{12}	0.356	Y_f (MPa)	10	μ (MPa)	7
v_{23}	0.49	G_1 (J/m²)	100	η_G	0.4
G_{12} (GPa)	4.5	G_2 (J/m²)	300	c	50000
				σ_0 (MPa)	23

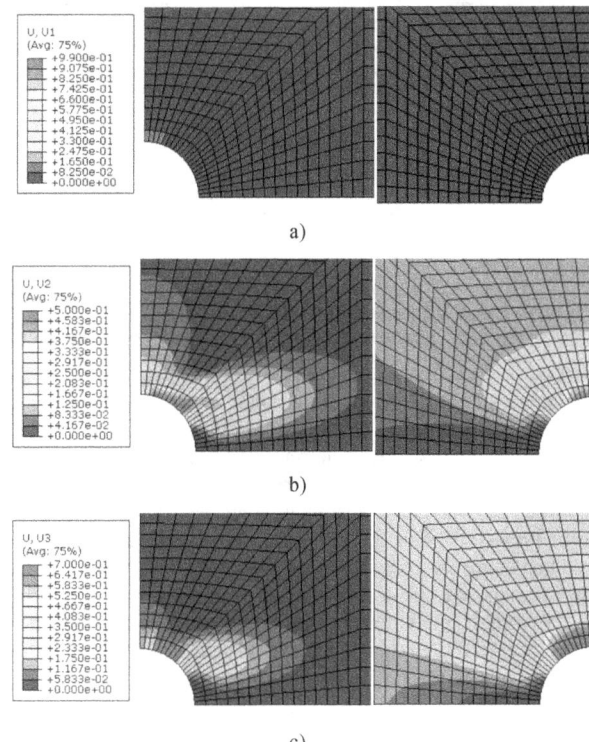

Figure 4. Damage calculation results for [0/90]ₛ composite: left figures show 0-layer damage mechanisms, right ones 90-layer; a) fibre damage, b) matrix damage, c) transverse cracking.

No fibre damage occurs in 90° - ply as shown in Fig. 4a at the left side. High concentration of matrix damage and transverse cracking located on upper zone under the hole (Fig. 4b, 4c at left). Most of whole ply is affected by damage except lower zone.

5. Conclusions

This paper discuss a new computational method based on continuous damage mechanics and plasticity for fibre reinforcement polymers structures which are subjected to high cycles fatigue loading. Evolution laws of residual strains are governed by plastic and damage potential. Modified Raghava yield function with hardening has been used to evaluate plastic strain. Plasticity yield function implies hydrostatic sensitivity of UD composites. Damage surface was described in terms of thermodynamic forces conjugated with damage.

Constitutive model based on micromechanics of composites takes into account mechanisms of degradation material properties as damage fibre, damage matrix and transverse cracking. Evolution laws are based on damage energy criteria.

A direct numerical procedure based on the calculation of the residual strain per stabilized cycle allows to predict life of a composite structure. The developed method has been verified using a FRP composite plate with UD plies under tensile-tensile cycle. The obtained results are consistent with published in literature. A significant reduction in the calculation time has achieved using a direct computational algorithm.

References

Abdelal, G.F., Caceres, A. and Barbero, E.J. (2002). A micro-mechanics damage approach for fatigue of composite materials, *Composite Structures*, Vol. 56, No 4, pp. 413-422.

Abu Al-Rub, R.K., and Voyiadjis, G.Z. (2003). On the coupling of anisotropic damage and plasticity models for ductile materials. *International Journal of Solids and Structures*, Vol. 40, No 11, pp. 2611-2643.

Chow, C. L. and Chen, X. F. (1992). An anisotropic model of damage mechanics based on endochronic theory of plasticity, *International Journal of Fracture*, Vol. 55, No 2, pp 115-130.

Hassan, N.M., and Batra, R.C. (2008). Modeling damage in polymeric composites, *Composites Part B: Engineering*, Vol. 39, No 1, pp. 66-82.

Jabbado, M. and Maitournam, M.H. (2007). A high-cycle fatigue life model for variable amplitude multiaxial loading, *Fatigue Fracture Engineer Material Structure*, Vol. 31, pp. 67-75.

Lubineau, G. and Ladevèze, P. (2008). Construction of a micromechanics-based intralaminar mesomodel, and illustrations in ABAQUS/ Standard, *Computational Materials Science*, Vol. 43, No 1, pp. 137-145.

Maire, J.F. and Chaboche J.L. (1997). A new formulation of continuum damage mechanics (CDM) for composite materials, *Aerospace Science and Technology*, Vol. 1, No 4, pp. 247-257.

Michel, S.A., Kieselbach, R. and Martens H.J. (2006). Fatigue strength of carbon fibre composites up to the gigacycle regime (gigacycle-composites) *International Journal of Fatigue*, Vol. 28, No 3, pp. 261-270.

Staat, M. and Heitzer, M. (2002). *Numerical Methods for Limit and Shakedown Analysis*, Report, NIC Series.

Voyiadjis, G. Z. and Park, T. (1997). Anisotropic damage effect tensors for the symmetrization of the effective stress sensor, *Journal of Applied Mechanics, ASME*, Vol. 64, pp. 106-110.

Vyas, G.M., Pinho, S.T. and Robinson, P. (2011) Constitutive modelling of fibre-reinforced composites with unidirectional plies using a plasticity-based approach, *Composites Science and Technology*, Vol. 71, pp. 1068–1074.

Raghavan, P. and Ghosh, S. Li S. (2004). Two scale response and damage modeling of composite materials, *Finite Elements in Analysis and Design*, Vol. 40, No 12, pp. 1619-1640

Large Eddy Simulation (LES) of Glass Fibre Dispersion in an Internally Spout-Fluidised Bed for Thermoplastic Composite Processing

Xiaogang Yang[1], Xiaobing Huang[1], Yuan Zong[2], Gance Dai[2]

[1] Institute for Art, Science and Technology, Glyndŵr University, Plas Coch, Mold Road, Wrexham, LL11 2AW, UK
[2] State Key Laboratory of Chemical Engineering, East China University of Science and Technology, Shanghai, 200237, P.R.China

Abstract: Large eddy simulation (LES) has been conducted to investigate glass fibre dispersion in an internally spout-fluidised bed with draft tube and disk-baffle, which was used in the manufacture of long glass fibre reinforced thermoplastic composites. The LES results have demonstrated that the internally spout-fluidised bed with draft tube and disk-baffle can remarkably improve its hydrodynamic behaviour, which can effectively disperse fibre bundles and promote pre-impregnation with resin powder in manufacturing fibre reinforced thermoplastics. The hydrodynamics of the spout-fluidised bed has been investigated and reported in a previous paper (Hosseini et al., 2009). This study attempts to reveal important features of fibre dispersion and correlations between the fibre dispersion and the characteristics of turbulence in the internally spout-fluidised bed using the LES modelling, focusing on the likely hydrodynamic impact on fibre dispersion. The simulation has clearly indicated that there exists a strong interaction between the turbulent shear flow and transported fibres in the spout-fluidised bed. Fibre entrainment is strongly correlated with the local vorticity distribution. The dispersion of fibres was modelled by a species transport equation in the LES simulation. The turbulent kinetic energy, Reynolds stress and strain rate were obtained by statistical analysis of the LES results. The LES results also clearly show that addition of the internals in the spout-fluidised bed can significantly change the turbulent flow features and local vorticity distribution, enhancing the capacity and efficiency of fibre flocs dispersion.

Key Words: Spout-fluidied bed; Fibre flocs; Dispersion; Turbulent flow; LES.

1. Introduction

Glass fibre-reinforced thermoplastics have been widely used in different industrial applications. Examples include automotive, airplane, pressure vessels where the use of non-reinforced thermoplastics is unable to provide the specified mechanical and thermal performance. However, high viscosity of thermoplastic and poor dispersion of long fibres imposes many problems in the manufacturing process of thermoplastic composites. A lot of efforts have been made in order to overcome these problems. In order to improve the impregnation with thermoplastic resins, the most attractive techniques that have been developed is to minimise the distance that resin must flow to penetrate the reinforcement. This observation has led to the development of commingled yarns or powder impregnation. The realisation of powder impregnation can be easily achieved using a dry process in which the polymer powders are dispersed through a fluidized bed and are deposited onto the fibre surface by additional force (Padaki and Drzal, 1999). To attain the effective fibre dispersion, additional processing equipment, such as pneumatic spreaders, is usually employed with an increase in production cost.

Dai et al. (2001) proposed the use of the modified spout-fluidised bed to disperse fibre tows and to realise powder impregnation. Using this technology, they have successfully manufactured a new fibre reinforced polymer composite in the laboratory. The aim of this paper is to explore the dispersion mechanism of the fibre flocs in such modified spout-fluidised bed by means of numerical simulation approach (LES) so that the results can provide an essential guidance for optimisation of fibre floc dispersion process. Turbulent flow behaviours in the spout-fluidised bed with different allocations of internals were

investigated using large eddy simulation (LES). The correlations between dispersion of fibre flocs and characteristics of turbulent shear flows were assessed.

Fundamental studies focusing on the conversion of fibre bundles to suspended individual fibres are still rarely reported so far. This may be attributed to the following factors; one is the inherent complexity of fibre flocs, affected by their fragility, coating and physical properties and the other is the complex rupture mode of the fibre flocs (Kuroda and Scott, 2002) caused by surface erosion (gradual shearing off of small fragments from the surface) and large-scale splitting (breakup into fragments of comparable size). There were some applications of fragmentation of flocs in chemical and biological engineering processes, such as protein precipitates (Zumaeta et al., 2007), inorganic nanoparticles (Wengeler and Nirschl, 2007) and colloid particles (Teung et al., 1997) but these fragmentation processes are primarily to depose flocs in turbulent flows using agitated vessels or other process devices. The effect of the turbulent flow field on the fragmented sizes of the flocs has been investigated by Bouyer et al. (2005); Shamlou et al. (1996); Tambo and Hoaumi (1979). These studies have clearly demonstrated that the size of the flocs depends strongly on the interaction between of the flocs cohesion and the shear stress exerted by the turbulent flow on the fibre flocs. These studies also revealed that the fragmentation occurs when the force due to shear stresses is strong enough to overcome the floc cohesion while the flocs withstand when their strength surpasses the shear stresses. It has been widely accepted that the cohesion depends on the physicochemical property of flocs while the features of turbulent flows are mainly controlled by the adopted reactor geometry and operation conditions. As the shear stresses are related to the velocity gradient and turbulence fluctuations, investigations also revealed that the un-fragmented maximum floc diameter can be correlated to average turbulence dissipation rate or local shear gradient, i.e. $d \propto \bar{\varepsilon}^{-1/4}$ or $d \propto G_v^{-1/2}$ (Yeung et al., 1997; Bouyer et al., 2005; Coufourt et al., 2005). Although these empirical relations can be used to qualitatively describe the relationships between the floc size and the hydrodynamics, they cannot provide physical insight into the complex phenomena of fibre floc de-agglomeration or break-up in the turbulent flows. In fact, the use of an average turbulence dissipation rate e to characterise the turbulent shear flow will lead to omission of important details of the local turbulence which is usually crucial for dispersion of fibre flocs. Considering this factor, Ducoste et al. (1997) proposed to correlate the floc size to local shear stress and strain rate. Because the turbulence kinetic energy and dissipation are closely related to shear velocity gradient, it will be reasonable to postulate the floc size to directly correlate to the local shear velocity gradient or vorticity strength. It is evident from the previous studies that fibre floc breakage is likely caused by shear-induced turbulence stresses in turbulent flows (Hosseini et al., 2009), and the final breakage of fibre flocs is dependant upon whether or not the exerted shear stress force is greater than the floc cohesion force.

As turbulent shear stresses have significant impact on dispersion of fibre flocs, deliberate application of turbulence modulation may be effective for controlling dispersion of glass fibre flocs. Spout-fluidised bed has been recognised as an effective means for gas-solid mixing and has been extensively employed in solid drying, coating, blending. Many modifications for spout-fluidised beds have been proposed for improvement of heat transfer and for enhancement of fluid-solid mixing efficiency (Zhou et al., 2004; Zhong et al., 2006) but the use of modified spout-fluidised bed for dispersion of fibre flocs is rarely reported in the open literature. Dai et al. (2001) have proposed the use of modified spout-fluidised bed with a draft tube and a disk-baffle to disperse fibre flocs in the preparation process for manufacturing fibre reinforced thermoplastic composites. It was found that introduction of the internals changes the flow patterns in the modified spout-fluidised bed, resulting in several "sub-flows", which may be classified as turbulent jet, impinging jet and wall jet. Though these typical flows have been extensively studied and well understood in many previous studies, the superposition of these flows has not been fully investigated. Thus, a better understanding of the hydrodynamics and fibre dispersion in such modified spout-fluidised bed will be beneficial to the control and optimisation of fibre floc dispersion process. Zhong et al. (2006) have adopted Eulerian-Lagrangian CFD modelling approach to study the gas-solid turbulent flow in a spout-fluidised bed. The particle motion was modelled using discrete element method (DEM) while the carrier gas flow was modelled using k-ε turbulence model. Wu and Arun (2008) employed Eulerian-Eulerian two fluid model in CFD modelling to simulate the gas-particle flow behaviour of spouted beds and the simulation has quite well predicted the overall flow patterns. Zhao et al. (2008) investigated the dynamics of particulate materials in two-dimensional spouted beds with draft plates by adopting the DEM for describing particle motion and by using the low Reynolds k-ε turbulence model for solving for fluid flow. Because the k-ε turbu-

lence model was employed in these studies, one of apparent disadvantages is that the details of turbulence structures in the flow were not able to be caught in the simulations. A successful prediction of dispersion of glass fibre flocs requires the details of turbulence structures while it is now generally accepted that large-eddy simulation (LES) can provide the detailed turbulence information except for those dissipated eddies. Thus, the present study has employed the LES to acquire the fluid flow details in the internally spout-fluidised bed. Since fibres with large aspect ratio tend to aggregate and the volume concentration of the glass fibres in the spout-fluidise bed is lower than 0.001% in the operation, the effect of glass fibres on the flow can be neglected, i.e. one-way coupling is assumed. This paper will only consider the effect of the turbulent flow on glass fibre dispersion in the spout-fluidised bed and dispersion of glass fibres is described using scalar transport model, focusing on quantification of the fibre floc dispersion in the spout-fluidised bed.

This paper is organised as follows. Section 2 describes mathematical models and numerical details of the LES employed for simulation of glass fibre dispersion in the internally spout-fluid bed. Section 3 presents the LES simulation results with detailed discussion. Finally, some important conclusions drawn from the present work are given in Section 4.

2. Mathematical models and numerical simulation

2.1. Flow field and fibre transport descriptions

In LES, the flow variables are decomposed into resolved scales, associated with the larger eddies, and the modelled sub-grid scales, related to the more universal smaller eddies. The resolved scale $\bar{\phi}$ is obtained utilising a filtering:

$$\bar{\phi}(x) = \int_D \phi(x')G(x, x')dx' \qquad (1)$$

where D is the computational domain, G the filter function, and x and x' represent the vector positions. The filtered continuity and Navier–Stokes equations on using the previous decomposition and filtering procedure can be written as:

$$\frac{\partial \bar{u}_i}{\partial x_i} = 0 \qquad (2)$$

$$\frac{\partial \bar{u}_i}{\partial t} + \frac{\partial}{\partial x_j}\left(\bar{u}_i\bar{u}_j\right) = -\frac{1}{\rho}\frac{\partial \bar{p}}{\partial x_i} + \nu\frac{\partial}{\partial x_j}\left(\frac{\partial \bar{u}_i}{\partial x_j} + \frac{\partial \bar{u}_j}{\partial x_i}\right) - \frac{\partial \tau_{ij}}{\partial x_j} \quad (3)$$

where r is the fluid density, $\tau_{ij} = \overline{u_i u_j} - \bar{u}_i\bar{u}_j$ is the SGS (sub-grid scale) stress, representing the interaction between small and large scales. Smagorinsky (1963) postulated that the SGS stresses τ_{ij} can be expressed as

$$\tau_{ij} = 2\nu_T \bar{S}_{ij} \qquad (4)$$

where the eddy viscosity n_T and the strain rate in the resolved velocity field are estimated by

$$\nu_T = C\bar{\Delta}^2 \left|\bar{S}\right| \quad \bar{S}_{ij} = \frac{\partial \bar{u}_i}{\partial x_j} + \frac{\partial \bar{u}_j}{\partial x_i} \qquad (5)$$

where Δ is a length scale associated with the filter width (or mesh size) and C is a constant (Smagorinsky's constant). In the current study, $\overline{\Delta} = \left(\overline{\Delta}_x \overline{\Delta}_y \overline{\Delta}_z \right)^{1/3}$ and the filter width $\overline{\Delta}_{x(y,z)}$ is taken the same as the mesh size in the x (y, z) direction. The real flow behaviour of eddies is highly complicated in the internally spout-fluidised bed and the value of C may be dependent on the local flow behaviour. For simplicity, we have taken $C = 0.1$ for homogeneous and isotropic turbulence as in Yeh and Lei (1991) and found that this value is reasonable.

In order to describe glass fiber mass transport and dispersion in the spout-fluidised bed, the filtering as defined in equation (1) is applied to the species transport equation used to describe the glass fiber mass concentration, which yields

$$\frac{\partial \rho \tilde{C}}{\partial t} + \frac{\partial}{\partial x_j}(\rho \overline{Cu_j}) = \frac{\partial}{\partial x_j} \left(\overline{\rho D_f \frac{\partial C}{\partial x_j}} \right) \qquad (6)$$

where $\tilde{C}$ is the LES grid-resolved species mass. Furthermore, the filtering of the second term in the left side of equation (6) is assumed to take the following form:

$$\rho \overline{Cu_j} = \rho \tilde{C}\tilde{u}_j + \left(\rho \overline{Cu_j} - \rho \tilde{C}\tilde{u}_j \right) \qquad (7)$$

The first term on the right hand of equation (7) represents the resolved convective flux of glass fibre mass concentration and the second term can be interpreted as the subgrid scale convective flux (SGS). When adopting the approximation

$$\overline{\rho D_f \frac{\partial C}{\partial x_j}} \approx \rho \tilde{D}_f \frac{\partial \tilde{C}}{\partial x_j}$$

and assuming that the filtered dispersion coefficient is proportional to the carrier fluid viscosity, i.e.

$$\tilde{D}_f = D_f(\tilde{T}, \rho) = \frac{\mu(\tilde{T})}{\rho Sc_f}$$

the filtered glass fibre mass concentration transport equation can be written as

$$\frac{\partial \rho \tilde{C}}{\partial t} + \frac{\partial}{\partial x_j}(\rho \tilde{C}\tilde{u}_j) = -\frac{\partial \lambda_f}{\partial x_j} + \frac{\partial}{\partial x_j} \left(\rho \tilde{D}_f \frac{\partial \tilde{C}}{\partial x_j} \right) \qquad (8)$$

where $\lambda_f = \rho \overline{Cu_j} - \rho \tilde{C}\tilde{u}_j$. Solving equation (8) requires a closure model for l_f. By analogy to decomposition of the flow variables into resolved components and subgrid scale components employed in the large-eddy simulation, the effect of the subgrid scale l_f is accounted for similar to the Smagorinsky-Lilly model. Since small scales tend to be isotropic, the subgrid scale stress models are based on the eddy-viscosity assumption. Thus, the subgrid term in species glass fibre mass can be approximated by the gradient-transport model, which yields

$$\lambda_f = -\rho \frac{\nu_t}{Sc_t} \frac{\partial \tilde{C}}{\partial x_j} \qquad (9)$$

where Sc_t is the turbulent Schmidt number and n_t is the Smagorinsky viscosity.

It has been found from the LES that the gradients of the mean mass fraction C in the spout-fluidised bed are small and direct solution of equation (8) yields little information concerning the glass fibre mass variability responsible for flocculation. Of more interest, is the flocculation intensity, as defined by Raghem Moated (1999),

$$Fl = \frac{\overline{\left(\tilde{C} - \overline{C} \right)^2}}{C_m^2} \qquad (10)$$

where $\overline{C}$ is the time averaged local fibre concentration and C_m is the average fibre concentration in the whole flow field. The flocculation intensity indicates the level of local mass variability. If there are flocs, the flocculation intensity will be high since glass fibre mass is concentrated in the flocs rather than is distributed between them.

2.2. The geometry of the spout-fluid bed and numerical detail

Four different configurations of the spout-fluid bed were considered in the simulation as shown in Fig. 1. The diameter of the bed is 280 mm and its height is 1250 mm. A cone bottom with expansion angle of 60° is mounted to avoid detaining the pockets of stagnated particles. A concentric circular draft tube with internal diameter of 90 mm and length of 700 mm long was fitted to convert the set-up from case A to case B for which the distance between the spout nozzle and the draft tube was kept to 217 mm. When a disk-baffle was installed above the draft tube, the set-up was converted to either case C or case D. The gap between the draft tube and the disk-baffle was kept to 100 mm. The only difference between case C and case D is to consider the effect of rotation. Spouting air directly enters the bed through the spout nozzle while fluidizing air is introduced into the bed via the orifices on the gas distributor. The volumetric flow rate of the spouting gas used in the simulation was 160 m³/h and that of the fluidizing gas was 40 m³/h, the same as those used in the experiments. All four configurations were simulated so as to understand the effect of the internals on the flow behaviour in the bed. For simulation of case D, the disk-baffle was assumed to rotate at a speed of 600 rpm.

In the simulations no-slip boundary condition was imposed to the bed walls. The exits of the bed were specified as pressure outlets. The spout nozzle and fluidizing gas distributor were defined as velocity inlets with the specified velocities for spouting gas and fluidizing gas respectively. The disk-baffle in case C was treated as a stationary wall while that in case D was specified as a rotational wall at an angular speed of 62.83 rad/s. The pressure-velocity coupling was obtained using SIMPLE algorithm and discretisation scheme for pressure was second order and the scheme for momentum was bounded central differencing. The time step for all cases was kept 0.0001s. The convergence criterions are that for all parameters the residuals were less than 1×10^{-4}. The average grid size was about 3 mm in the simulations. However, a refined mesh with $Y^+ \approx 5$ in the vicinity of the walls, in consistent with the mesh size requirement for LES, was adopted. A grid dependence check has been conducted with the number of mesh cells doubles compared to the refined mesh but the mesh is uniformly distributed. The trial simulation has indicated that grid invariance of computed results has been ensured. Thus, a mesh system having the refined mesh with $Y^+ \approx 5$ close to the walls and the average grid size of 3 mm is adopted in all our LES simulations. All the simulations were conducted using the commercial CFD Software Fluent 6.3.26.

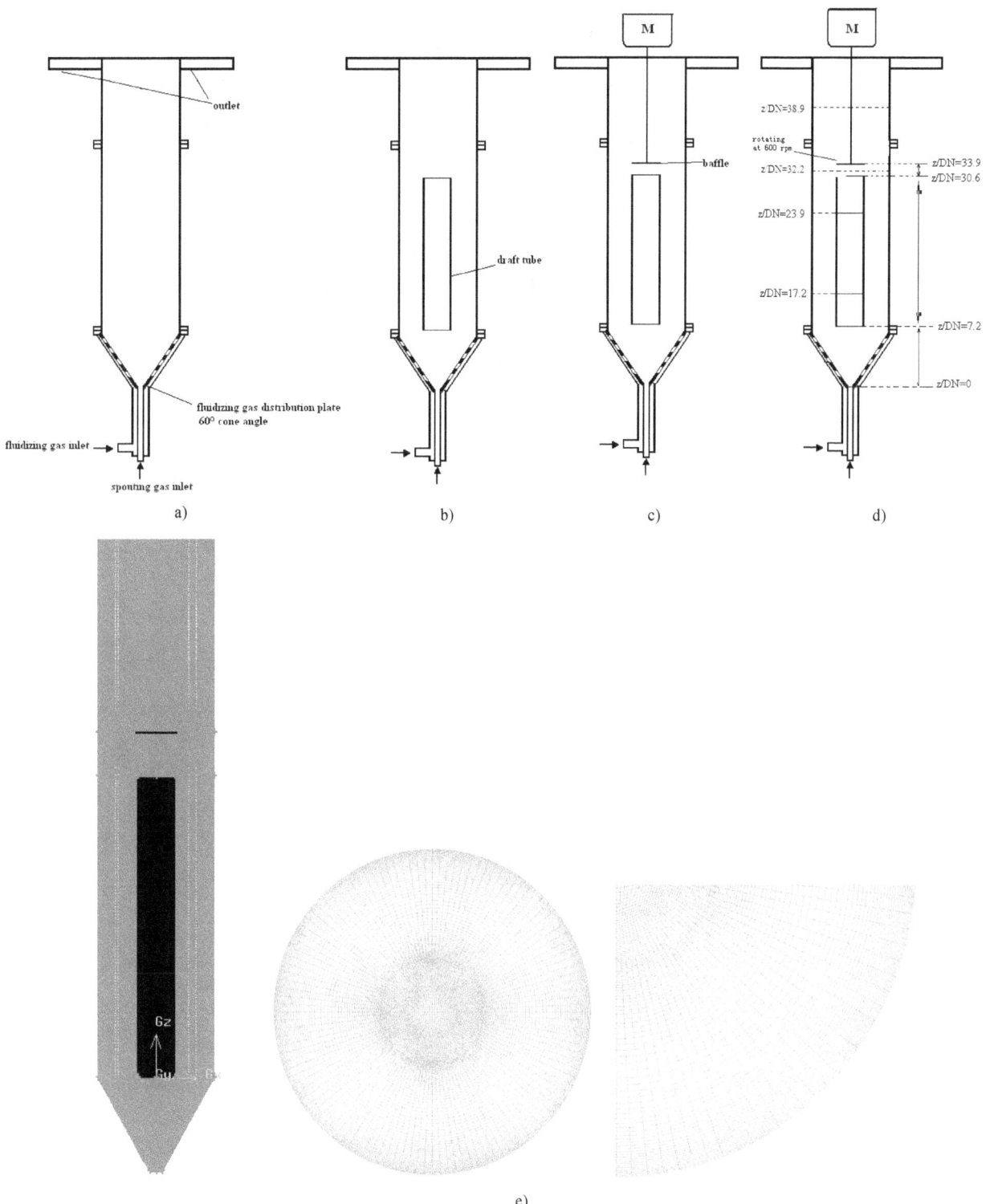

Figure 1. Schematic diagram of four configurations of the spout-fluidised bed used in the simulation:
(a) case A; (b) case B; (c) case C; (d) case D; (e) mesh set-up for CFD modelling.

3. Results and discussion

3.1. Flow pattern in the spout-fluid beds

Since the volume concentration of the fibres in the spout-fluid bed is lower than 0.001% in the current study, the effect of the glass fibres on the entire flow can be neglected, i.e. only one-way coupling is considered. The LES simulations have been carried out to reveal the turbulent shear flow features in the spout-fluidised bed, in par-ticular with addition of the draft tube and disk-baffle. In contrast to the previous study on a spouted bed (Wu and Arun, 2008) in which the flow field was characterised to consist of three typical regions, a central spout, an annulus and a fountain region, the current LES results indicate that the turbulent flow in the internally spout-fluidised bed can be dynamically and geometrically characterised with 4 regions. These regions are: (1) the entry region below the draft tube; (2) the annulus located between the draft tube and the walls; (3) the impinging region between

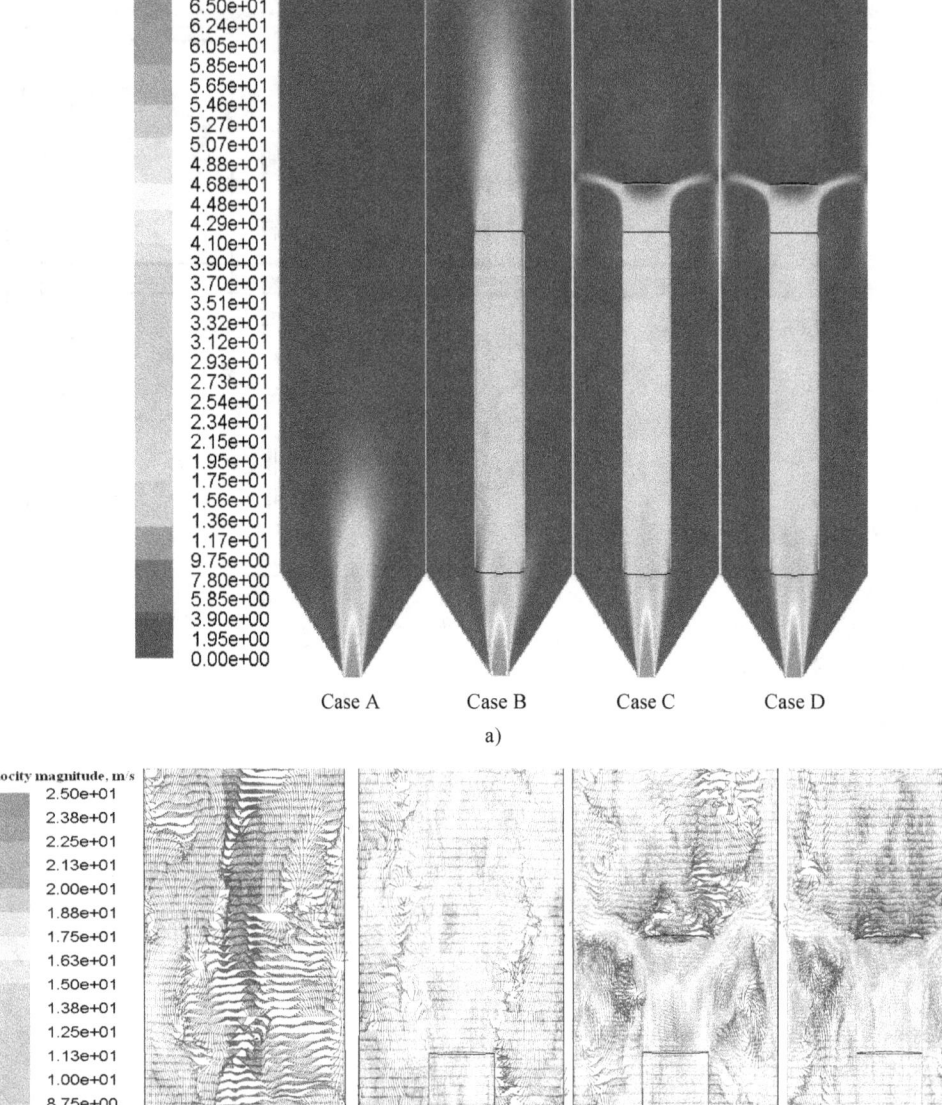

Figure 2. (a) Distribution of the mean flow for all cases; (b) Velocity vector distribution for the impinging and exit region for all cases. U_0=65.0 m/s.

the draft-tube and disk-baffle; and (4) the exit zone above the baffle. Fig. 2(a) showed the time-averaged velocity distribution contours for four different cases. Instantaneous velocity fields in these flow regions are shown in Fig. 2(b). It can be seen clearly from the figure that addition of the internals has a significant impact on the flow behaviour.

3.1.1 The entry region and the draft tube

The flow in the entry region of the spout-fluidised bed has typical characteristics of a nozzle jet flow. Introduction of the draft tube has redistributed and modulated the fluid flow, resulting in the presence of relatively regular flow pattern similar to the turbulent shear flow in a circular

tube. Fig. 3 displays the profiles of the axial mean velocity along the centreline and radial mean velocity distributions at different cross-sections for different cases. For case A without the draft tube, the flow can be characterised by jet flow in a confined closure (Moated, 1999). In the early stage of the flow, the jet decays quite quickly and the axial mean velocity on centreline reduces with increase of the axial distance. The length of the potential core of the jet is relatively short. The decay coefficient of the centreline axial mean velocity was found to be 6.94. When draft tube was added, the axial mean velocity decays faster than case A with a decay coefficient of 9.11. This phenomenon was also reported in the previous studies for confined impinging jets, e.g. Ashforth-Frost et al. (1997) and Baydar and Ozmen (2006), which can be at-

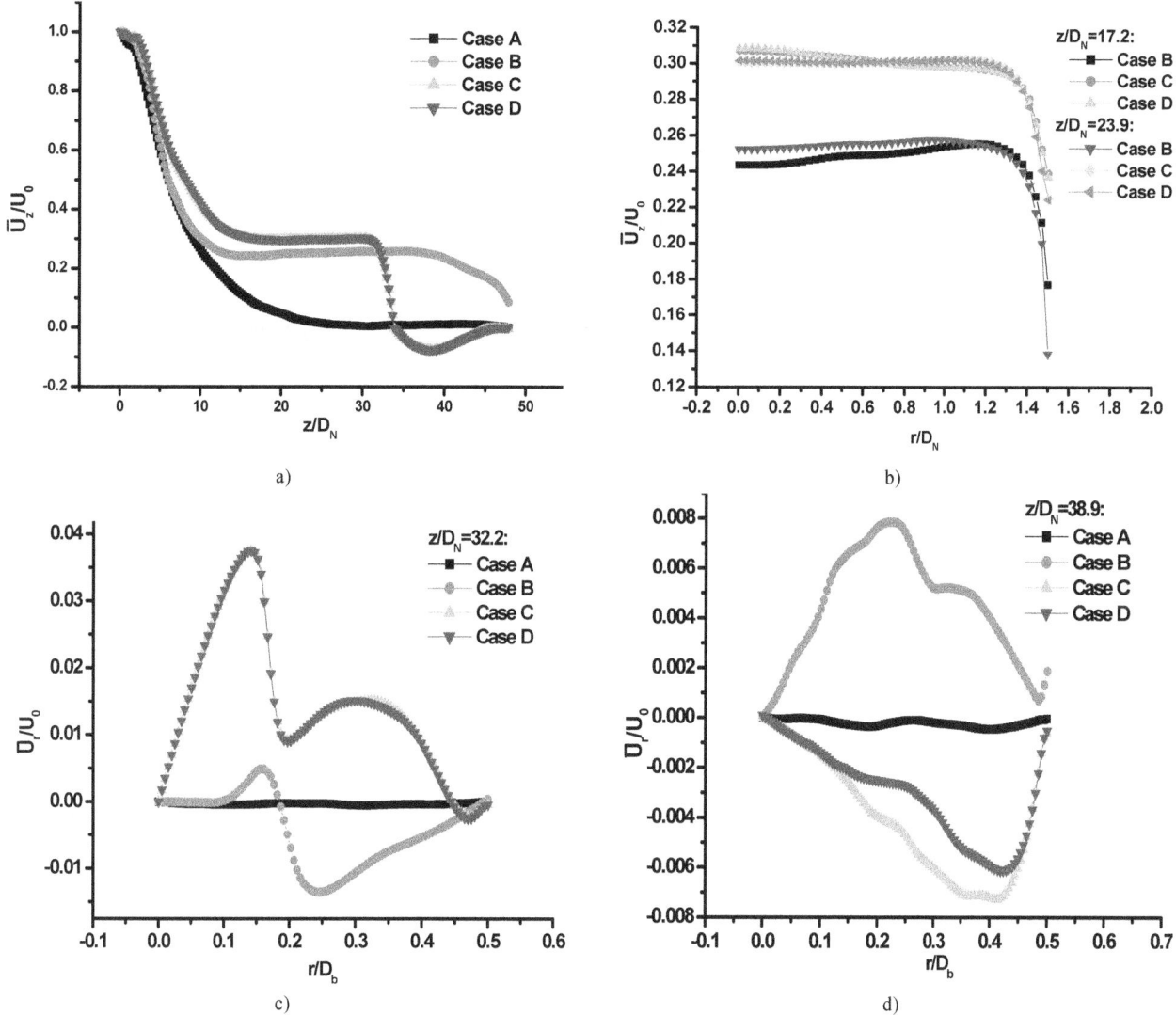

Figure 3. Mean velocity distribution (a) axial mean velocity along the centreline; (b) radial mean velocity distribution at z/D_N = 17.2 and z/D_N = 23.9; (c) radial mean velocity distribution at z/D_N = 32.2; (d) radial mean velocity at z/D_N = 38.9. U_0 = 65.0 m/s.

tributed to the confinement on the entrainment and spread of the jet due to the draft tube. The spread of jet flow for case A is only affected by the size of the enclosure while the spread of jet flow for other cases with the draft tube is confined by the draft tube, resulting in the flow to behave similar to the flow through a pipe so that the mean axial velocity remains almost unchanged until the flow leaves the draft tube as shown in Fig. 3(a).

A careful observation on Fig. 3(a) reveals that though the axial mean velocity when the jet flow is led into the draft tube remains unchanged for case B, case C and case D, the axial mean velocity in the draft tube for case B is lower than that for case C and case D. An explanation may be that the presence of the disk-baffle diversifies the outflow from the draft tube and part of the flow has been circulated from the impinging region and entrained into the entry jet, giving rise to a slightly higher axial mean velocity.

3.1.2 The annulus and impinging region

The impinging region was defined as the region where the jet flow touches the wall of the spout-fluidised bed. For case A, the jet flow develops initially but forms a pipe flow when the jet flow spreads to touch the wall. For case

B, the turbulent shear flow exiting from the draft tube was forced to change the flow direction due to the disk baffle to form a circular spread jet flow, causing the axial mean velocity variation and sudden change in radial mean velocity distribution at z/D_N = 32.2 (see Fig. 3(a) and (c)). The axial mean velocity distribution on centreline exhibits such behaviour, i.e. keeping unchanged in the draft tube and gradually reducing after the secondary jet flow is generated. Due to the effect of the entrainment of the secondary jet flow, turbulence intensity in the annulus region is enhanced and the axial velocity of the flow in this region increases in comparison to case A.

It was also observed from the simulation that the circular spread impinging jet simultaneously affects the flow behaviour in the annulus region and exit region. As can be seen from Fig. 3(a) that the axial mean velocity reduces steeply. The radial mean velocity increases rapidly from zero at the stagnation point to the maximum at position $r/D_b \approx 0.15$ and then decreases to a local minimum at $r/D_b \approx 0.2$. It increases and attains the second maximum at $r/D_b \approx 0.3$ and then decreases towards the wall. It is interesting to note that the position of appearance of the first maximum radial mean velocity corresponds to the outer fringe of the disk-baffle. In the formation of the circular spread impinge jet, large-scale toroidal vortices are gener-

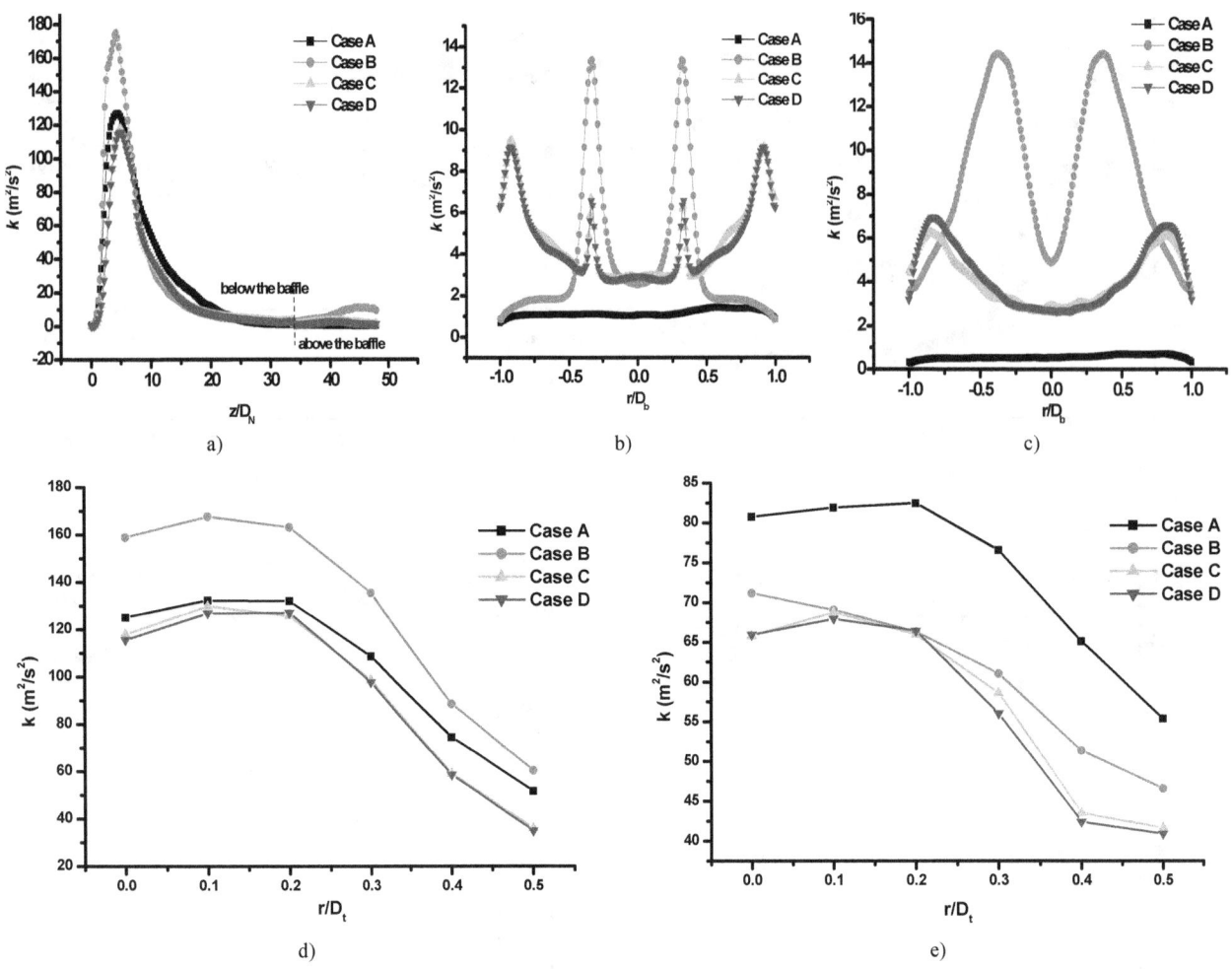

Figure 4. Profile of turbulent kinetic energy at different positions:
(a) along the centerline; (b) at $z/D_N = 32.2$; (c) at $z/D_N = 38.9$; (d) at $z/D_N = 5.0$; (e) at $z/D_N = 7.7$.

ated respectively on both above and beneath the impinge jet. These two toroidal vortices squeeze the impinge jet so as to form an initial accelerated and then decelerated jet flow, which results in a local maximum radial mean velocity, corresponding to the position of $r/D_b \approx 0.3$ as can be seen from Fig. 2(b) and Fig. 3(c). Popiel and Trass (1991) indicated that the ring-shaped toroidal vortices formed on the impingement surface would separate from the wall boundary between the stagnation region and wall jet region. Our results seem to be consistent with their finding.

The circular spread impinge jet when hitting the wall generates strong shear both upwards and downwards while the downward shear flow entrains the gas into the annulus and exit regions, intensifying the local turbulence. It was also found that the maximum velocity in the annulus for cases C and D are enhanced to some extent, which is beneficial to fibre floc suspension.

3.1.3 The exit region

The strong upward shear flow near the wall in the exit region due to the circular spread impinge jet redistributes and adjusts itself behind the disk baffle. Fig. 3(d) illustrates the profiles of radial mean velocity for all cases in the middle of the exit region ($z/D_N = 38.9$). It can be seen from the figure that the flow for case A in this region has a typical characteristics similar to the fully developed turbulent flow in a circular pipe, i.e. the radial component of the velocity is almost negligibly small. While for case B, the secondary jet flow from the draft tube has developed quite well at this position and the radial mean velocity profile exhibits the feature of self-preserving which was also confirmed in the previous studies for the confined jet flow. For cases C and D, the upward shear flow (wall jet) along the bed wall is still developing in this region and the flow expands greatly towards the centre of the bed, thus having negative radial velocity component as shown in Fig. 3(d). The upward shear flow also induces huge entrainment and causes a big re-circulation vortex behind the disk-baffle, as can be observed by the existence of two apparent re-circulation vortices almost symmetrical to the centreline.

3.2. Turbulence characterisation

Dispersion of flocs depends on interaction between turbulent stresses and floc cohesion force. When turbulence stresses exceed the force for floc cohesion, flocs become disintegrated and dispersed. In the present study, the carrier fluid is air and its viscosity is small so that it is impossible to disperse fibre bundle by viscous shear force. The only feasible way to breakup fibre bundles is to take advantage of the features of turbulent shear flow, i.e. significant enhancement of Reynolds stresses. As the cohesion strength the fibre bundle was obtained by using the

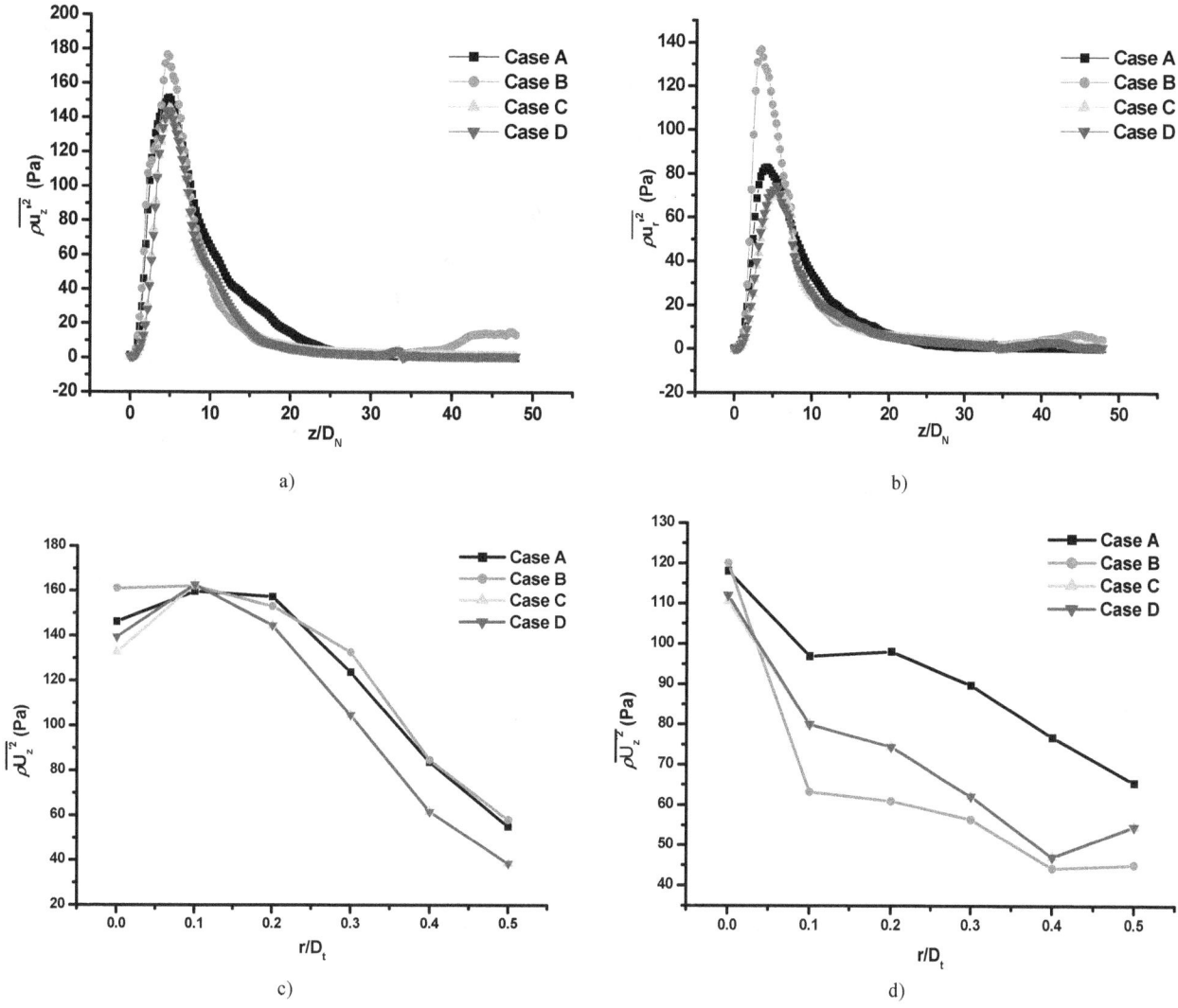

a)

b)

c)

d)

Figure 5. Profiles of normal Reynolds stress variation along the centreline (a) $\overline{\rho u_z'^2}$; (b) $\overline{\rho u_r'^2}$.
Profiles of axial normal Reynolds stress $\overline{\rho U_z'^2}$ at two different positions (c) at the entry region $z/D_N=5.0$; (d) in the draft tube $z/D_N=7.7$.

rheometer (Zong et al., 2011), it becomes essential to explore whether or not the turbulent stress generated in the turbulent flow in the modified spout-fluidised beds is strong enough to subdue the cohesion strength of the fibre bundle so that fibre flocs can disperse during the flow.

Fig. 4 illustrates the variations of turbulent kinetic energy along the centreline and radial distributions of turbulence kinetic energy at two different cross-sections ($z/D_N = 32.2$ and $z/D_N = 38.9$) of the bed. Since addition of the internals alters the flow patterns in the bed, it can be seen from the figure that the effect of the internals is remarkable, in particular the draft tube. Without the draft tube, the jet flows from the spouting nozzle, developing strong shear flow in the downstream of the fringe of the nozzle. The shear rate reaches the maximum at $z/D_N \approx 5$. This dynamic behaviour can be visualised from Fig. 2 as the ring-shaped vortices generated around the periphery of the nozzle are shedding and spreading, eventually merging to form large scale eddies. Addition of the draft tube confines the spread of shear flow but the shear rate is enhanced as a result of the shear flow to be reinforced in the draft tube. Such turbulence modulation has resulted in a remarkable increase in turbulence kinetic energy as can be seen from Fig. 5. However, introduction of the disk-baffle will restrain the turbulence kinetic energy in the

area between the exit of the draft tube and the disk-baffle, partly associated with a reduction in shear rate, i.e. $\partial Vz/\partial r$. For cases C and D, development of the first jet flow from the spouting nozzle is further confined by the circulation formed in the annulus region. Consequently, the turbulent kinetic energy in the entry region slightly decreases as compared with case A (see Fig. 5(a)). It can be also seen from Fig. 5 that the turbulence kinetic energy intensity overall becomes small for case C and case D behind the disk baffle, but the turbulence kinetic energy for case B increases a bit and then decreases. A likely explanation is that the turbulent flow tends to be more uniform downstream of the disk-baffle, where the shear flow of the wall jet has experienced re-development and fewer large eddies are presented as indicated in Fig. 2(b). It is interesting to note that the radial distributions of turbulence kinetic energy at $z/D_N = 32.2$ for case C and D present bimodal peaks as shown in Fig. 5(b). This behaviour is obviously associated with the shear flow features in the impinge region, where the secondary jet flow exiting from the draft tube is converted to a circular radial impinge jet. The positions of these peaks correspond to the locations with the larger local shear rates. Fig. 5(c) shows that except for case A, all other cases present bimodal peak distribution of the radial turbulence kinetic

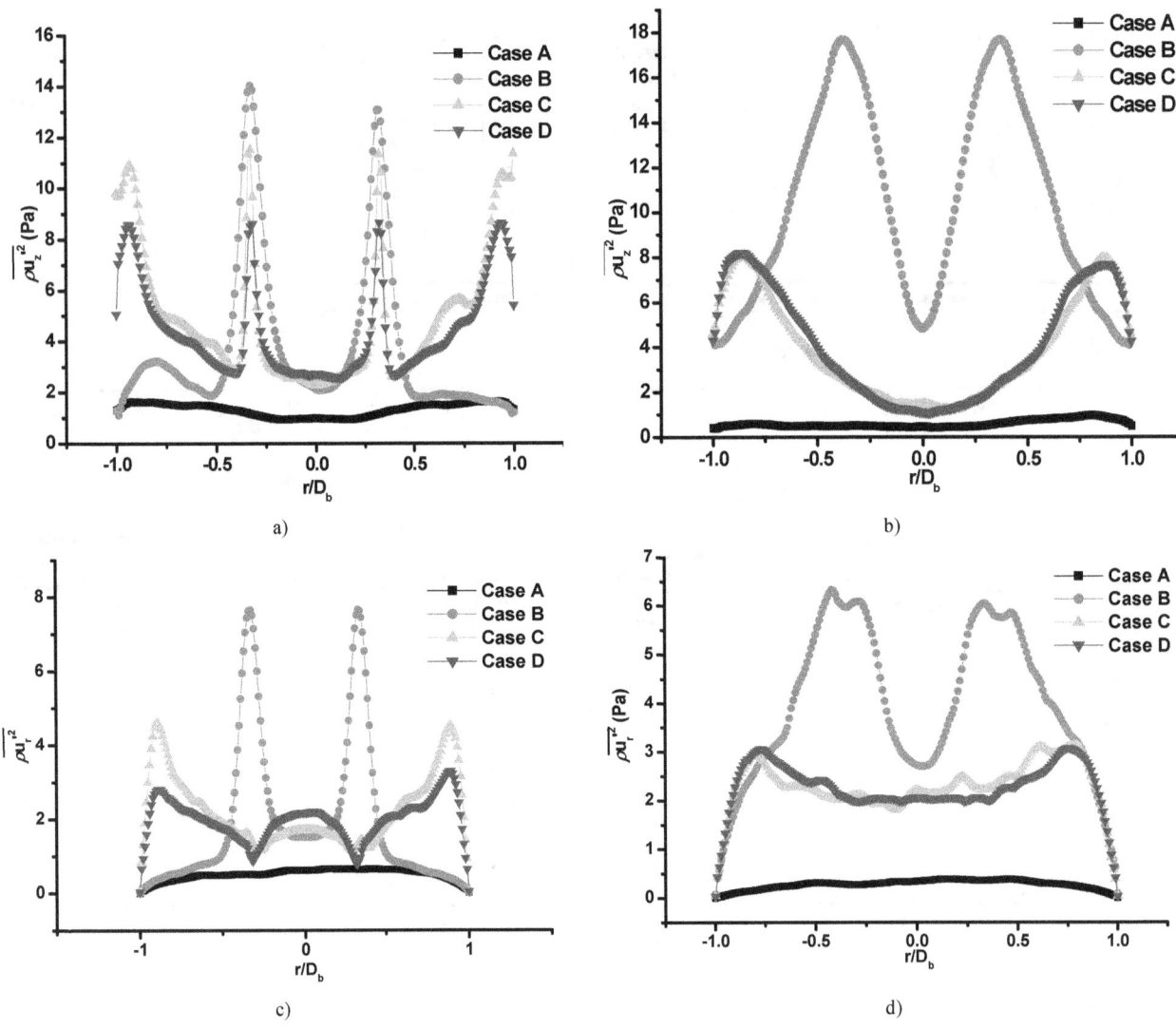

Figure 6. Profiles of normal Reynolds stress: (a) $\overline{\rho u_z'^2}$ at $z/D_N = 32.2$; (b) $\overline{\rho u_z'^2}$ at $z/D_N = 38.9$; (c) $\overline{\rho u_r'^2}$ at $z/D_N = 32.2$; (d) $\overline{\rho u_r'^2}$ at $z/D_N = 38.9$.

energy at $z/D_N = 38.9$. Among cases with draft tubes, case B exhibits the peak values of turbulence kinetic energy twice times of that for case C and D as a result of strong shear layers formed due to the straightforward secondary jet flow. Due to redistribution of the flow field behind the disk-baffle, the peaks in the turbulence kinetic energy profiles for case C and D have moved towards the wall and their values have been significantly reduced in comparison to case B. This indicates that the shear rates generated by the wall jet (caused by the impinge jet onto the bed wall) in case C and D are smaller. Since turbulence generation is strongly associated with the shear, it implies that the overall shear rate in case B is strongest but this may be harmful for long fibre suspension and transport because the fibres will strongly interact with large eddies.

Fig. 6 illustrates the axial and normal Reynolds stress distributions at cross-sections $z/D_N = 32.2$ and $z/D_N = 38.9$ for all four cases while the shear stress $\overline{\rho u_r' u_z'}$ contours for all cases are shown in Fig. 7. It can be seen from Fig. 6 that the normal Reynolds stress component $\overline{\rho u_z'^2}$ is much greater than the other Reynolds stress components for all cases with the maximum falling into the entry region. The maximum values achieved in all cases are greater than 140 Pa, which is much higher than the critical

cohesion strength of fibre bundles. This implies that most of fibre bundles will be disintegrated by the turbulent shear flow in the draft tube. For cases C and D, the Reynolds stresses have similar distributions to the turbulence kinetic energy distributions but there exist tiny difference among the profiles. This may be resulted from the effect of rotation of the disk-baffle. At position $z/D_N = 38.9$, the profiles of the normal Reynolds stresses $\overline{\rho u_z'^2}$ and $\overline{\rho u_r'^2}$ for both cases are highly similar, i.e. the normal Reynolds stresses increase towards the bed wall, apparently relating to the entrainment effect of the upflowing wall jet. By contrast, the radial normal Reynolds stress in case A is relatively small. Fig. 7 also indicates that the time-averaged resolved shear Reynolds stress distribution is roughly anti-symmetric about the centreline due to the changes of the mean shear rate from positive to negative, which is consistent with the results as reported by Webster et al. (2001) and Beaubert and Viazzo (2003). The sign of the shear stress agrees with the net transport of high momentum away from the centreline. It should be noted here as can be seen from Fig. 7 that the overall magnitudes of the shear stress in case C and case D are lower than that of case B, which may be beneficial to the formation of a relatively uniform fibre floc suspension in the upper part of the spout-fluidised bed.

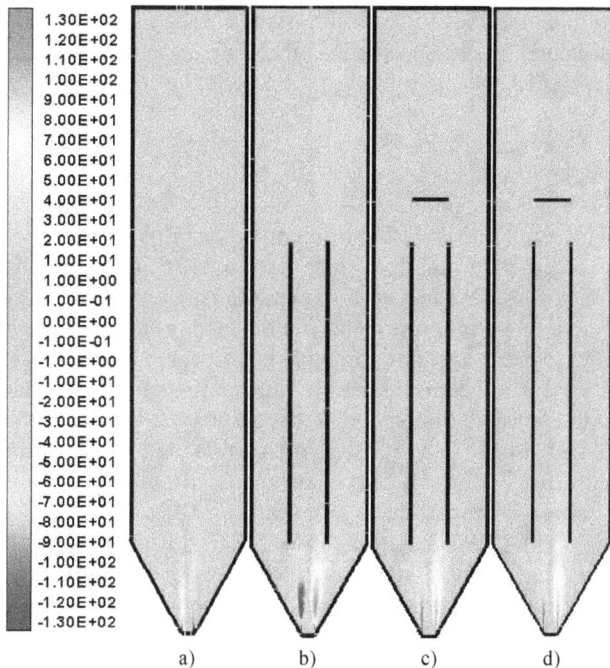

Figure 7. Resolved shear component of Reynolds stress $\overline{\rho u_r' u_z'}$.
(a) case A; (b) case B; (c) case C; (d) case D.

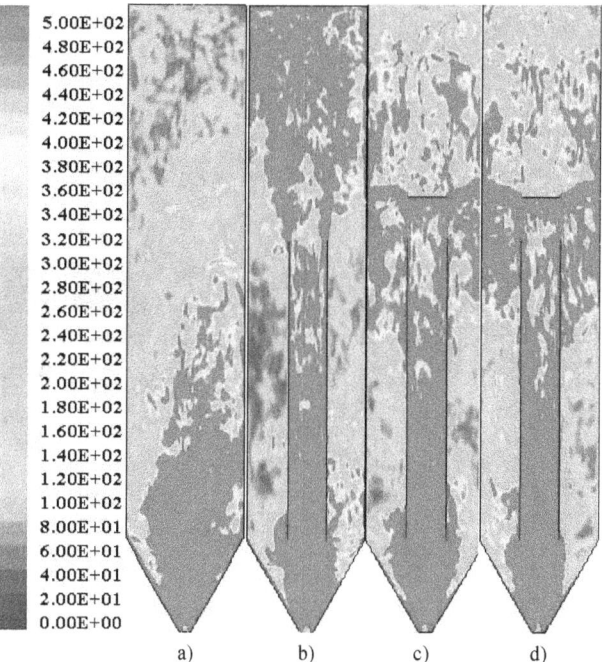

Figure 8. Instantaneous contours of shear strain rate for all cases at
t = 10 s: (a) case A; (b) case B; (c) case C; (d) case D.
(The shear strain rates ranging from 0 to 500 s⁻¹ are displayed only and
red represents those values equal to or higher than 500 s⁻¹).

3.3. Correlation between fibre dispersion and local turbulent eddies

It is expected that there exists a strong correlation between glass fibres and turbulent large eddies because turbulent kinetic energy generation is closely related to turbulent eddies. Fig. 9 illustrates the distributions of fibre mass concentration and shears rates in the spout-fluidised bed. It can be seen clearly that those locations with relatively high fibre mass concentration correspond to high vorticity, implying local high kinetic energy generation. The glass fibres are entrained by large turbulent eddies. In order to quantify such correlation, the following correlation is defined to explore this behaviour.

$$\gamma = \frac{\overline{(\tilde{\omega}_i - \overline{\tilde{\omega}}_i)(\tilde{C} - \overline{\tilde{C}})}}{\sqrt{\overline{(\tilde{\omega}_i - \overline{\tilde{\omega}}_i)^2}}\sqrt{\overline{(\tilde{C} - \overline{\tilde{C}})^2}}} \quad (11)$$

Fig. 10 shows the correlation coefficient for fibre mass concentration and local vorticity. It was demonstrated that the correlation coefficient is higher in the regions where the local vorticity is high, as observed in the impinging region in contrast to the entry and exit regions. The maximum |g| appears in the vicinity of the bed wall at the exit of the disk baffle. The circular spread jet flow generates strong vortices when hitting on the bed wall so that the flocs are entrapped by such eddies and high floc concentration is found in this region (Lin, 2008). The radial distribution of the correlation coefficient shown in Fig. 10(b) further supports this argument.

Fig. 11(a) shows the profiles of Fl along the axis centerline. It can be seen from the figure that Fl attains the maximum at inlet of the draft tube ($z/D_N \approx 4.8$) and maintains a relatively greater value along the draft tube. This may be attributed to a strong shear generated by the spouting nozzle, forming large turbulent eddies which entrain the fibres into the draft tube. The fibre flocs/

buddles are apparently influenced by this shear in this stage which is beneficial to floc dispersion. When the flow carries the flocs through the draft tube, dispersion of the fibre flocs/buddles is reduced but Fl remains almost unchanged until fibres leave the draft tube. Fig. 11(b) illustrates the radial distributions of Fl at $z/D_N = 32.23$ and $z/D_N = 38.9$, which correspond to the locations of middle impinging jet and exit region respectively. In the impinging region, Fl reaches the maximum at the centre and the minimum at the location where the circular spread jet just forms. Fl increases towards the bed wall, indicating the fibres are entrapped by the large eddies formed on the wall. Xu and Aidun (2005) indicated that the size of the flocs directly depends on the interaction between flocs and turbulence intensities in the flow field. When turbulent Reynolds stresses is greater than the cohesion strength of the floc network, disintegration of the flocs is occurring. Otherwise, the flocs may gradually grow. Therefore, the relatively weak turbulence in the jet regime easily causes the formation of the floc while local high vorticity and strong interaction between glass fibres and eddies enhance fibre floc dispersion. In the exit region, the wall jet flow experiences a redistribution process due to the effect of the disk-baffle. The upward shear flow is developing along the bed wall. Owing to formation of vortices with high vorticity, the flocs are entrained these eddies and the local floc mass concentration increases. Thus, how to reasonably control the local vorticity of the turbulent large eddies requires further investigation.

It should be noted here that addition of internals into the spout-fluidised beds is in fact to partition the flow into several sub-flow regions with the hydrodynamic characteristics of either flow contraction or expansion. Blaster (2000) and Kuroda et al. (2002b) have investigated fibre floc deformation in shear and strain flow and their results suggested that the critical mechanical stress required for

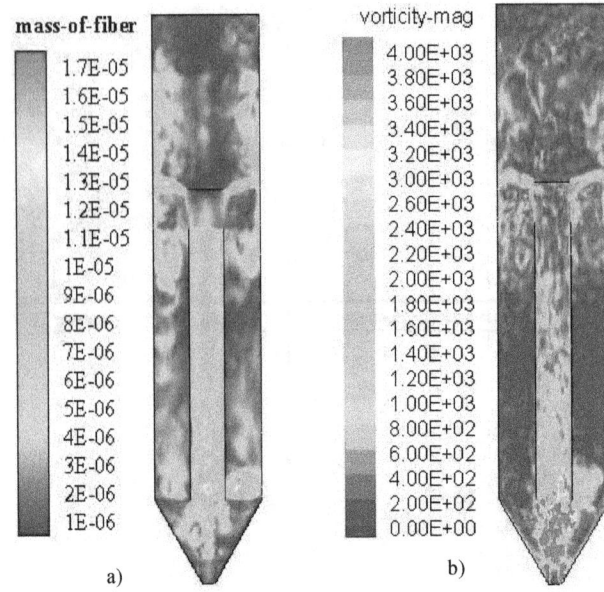

a)   b)

Figure 9. (a) Local fiber mass concentration; (b) local vorticity magnitude ($|\Omega|^3$ 4000s^{-1} highlighted by 'red').

breaking up the fibre floc depends on types of flows applied and how the fibre flocs are elongated in the flow. A strong shear flow having a local high vorticity is more effective in breaking up droplets and in agglomerating the fibre flocs than a simple shear flow. In this sense, formation of complex turbulent shear flow patterns in the spout-fluid bed by addition of the internals can effectively intensify the vortex stretching due to the formation of different

turbulent eddies so that fibre flocs can be deformed and stretched, further promoting fibre floc dispersion. This is crucial for the case of long fibre flocs to be concerned.

4. Conclusions

The hydrodynamic characteristics of a spout-fluid bed with addition of different internals for purpose of effective dispersion of fibre flocs were studied in this paper. Since fibre floc break-up in shear flows depends to great extent on both floc strength and the features of the applied flow field, fibre floc strength measurements were conducted in a rheometer and the LES simulation of the spout-fluid bed with the internals was carried out to obtain turbulent kinetic energy, Reynolds stress and shear strain rate distributions. Conclusions drawn from the present study are summarised as follows:

1) The LES simulation results have shown that addition of the internals (draft tube and disk-baffle) into the spout-fluidised bed can significantly alter the shear flow field, turbulence structures and local vorticity distributions.

2) Through controlling the shear flows and effectively enhancing the turbulence intensities, Reynolds stresses can be higher than the strength of fibre flocs coherence so that fibre flocs can be well dispersed.

3) Dispersion of fibre flocs benefits from turbulence characteristics in the spout-fluidised bed when internals are effectively utilised.

4) LES results clearly indicate existence of a strong correlation between the glass fibre concentration and local

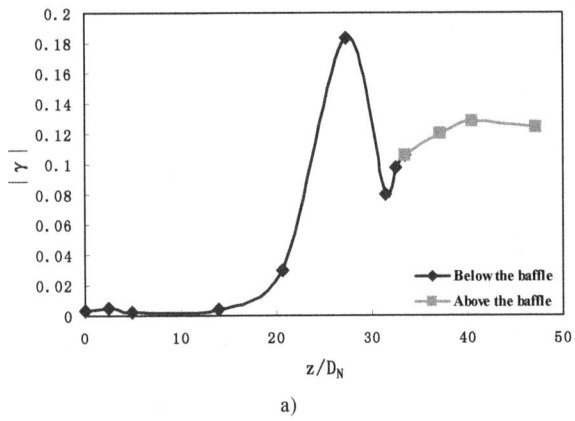

a)

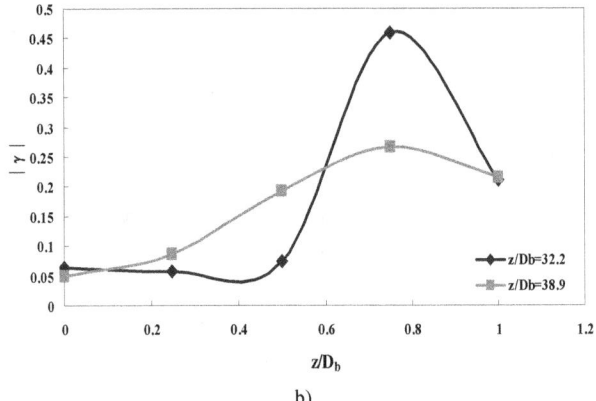

b)

Figure 10. Profile of correlation coefficient $|\gamma|$ between mass of fiber and local vorticity at different location.
(a) axial distribution of $|\gamma|$ along the centreline; (b) radial distribution of $|\gamma|$ at $z/D_b = 32.2$ and $z/D_b = 38.9$ respectively.

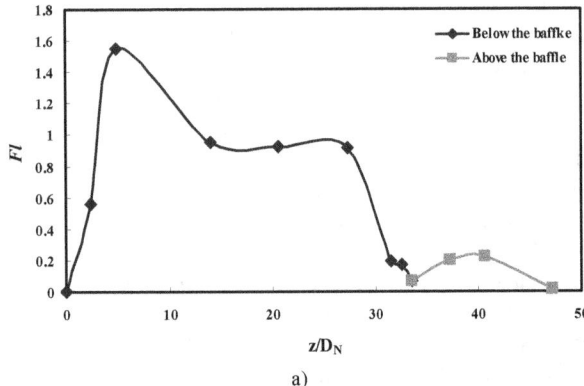

a)

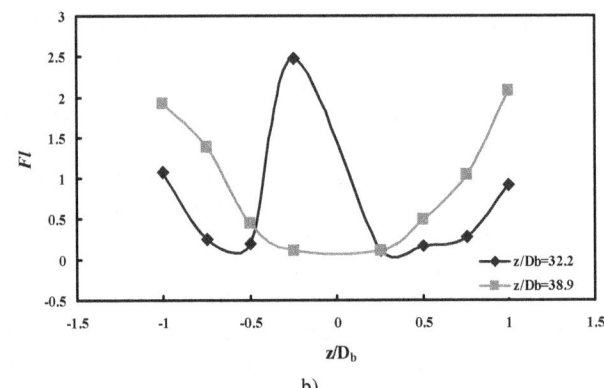

b)

Figure 11. Profile of flocculation intensity (Fl) at different location.
(a) axial distribution of Fl along the centreline; (b) radial distribution of Fl at $z/D_b = 32.2$ and $z/D_b = 38.9$ respectively.

vorticity. A local high vorticity corresponds a relatively high glass fibre concentration. The glass fibres are remarkably entrapped by the large eddies.

5. Acknowledgement

The authors gratefully acknowledge the support provided by the 111 Project (B08021). Yuan Zong would also like to acknowledge the support of State Key Laboratory of Chemical Reactors, East China University of Science and Technology for a Doctoral programme scholarship. Xiaogang Yang would especially like to thank the support of Glyndŵr University for a Research Fellowship.

Nomenclature

C	Smagorinsky's constant;
D	computation domain;
D_b	diameter of the spout-fluid bed, m;
D_d	diameter of the disk-baffle, m;
D_N	the diameter of the spouting nozzle, m;
D_t	diameter of draft tube, m;
G	the grid filter function;
G_v	local velocity gradient, s^{-1};
H	the distance between disk-baffle and outlet of draft tube, m;
h	bob height, mm;
M	torque, m·s
U	mean velocity, $m \cdot s^{-1}$;
U_j	velocity of the flow at the entrance of draft tube, $m \cdot s^{-1}$;
U_c	axial mean velocity along centreline, $m \cdot s^{-1}$;
U_0, U_N	flow velocity at the exit of spouting nozzle, $m \cdot s^{-1}$;
R	the radius of the spout-fluidised bed, m;
$\overline{S}_{ij}$	resolved scale strain tensor;
V_r, V_θ	radial component and circumferential component;
d	floc size, mm;
k	turbulent kinetic energy, $m^2 \cdot s^{-3}$;
r	radius of the baffle, m;
$\overline{u}_i, \overline{u}_j$	the filtered velocity components, $m \cdot s^{-1}$;
u', v'	fluctuation velocity, $m \cdot s^{-1}$;
X	axial distance from the nozzle, m;
x_0	distance of virtual origin from the nozzle, m.

Greek letters

Ω	angular velocity, $rad \cdot s^{-1}$;
a	volume fraction;
$\dot{\gamma}_{bc}$	average shear rate, s^{-1};
$\overline{\Delta}$	local mesh size, m;
e	turbulence dissipation rate, $m^2 \cdot s^{-3}$;
k	cup and bob radius ratio, $k = R_c/R_b$;
n_T	eddy viscosity, Pa·s;
r	density, $kg \cdot m^{-3}$;
s_{bc}	average shear stress, Pa;
t_{ij}	sub-grid scale stress, Pa;
ω	rotational speed, $rad \cdot s^{-1}$.

Subscripts

G	gas phase;
F	fibre phase.

References

Ashforth-Frost, S., Jambunathan, K. and Whitney, C.F. (1997). Velocity and turbulence characteristics of a semiconfined orthogonally impinging slot jet. *Experimental Thermal and Fluid Science*, Vol. 14, pp. 60-67.

Baydar, E. and Ozmen, Y. (2006). An experimental investigation on flow structures of confined and unconfined impinging air jets. *Heat Mass Transfer*, Vol. 42, pp. 338-346.

Beaubert, F. and Viazzo, S. (2003). Large eddy simulations of plane turbulent impinging jets at moderate Reynolds numbers. *International Journal of Heat and Fluid Flow*, Vol. 24, pp. 512-519.

Blaser, S. (2000). Flocs in shear and strain flows. *Journal of Colloid and Interface Science*, Vol. 225, pp. 273-284.

Bouyer, D., Coufort, C., Liné, A. and DoQuang, Z. (2005). Experimental analysis of floc size distributions in a 1-L jar under different hydrodynamics and physicochemical conditions. *Journal of Colloid and Interface Science*, Vol. 292, pp. 413-428.

Coufourt, D., Bouyer, D. and Liné, A. (2005). Flocculation related to local hydrodynamics in a Taylor-Couette reactor and in a jar. *Chemical Engineering Science*, Vol. 60, pp. 2179-2192.

Dai, G., Huang, J., Sun, B. and Zhou, X.D. (2001). Agitated-spouted fluidited bed and use in preparing fiber reinforced composite, in: China, 01112947.6 [P], 5-22-2001.

Ducoste, J.J., Clark, M.M. and Weetman, R.J. (1997). Turbulence in flocculators: Effects of tank size and impeller type. *AIChE J.*, Vol. 43, pp. 328-338.

Hosseini, S.H., Zivdar, M. and Rahimi, R. (2009). CFD simulation of gas-solid flow in a spout bed with a non-porous draft tube. *Chemical Engineering and Processing*, Vol. 48, pp. 1539-1548.

Kuroda, M.M.H. and Scott, C.E. (2002a). Initial dispersion mechanisms of chopped glass fibres in polystyrene. *Polymer Composites*, Vol. 23, pp. 395-405.

Kuroda, M.M.H. and Scott, C.E. (2002b). Blade geometry effects on initial dispersion of chopped glass fibres. *Polymer Composites*, Vol. 23, pp. 828-838.

Lin, J. (2008). *Multiphase Fluid Dynamics for Extraordinary Particle Flow-Two Phase Flow with Cylindrical Particles*. Beijing: Science Press.

Moated, R. (1999). *Characterization of Fiber Suspension Flows at Papermaking Consistencies*. PhD Thesis, University of Toronto.

Padaki, S. and Drzal, L.T. (1999). A simulation study on the effects of particle size on the consolidation of polymer powder impregnated tapes. *Composites: Part A*, Vol. 30, pp. 325-337.

Popiel, C.O. and Trass, O. (1991). Visualization of a free and impinging round jet. *Experimental Thermal and Fluid Science*, Vol. 4, pp. 253-264.

Shamlou, P.A., Gierczycki, A.T. and Titchener-Hooker, N.J. (1996). Breakage of flocs in liquid suspensions agitated by vibrating and rotating mixers. *The Chemical Engineering Journal*, Vol. 62, pp. 23-34.

Smagorinsky J. (1963). General circulation experiments with the primitive equations. I. The basic experiment. *Month. Wea. Rev.*, Vol. 91, pp. 99-164.

Tambo, N. and Hozumi, H. (1979). Physical characteristics of flocs - II. Strength of floc. *Water Research*, Vol. 13, pp. 421-427.

Webster, D.R., Roberts, P.J.W. and Ra'ad, L. (2001). Simultaneous DPTV/PLIF measurements of a turbulent jet. *Experments in Fluids*, Vol. 30, pp. 62-72.

Wengeler, R. and Nirschl, H. (2007). Turbulent hydrodynamic stress induced dispersion and fragmentation of nanoscale agglomerates. *Journal of Colloid and Interface Science*, Vol. 306, pp. 262-273.

Wu, Z. and Arun, S.M. (2008). CFD modelling of the gas–particle flow behavior in spouted beds. *Powder Technology*, Vol. 183, pp. 260-272.

Xu, H. and Aidun, C.K. (2005). Characteristics of fiber suspension flow in a rectangular channel, *International Journal of Multiphase Flow*, Vol. 31, No 3, pp. 318-336.

Yeh, F. and Lei, U. (1991). On the motion of small particles in a homogeneous isotropic turbulent flow. *Phys. Fluids A*, Vol. 3, pp. 2571-2586.

Yeung, A., Gibbs, A. and Pelton, R. (1997). Effect of shear on the strength of polymer-induced flocs. *Journal of Colloid and Interface Science*, Vol. 196, pp. 113-115.

Zhao, X.L., Li, S.Q., Liu, G.Q. and Yao, S.Q. (2008). Flow patterns of solids in a two-dimensional spouted bed with draft plates: PIV measurement and DEM simulations. *Powder Tech.*, Vol. 183, pp. 79-87.

Zhong, W., Xiong, Y., Yuan, Z. and Zhang, M. (2006). DEM simulation of gas–solid flow behaviors in spout-fluid bed. *Chemical Engineering Science*, Vol. 61, pp. 1571-1584.

Zhou, H.S., Flamant, G. and Gauthier D. (2004). DEM-LES of coal combustion in a bubbling fluidized bed. Part I: Gas-particle turbulent flow structure. *Chemical Engineering Science*, Vol. 59, No 20, pp. 4193–4203

Zong, Y., Yang, X. and Dai, G. (2011). Design simulation of glass-fiber-loaded flow in an internally spout-fluidized bed for processing of thermoplastic composites. I. flow characterization. *Industrial and Engineering Chemistry Research*, Vol. 50, pp. 9181-9196.

Zumaeta, N., Byrne, E.P. and Fitzpatrick, J.J. (2007). Predicting precipitate breakage during turbulent flow through different flow geometries. *Colloids and Surfaces A: Physicochemical and Engineering Aspects*, Vol. 292, pp. 251-263.

Three-Level Design of Composite Structures

Oleg Tatarnikov[1,2], Maria Gaigarova[1]

[1] Rocket and Spacecraft Composite Structures Department, Faculty of Special Machinery, Bauman Moscow State Technical University, 5 2nd Baumanskaya Street, Moscow, 105005, Russia
[2] Higher Mathematics Department, Plehanov Russian University of Economics, 36 Stremyanny Per., Moscow, 117997, Russia

Abstract: A composite can be considered at three scale levels: fibre-matrix, structural element and a composite volume where averaging of properties is allowed. At each level of consideration a certain set of initial data has to be used. These data determine composite component properties, the characteristics of a preform structure and finally the construction parameters. The set of data has to be changed in the transition from one scale level to another following the simulation model transformation. Using such an approach it is necessary to establish a link between the different composite models. The necessary set of relations can be found from the finite element simulation of a composite under mechanical and thermal loading.

Key Words: Composite component, Perform structure, Scale level of composite, Finite element simulation, Strain-stress analysis.

1. Introduction

The process of composite construction design includes several steps. At each of those a set of parameters is used. These characterize the material and component properties, the construction elements and manufacturing process features. In the transition from one level to another the set of parameters is changes according to simulation model which is currently used. This is why it is necessary to define the relation between data that is used on different levels of composite modelling.

This problem can be solved using information technologies such as databases of composite component properties, software modules for calculation of composite characteristics and finite element analysis programs, permitting simulating composite construction while a modal or full-scale test.

Some of the results, concerning spatially-reinforced composites, which have been obtained using a three level approach, are presented below.

2. Scale levels of spatially-reinforced composite

At the present time spatially reinforced composites with ceramic, carbide and carbon components such as AL2O3, SiO2, Si3N4, SiC, graphite, pyrocarbon etc. take up a special place among materials, which are applied in advanced constructions operating under extreme conditions. Such types of composites can be used at working temperatures exceeding 1200°C. For example, heaters for industrial furnaces, pump components that are used for transferring of aggressive liquids and gases, aerospace parts, high speed train and heavy truck brakes, metallurgy crucibles and electrodes and many other items may be manufactured from these materials.

Efforts directed to creation of an adequate mathematical model of a spatially reinforced composite which is characterized by complicated anisotropy and high level of nonuniformity are necessary to bring an understanding of composites on the different scale levels, such as:

- Microlevel – single reinforcement (Fig. 1);
- Mesolevel – representative volume of a composite, consisting of a number of structural cells (Fig. 2);
- Macrolevel – composite construction (Fig. 3).

On the microlevel a two-phase structure which plays the role of the composite is considered. The components being elementary fibres and a matrix which fills interspaces between them. The physical and mechanical characteristics of the components as well as geometrical parameters of the reinforcement are used as the initial data here. In the resulting model the composite, consisting of matrix and several thousand elementary fibres, is transformed to an uniform anisotropic solid

The representative volume of a composite . (Christensen, 1979) which appears as an assemblage of structural cells is considered on the mesolevel. This is a two-phase structure as well but in contrast to first level it consists of uniform anisotropic reinforcements and matrix filled interspaces. Generally this type of matrix is different compared to the one situated inside the reinforcement. Hereinafter the matrix inside the reinforcement will be indicated as matrix I, the other one – matrix II. The aver-

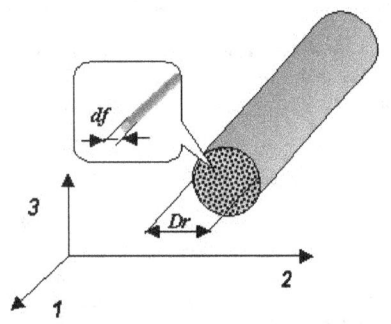

Figure 1. Single reinforcement.

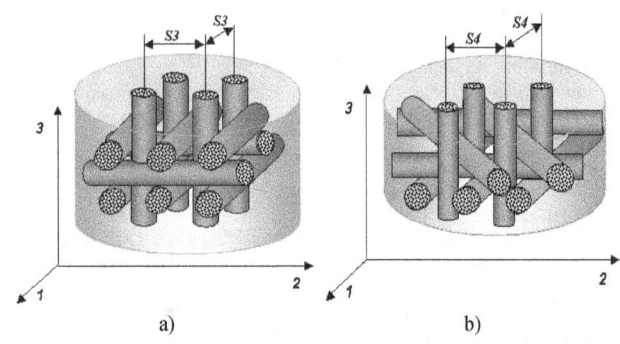

a)   b)

Figure 2. Representative volume of a composite: a) 3D, b) 4D.

Figure 3. Composite construction.

age physical and mechanical characteristics of the reinforcement, properties of matrix II and geometrical parameters of the elementary cell are used as initial data on the second structural level of a composite. In the result the representative volume of the composite is changed to a uniform anisotropic material.

Finally a composite construction represents the macrolevel of a spatially reinforced composite. Here the composite is assumed to be an anisotropic homogeneous material having average properties and composite construction in the context of mechanics of solid media.

2. Micro- and mesolevel of composite

As mentioned above with each transition step from one scale level to another the conversion from heterogeneous to homogeneous model takes place. There are two basic approaches, which create these relations. First one is Voigt's approach, the second - Reiss's one. In Voigt's approach uniformity of strain is supposed, in opposite within Reiss's approach stresses are assumed to be uniform. In addition it is usually assumed the fibres introduce anisotropy (Christensen, 1979; Gay et al., 2002). The last assumption of fibre isotropy is correct in the case of glass and metal ones as contrasted with high level of carbon and organic fibres anisotropy (Simamura, 1984; McAllister and Lachman, 1983). Besides, in some cases of loading the assumption of strain or stress uniformity is not correct in the averaging volume of the material. In

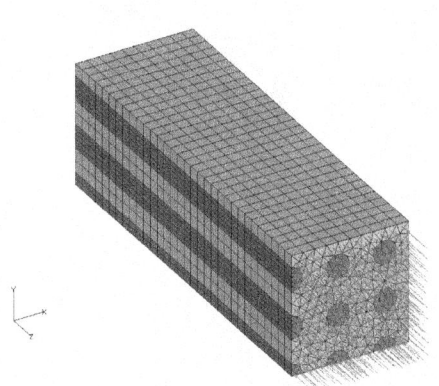

Figure 4. Finite element model of 1D composite specimen

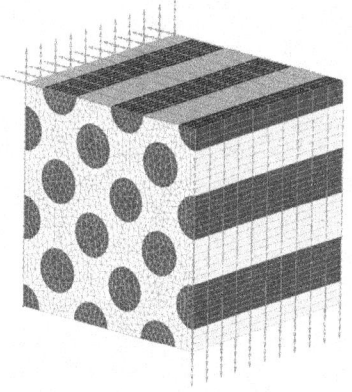

Figure 7. Model of specimen for share test

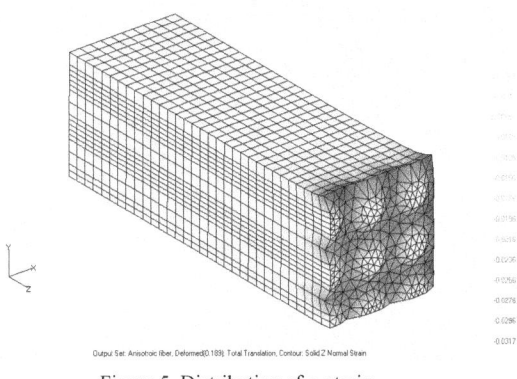

Figure 5. Distribution of ε_z strain

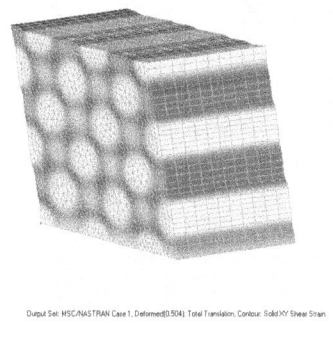

Figure 8. Distribution of γ_{yz} strain

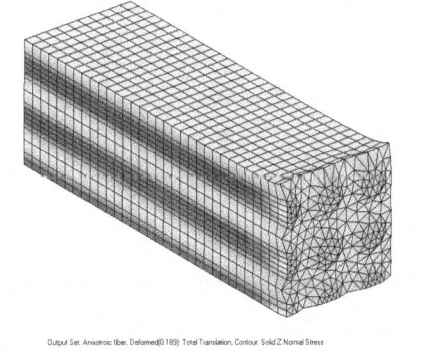

Figure 6. Distribution of σ_z stress

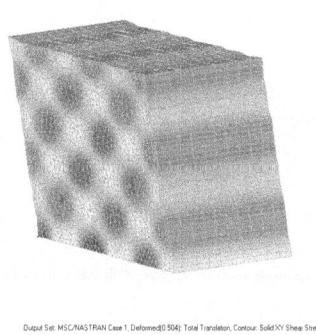

Figure 9. Distribution t_{yz} stress

Table 1
Elastic properties of 1D composite

Fiber contents	$C_f = 0.2$				$C_f = 0.4$			
Modulus	E_z	E_y	G_{xy}	G_{zx}	E_z	E_y	G_{xy}	G_{zx}
Calculated data, GPa	44.7	4.27	1.45	1.75	93.5	5.05	2.15	2.68
Voigt approach, GPa	46.1	5.35	2.12	-	91.5	7.68	3.16	-
Reiss approach, GPa	-	3.57	1.36	1.42	-	4.36	1.78	1.86

such cases it makes a sense to use an approach based on numerical simulation to determine averaged properties of a composite.

Let us consider some results of finite element modelling.

In Fig. 4 the finite element model of the 1D composite specimen under tensile loading along fibre direction is presented. Fibre contents vary from $C_f = 0.2$ up to $C_f = 0.4$.

The carbon fibre is considered as a transversal isotropic material. It's mechanical properties are taken to be as follows:

- Longitudinal Young's modulus $E_z = 230$ GPa;
- Transversal Young's modulus $E_x = 15$ GPa;
- Poisson ratio $v_{xz} = 0.001$;
- Poisson ratio $v_{yz} = 0.3$;
- Shear modulus $G_{zx} = 50$ GPa.

The matrix of the composite is considered as an isotropic material with the following mechanical properties:

- Young's modulus of matrix $E_m = 3$ GPa;
- Poisson ratio $v_m = 0.3$,

In Fig.5 and 6 strain and stress distributions in the specimen are shown. These figures illustrate respectively uniform distribution of strain within the specimen and a rather non-uniform stresses distribution. This fact confirms a good correspondence of Voigt approach in case of determining of Young's modulus for 1D composite with anisotropic fibre. The difference between averaged and calculated value is equal to 4%.

Fig. 7 illustrates the finite element model of a 1D composite specimen under share loading in the plane YZ. Shear strain and stress are presented in Fig. 8 and 9.

The composite shear modulus was calculated as a result of dividing of volume averaged shear stress by volume averaged shear strain. In this case both the Voigt and Reiss approach do not give an acceptable prediction. The difference between calculated shear modulus and Voigt averaged value is as high as 46%; Reiss averaging gave a difference in the values of value 17%. Some other results of the calculation of elastic properties of 1D carbon composite are presented in Table 1.

The following conclusions may be made after analyzing the data:

1) All the results obtained from numerical experiment are located within so called Hill's fork.
2) Voigt approach gives an acceptable result for the longitudinal modulus only.
3) Reiss approach gives an acceptable result for the shear modulus G_{xy} in case of low fibre contents.

A number of structural cells of a composite are considered on the mesolevel. Fig. 10 – 12 present the finite element model, strain and stresses in a 3D composite specimen under compressive loading.

In the case of 3D composites the empirical relation between properties of components and longitudinal Young's modulus E_L of the composite was obtained (on

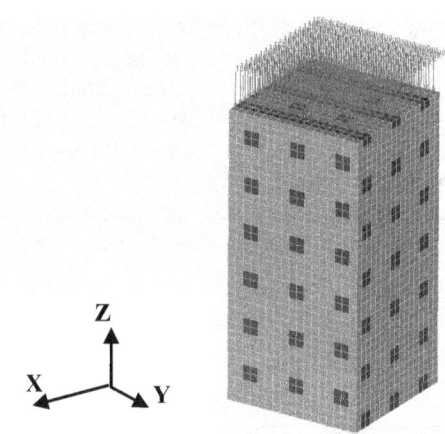

Figure 10. Finite element model of 3D composite specimen

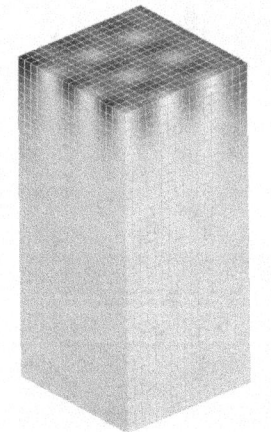

Output Set: NX NASTRAN Case 1, Contour: Solid Z Normal Strain

Figure 11. Distribution of ε_z strain

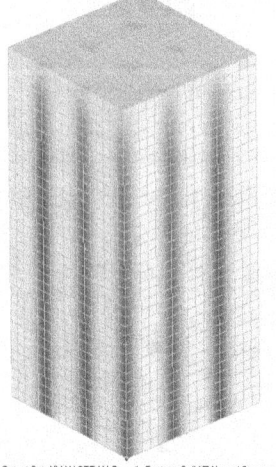

Output Set: NX NASTRAN Case 1, Contour: Solid Z Normal Stress

Figure 12. Distribution of σ_z stresses

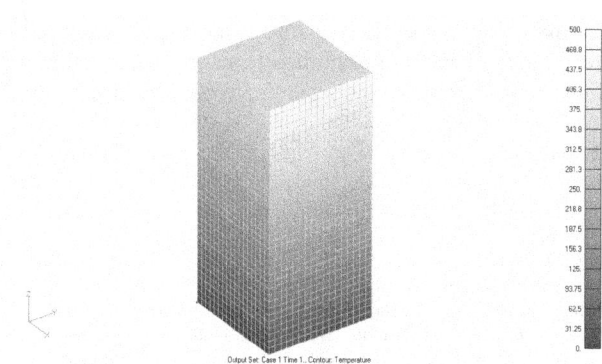

Figure 13. Temperature distribution in 3D composite specimen

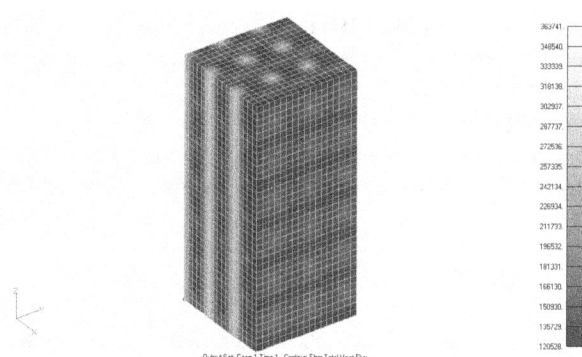

Figure 14. Heat flux distribution in 3D composite specimen

Table 2
Elastic properties of 3D composite

Fiber contents	$C_f = 0.33$	$C_f = 0.75$
E_l, GPa numerical result	16.4	33.1
E_l, GPa according (1)	15.9	32.6

the assumption that reinforcements in x and y directions are equivalent):

$$E = c_z E_z + \frac{1}{2}\left(\frac{E_x E_m}{E_x c_m + 2 E_m c_x} + \frac{E_y E_m}{E_y c_m + 2 E_m c_y} \right)(1 - c_z) \quad (1)$$

where E_m is Young's modulus of matrix, E_z is longitudinal Young's modulus of z – reinforcement; E_x and E_y are transversal modulus of reinforcements in x and y directions respectively.

The data presented in Table 2 illustrates an acceptable agreement between the results received from the numerical analysis and calculations of elastic modulus by means of relation (1). Elastic properties of components were taken as follows:

$E_m = 3$ GPa;
$E_z = 114$ GPa;
$E_x = E_y = 4,93$ GPa.

Numerical modelling of thermal conductivity coefficient l_z is illustrated in Fig. 13 and 14. These two figures illustrate the distributions of temperature and heat flux within the 3D composite specimen.

In a similar way the other characteristics necessary for thermal and strain-stress analysis can be obtained.

3. Macrolevel of composite construction

On the macrolevel a composite is considered as an anisotropic homogeneous material with properties averaged by a representative volume.

Let us consider some of the results, which were obtained from sequential transition from the microlevel – elementary fibre and matrix I, up to macrolevel – composite construction. As an example we take up thermal and

Table 3

Composite type	Maximal tensile stress s_θ, MPa	Maximal compressive stress s_θ, MPa
3D	17.73	-35.35
4D	33.54	-53.82

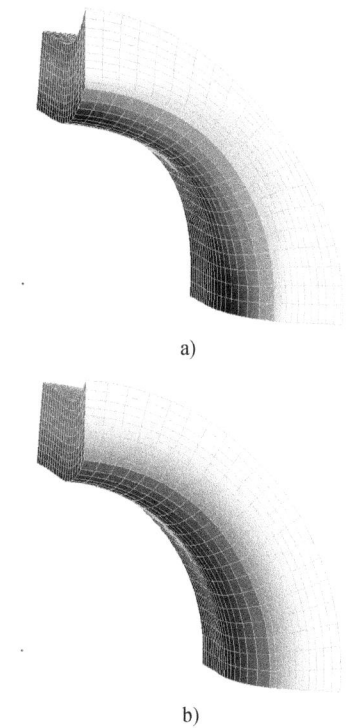

a)

b)

Figure 15. Elastic displacements in 3D (a) and 4D (b) cylinders

elastic conditions of a 3D (Fig. 2a) and 4D (Fig. 2b) composites having the same reinforcements in all directions. The same type of elementary fibre and matrix is used for both 3D and 4D cylinders. The fibre content is considered to be maximal for appropriate reinforcement structure. To avoid the influence of geometry factors in the construction a hollow cylinder was chosen.

So, the two materials under consideration have only one basic difference namely in the number of reinforcement directions. This difference is gives rise to two points:

1) type of anisotropy, 3D composite is being an orthotropic material, 4D – transversal isotropic one;
2) maximal value of reinforcement content.

External diameter for both cylinders is equal to 0.5m, internal diameter – 0.35m and height – 0.4m. Mechanical properties of the materials were defined on composite mesolevel according to section 2.

The internal surface of cylinders is subjected to thermal flux $q = 7.5$ MWt/m^2. The duration of heating is 40 seconds.

The deformed condition of both cylinders has quantitative and qualitative differences. The cylinder with 4D reinforcement structure has axisymmetric distribution of strains whereas the 3D cylinder is deformed in non-axisymmetric way. Elastic displacements in cylinders are shown in Fig.15.

Comparison of the thermal stresses gives the result that maximal level of tangential stress σ_θ in 4D cylinder is higher compared with that in 3D cylinder. The data are presented in Table 3.

4. Conclusions

The proposed approach allows thermal and stress-strain analysis of the composite construction taking into account the main features of its structure and components properties. That gives the possibility to carry out the optimal concurrent design of composite materials and structures.

References

Christensen, R. (1979). *Mechanics of Composite Materials*. New York: Willey.

Gay, D., Hoa, S. and Tsai, S. (2002). Composite *Materials: Design and Application*. Boca Raton: CRC Press.

Simamura, S. (1984). *Carbon Fibres*. Tokyo: Omsya.

McAllister, L. and Lachman, W. (1983). *Handbook of Composites: Vol.4, Fabrication of Composites*. Amsterdam: North-Holland.

Three Dimensional Elastic-plastic Finite Element Analysis of Tensile Properties of Carbon Fibre Reinforced Aluminium Alloy Laminates

Zhang Jiazhen[1,3], Song Xin[1,2], He Xiaodong[3]

[1] Beijing Aeronautical Science and Technology Research Institute, Commercial Aircraft Corporation of China - COMAC, Beijing, 100083, P.R.China
[2] College of Mechanical and Power Engineering, Harbin University of Science and Technology, Harbin, 150080, P.R.China
[3] School of Aeronautics, Harbin Institute of Technology, 92 West Dazhi Street, Nan Gang District, Harbin, 150001, P.R.China

Abstract: Recently, fibre reinforced metal laminates (FMLs) are used in the aeroplane structures. For the outstanding performance, it is interested by the aviation researchers. A three dimensional elastic- plastic finite element model of carbon fibre reinforced aluminium alloy laminates has been setup by ABAQUS in this study, and the fracture behaviours of laminate has been simulated with the model under quasi-static loading conditions. The tensile strength and elastic modulus of laminate have been calculated by the model and good accordance has been achieved in according to the experimental results.

Key Words: Fibre reinforced metal laminate; FMLs; Tensile properties; Finite element analysis.

1. Introduction

Fiber metal laminates (FMLs), as a kind of interplay hybrid composite, showed in the Fig. 1, are composed of metal lamina and reinforced fibers which are bonded layer by layer, and then, solidifying under proper temperature and pressure (Wu and Guo, 1999). They have the advantages of metal materials and fiber composite materials simultaneously, such as, light weight, high anti-fatigue properties and vibrations resistance, high specific strength and specific stiffness, like traditional fiber composite, also, good toughness and machinability, like metal materials. These superior properties make it very suitable for the new generation airplane structures (Binienda and Qiao, 2005) and applied on a several in-service airplane style (Cao, 2007). Furthermore, due to the high strengthen alloy promoted, and with the study progressing (Selective Reinforce Panel, 2005), the study of FMLs properties, which are composed of different new materials, have been developed (Boscolo, et al., 2008).

It is very difficult to achieve the accurate solutions of tensile properties of new style FMLs by the analytic expression. So, a large amount of test must be done to determine the values of tensile properties, such as tensile strength, elastic modulus. With the computer capacity enhanced and numerical simulation techniques developed,

it is an efficient method to cut down the amount of new style FMLs tensile properties experiment and shorten the research period by the use of finite element analysis method (FEAM). The results in the study of Grassi and Zhang (2003) and Liao et al. (2006) showed that FEAM is very useful in the research of new style FMLs.

2. FMLs specimen preparation

Composed materials of FMLs specimens are, in this paper, LY12M aluminium alloy layers with 1mm thickness, T700 carbon fiber prepreg layers, the tensile properties are listed in Table 1, and bonded by J272 middle-temperature-setting-glue. The interlayer of aluminium alloy are bonded with fiber on double side, outer layer of aluminium alloy are bonded on single side which toward inner. The surface of aluminium alloy layers should be treated by the method of phosphoric acid anodizing according to the standard of HB/Z197-91, bonded with carbon fiber, after middle temperature solidifying with 0.3 MPa pressure, the carbon fiber reinforced aluminium alloy laminate is prepared. Solidifying time vs. temperature curve is showed in the Fig. 2.

After solidification process, the quality of FMLs must be checked by the non-destructive testing (NDT) technology. In this study, ultrasonic C-scan system is used. Different solidifying time and temperature affect the performance of FMLs significantly. Fig. 3 showed the NDT,

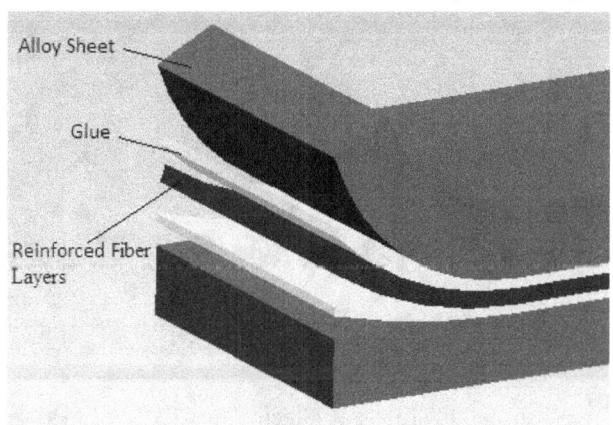

Figure 1. Schematic diagram of FMLs

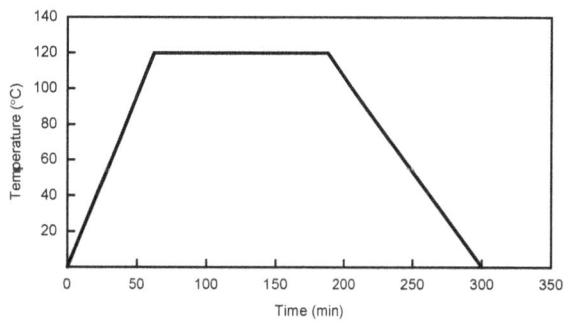

Figure 2. The solidifying time vs. temperature curve of laminate

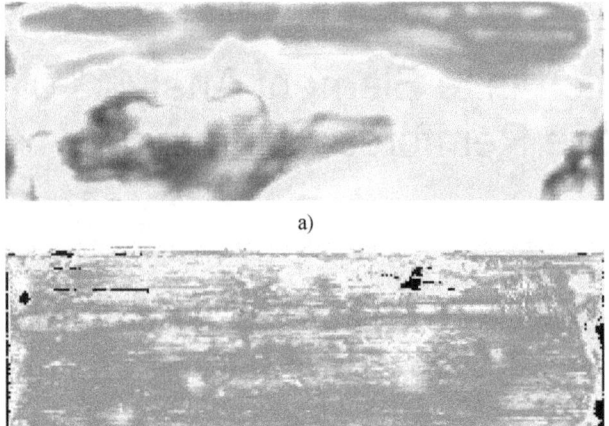

a)

b)

Figure 3. Different solidifying time vs. temperature curve of FMLs thermo-compression solidification process:
a) weak solidifying quality; b) good solidifying quality according to the process showed in Fig. 2.

C-scan, results of different process, green and blue-black area are defective region, most part of Fig. 3a, red and yellow area are the qualified solidifying portion of FMLs, most part of Fig. 3b.

Fig. 4 and Fig. 5 are the fractography analysis of FMLs with weak and good solidifying quality, respectively. In green and blue-black area, delaminating (Fig. 4b) and bubbles (Fig. 4c) may be exist.

The solidifying process will vary probably when the composed materials are different. It also needs to test expensively to gain the best results. So, reasonable test plan is a critical factor before a new style FMLs was studied. Here, finite element analysis also can play a key role in the research.

According to the ASTM E8/E8M-2008, the specimen is prepared as Fig. 6 showed. Loading mode is displacement load, speed is 2mm/sec., and material properties are list in the Table 1.

3. Finite element model

After thermo-compression solidifying process, each layer performance of FMLs is stable relatively. In this papers, Cohesive element of ABAQUS is used to simulate the bonding of each single layers, interface performance, the initiation and propagation of interface damage are defined by the parameter value of tensile-compression and shear orientation of cohesive element, the built-in criterions, which are quadratic exponent energy facture criterion and linear stiffness degradation criterion, are used to determine the bonding interface damage, but, the contact behaviour between the damaged interface is neglected for the reason that the test loading regime is quasi-static and monotonic. The reinforced fiber layers is simulated by continuous shell element, anisotropic attribute defined directly to the element, ABAQUS built-in criterions, which are Hashin criterion and linear stiffness degradation criterion, are used to determine the fiber damage. In the displacement loading mode, load is applied at reference point (RP) which has specified in advance, loading direction is along to the fiber orientation, and reacting

a)

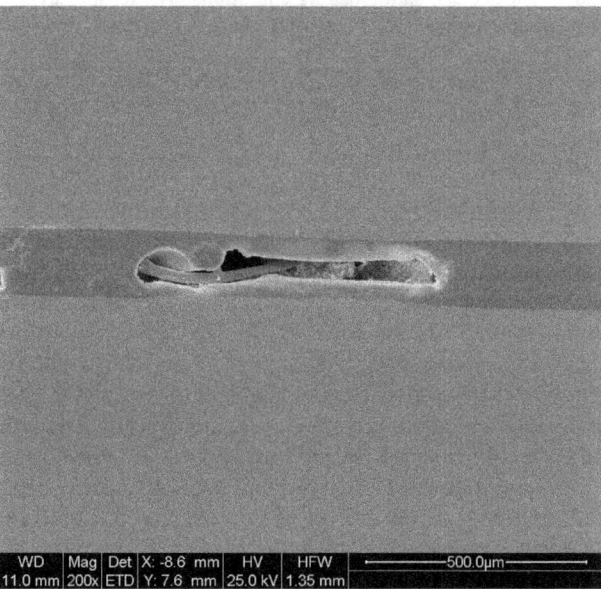

b)

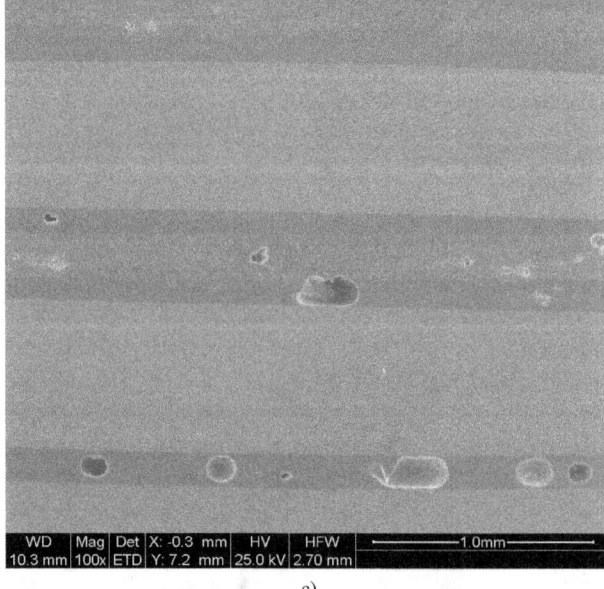

c)

Figure 4. Fractography analysis of weak solidifying quality:
a) incision surface; b) delaminating in 200X magnifying;
c) the composites bubbles in 100X magnifying.

force can output from RP directly, then, Mises stress is calculated by the RP reacting force.

a)

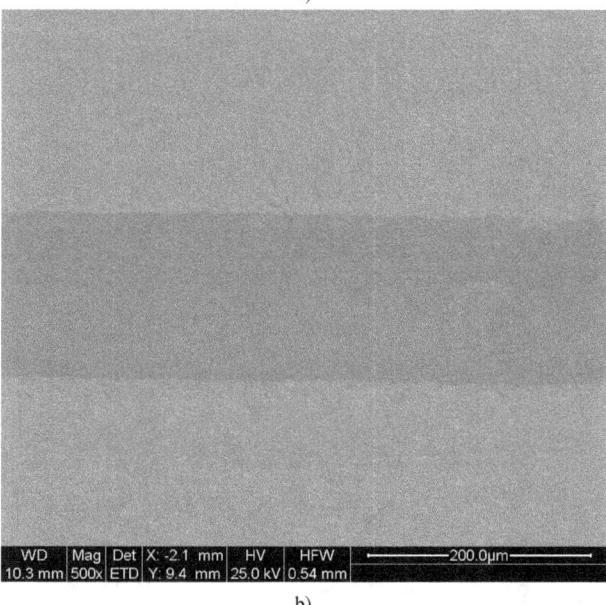

| WD | Mag | Det | X: -2.1 mm | HV | HFW | 200.0μm |
| 10.3 mm | 500x | ETD | Y: 9.4 mm | 25.0 kV | 0.54 mm | |

b)

Figure 5. Fractography analysis of good solidifying quality:
a) incision surface; b) fractography in 500X magnifying.

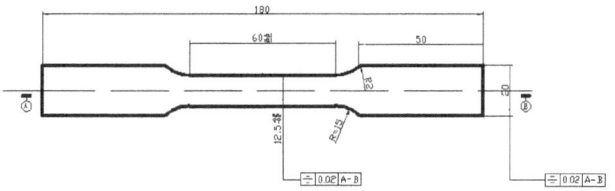

Figure 6. The configuration of tensile test specimen

Table 1
Material properties LY12M and T700 (Zhang Guoteng, ea al., 2009)

Materials	Tensile strength (MPa)	Elastic modulus (GPa)	Poisson's ratio
LY12M	220	70	0.3
T700	1831	99	0.311

The overall performances of FMLs are embodied by the interaction between each layer. So, three-dimensional (3D) finite element model must be setup to present interaction of each layer, moreover, because of the non-linear material attributes of metal, refining model is necessary to obtain convergence solution, and the amount of elements will increase sharply. So, it is necessary to use a simplified model with high accuracy to calculate.

Symmetrical specimen can be modelled in 1/2, 1/4 and 1/8 simplified finite element model. The value of RP reacting force is the total force of all nodes on the upper surface. When the model is simplified, the amount of

nodes will vary accordingly. The calculation results are list in the Table 2 to show the difference of tensile strength between these simplified FEA models. It showed that the 1/8 simplified model is a proper choice according to the consideration both in accuracy and computation efficiency.

One eighth of FMLs specimen is modelled by the method described before (Song Xin, 2008, 2010), as showed in Fig. 7. In the ABAQUS element library, linear hexahedron reduced integration elements, C3D8R (three dimensional eight nodded continuum element) was used to mesh the model of aluminium sheet and SC8R shell element was used to mesh the carbon fiber layer. The co-

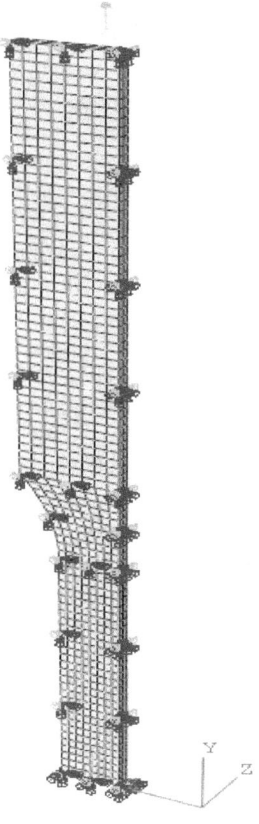

Figure 7. Simplified 3D finite element model of specimen.

Table 2
Comparison of tensile strength calculation results of simplified model to the whole model

| | Finite element calculation results of carbon fiber reinforced metal laminate | | | |
	whole	1/2	1/4	1/8
Tensile strength (MPa)	295	295	293	294
Error (%)	0	-0.047	-0.554	-0.301

Table 3
Property and damage model definition of cohesive element (MPa)

Elastic Property	Elastic Modulus in Normal	Elastic Modulus in Shear-1	Elastic Modulus in Shear-2
	13333MPa	5013MPa	5013 MPa
Quade Damage Initiation	Nominal Strain Normal-only Mode	Nominal Strain Shear-only Mode First Direction	Nominal Strain Normal-only Mode Second Direction
	0.003 mm	0.00798 mm	0.00798 mm

Table 4
Property and damage model definition of T700

Elastic Property (Type: Lamina) (MPa)		Damage Evolution (Type: Energy) (N/mm)		Hashin Damage Initiation (MPa)	
Longitudinal Tension Modulus	99220	Longitudinal Tension Fracture Energy	12.5	Longitudinal Tension Strength	1831
Longitudinal Compressive Modulus	97900	Longitudinal Compress Fracture Energy	12.5	Longitudinal Compress Strength	895
Shear Modulus in 12	5670	Transverse Tension Fracture Energy	1	Transverse Tension Strength	31
Shear Modulus in 13	5000	Transverse Compress Fracture Energy	1	Transverse Compress Strength	125
Shear Modulus in 23	5000			Longitudinal Shear Strength	72
				Transverse Shear Strength	72

hesive element is used to simulate the interlaminar behaviour of J272 middle-temperature-setting-glue; layer's thickness is 0.15mm, definition and damage model are showed in Table 3. T700 carbon fiber prepreg property definition and damage model are showed in Table 4, layer's thickness is 0.2 mm.

4. Analysis and validation

With room temperature, all specimens were tested on an electronic universal material testing machine, as showed in Fig. 8. FMLs test curve of stress-strain shows, in Fig. 9, like carbon fiber materials, an elastic performance. The ductility and cross-section reduction are smaller than the aluminium plate, showed in the Fig. 10. In the same loading conditions, strain of carbon fiber is far less than the strain of metal materials. Majority of load is endured by the carbon fiber before reached ultimate strength, a small part of load applied on the metal laminate with good adhesive conditions of each interface. When the fracture of carbon fiber occurred, the applied load is far beyond the tensile limits which the aluminium sheets can afford. So, the metal laminates ruptured rapidly before plastic deformation occurred.

Fig. 11 present the stress-strain curve calculated by finite element. Within the displacement loading of 0mm-0.25mm, elastic behaviour of specimen appears clearly; within the 0.25mm-0.5mm, a little yield phenomenon due to interact between aluminium laminate and carbon fiber appears not obviously. Fracture occurs when displacement beyond 0.5mm. The trend is agreed with experiment results.

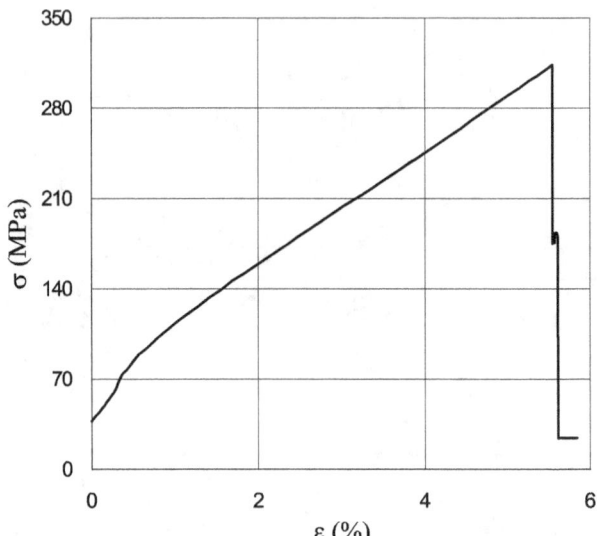

Figure 9. FMLs test curve of stress-strain

Figure 10. Fracture photo of two style specimen

Figure 8. Electronic universal material testing machine

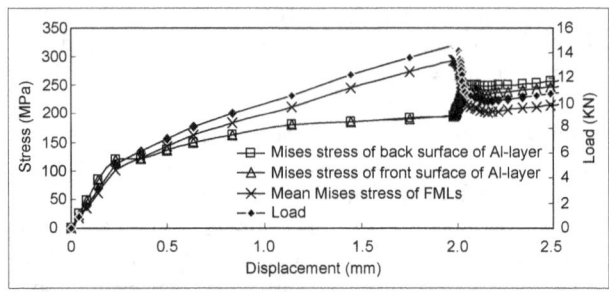

Figure 11. Calculation results of finite element model

Table 5
Tension properties comparisons

	Tensile strength (MPa)	Elastic modulus (GPa)	Specific elongation (%)
Experimental results	282	92	1.63
FEA calculation results	294	80	1.10
Error (%)	4.38	-13.03	-32.39

Table 6
Effect of tensile strength of aluminium laminate on the FMLs strength

Tensile strength of aluminium laminate (MPa)	264	317	370	423	476
Strength increment (%)	0	20	40	60	80
Variation of FMLs strength (%)	0	0	0	0	0

Another computation method of tensile strength is the expression derive from volume fraction theory of composites material, expressed as equation (1)

$$\sigma_{FMLs} = \mathrm{MVF} \cdot \sigma_{Metal} + \mathrm{FVF} \cdot \sigma_{Fibre} \qquad (1)$$

where MVF and FVF are represent the volume fraction of metal and fibre, respectively.

In Table 5, FEA calculation result of tensile strength larger than experimental results no more than about 5%, but, using equation (1), the calculation result is 354.51 MPa, lager than experimental results about 26%. So, FEA is a accurate method to assess the new style FMLs tensile strength.

The FEA result of elastic modulus and elongation are far different to experiment data. The main influences are: 1) the property dispersibility of composites is inevitable, even though in the same prepare process; 2) the value of simulation parameter in the finite element definition step need to optimize further up; 3) there are large and unavoidable measure error in the operation of tensile experiment when the specific elongation is calibrated by hand.

Furthermore, the effects of ultimate strength of metal laminate on the whole tensile strength of FMLs are studied by the 3D elastic-plastic finite element model described in section 3. The results are list in the Table 6. It shows that the strength of metal have a little influence on the FMLs strength.

5. Concluding remarks

Experiments validated that FMLs present elastic performance because of most load during quasi-static tensile loading is sustained by the reinforced fiber which have large elastic modulus. Calculated by the 3D elastic-plastic finite element model, a fairly accurate result of FMLs tensile strength obtained. It is an efficient method, not only in the new style FMLs tensile properties study, but also can be used to study the fracture mechanism of FMLs.

In the paper, conclude can be deduced that the effect of tensile strength of aluminium alloy on the whole tensile strength of FMLs is very small. For taking full advantage of super strength of reinforced fiber, in the new style FMLs design, elongation of materials should be considered that the elongation of metal must lager than the ultimate value of reinforced fiber.

References

Binienda, W.K. and Qiao, P. (2005). Advanced materials and structures: Analysis methods and results, *Journal of Aerospace Engineering*, Vol. 18, No 1, pp.1-2.

Boscolo, M., Zhang, X. and Allegri, G., (2008). Design and modelling of selective reinforcements to improve fail safety in integral aircraft structures, *AIAA Journal*, Vol. 46, pp. 2323-2331.

Cao, C. X. (2007). *New Materials Industry*, Vol.10, pp. 9-13.

Cao, Z. Q. (2006). *Aviation Manufacture Technology*, Vol. 8, pp. 60-62.

Grassi, M. and Zhang, X. (2003). Finite element analyses of mode I interlaminar delamination in z-fibre reinforced composite laminates, *Composites Science and Technology*, Vol. 63, pp. 1815-1832.

Liao, Y.-Q., Su, J.-H. and Ke, S.-L. (2006). Application of ANSYS in composite simulation analysis. *Fiber Composites*, Vol. 23, No 4, pp.63-66.

Selective Reinforce Panel. (2005). *International Committee on Aircraft Fatigue (ICAF)*, June, Hamburg, Germany, pp. 6-10.

Song, X. and Zhang, J. (2008). Analysis of a finite element model of 2D elastic-plastic crack, *Journal of Harbin University of Science and Technology*, Vol.13, No 5, pp.9-13.

Song, X., Liu, Z. and Zhang, J. (2010). Simulation study of fracture behavior in a FMLs specimen by 3D elastic plastic finite element method, *Thin Films 2010 and Compo 2010*, Harbin, China, June 11-14, pp. 1472-1476.

Wu, X. and Guo, Y. (1999). Fatigue life prediction of fiber reinforced metal laminates under variable amplitude loading. *Engineering Science*, Vol.1, No 3, pp. 36-39.

Young, J. B., Landry, J. G. N. and Cavoulacos, V. N. (1994). Crack growth and residual strength characteristics of two grades of glass-reinforced aluminium "Glare", *Composite Structures*, Vol. 27, pp. 457-469.

Zhang, G., Chen, W. and Yang, B. (2009). Testing research on mechanical properties of T700 carbon fiber/epoxy composites. *Fiber Composites*, Vol.49, No 2, pp. 49-52.

Modelling the Influence of Properties of a Matrix on the Strength of Unidirectional Fibrous Composite Materials

Peter Mikheev[1], Alexandre Berlin[2]

[1] Bauman Moscow State Technical University, 5 2nd Baumanskaya Street, Moscow, 105005, Russia
[2] Semenov Institute of Chemical Physics, Russian Academy of Sciences, 4 Kosygina Street, Moscow, 117977, Russia

Abstract: In this article a theoretical approach allowing prediction of deformation of fibrous or layered composite materials with changing of properties of fibres and matrices, taking into account the actual distribution of strength of fibres is proposed. The main difference of the proposed approach from the earlier ones is an assessment of the distribution of deformation to failure of fibres, instead of distribution of strength is used. The proposed approach allows consideration of the deformation behavious of fibres, distribution of their properties, properties of a polymeric matrix, and gives a method of graphically finding the failure deformation of the unidirectional composite material. The proposed model allows prediction of a change of deformation to failure of unidirectional composite materials on the basis of the chart of the deformation of the polymer matrix. It make also it possible to take into account temperature effects via temperature dependences of the modulus of elasticity, a yield stress limit and limiting deformation of the polymer matrix. Possible changes of the deformation properties of unidirectional composite materials were demonstrated for a few examples - for several types of epoxy matrix, and for a number of reinforcing fibres (carbon fibres and aramid fibres).

Key Words: Composite materials, Fibers, Deformation, Stress concentration, Elasticity module, Distribution of strength.

1. Introduction

In the creation of a unidirectional composite it is necessary to consider many factors. On the one hand, properties of a polymeric matrix shall be taken into account. On the other hand, distribution of strength of the reinforcing fibres, caused by features of their manufacturing process, shall be considered.

According to a known model of damage accumulation , proposed by Rosen, for the unidirectional fibrous composite materials, strength of the unidirectional composite can be estimated via strength of a bunch of reinforcing fibres whose length equals to doubled the critical (ineffective length) of δ (Nemez and Strelyaev, 1970) (Fig. 1). The ineffective length δ corresponds to the zone of stress concentration near the end of the fibre.

Weibull's two-parametrical distribution is usually applied to an assessment (Fudzy and Dzako, 1982) of distribution of strength:

$$P(\sigma, L) = 1 - \exp\left[-L\left(\frac{\sigma}{\sigma_0}\right)^m \right] \qquad (1)$$

where m and σ_0 are the parameters, characterizing the fibre strength distribution, L is fibre length.

For a constant value of modulus of elasticity of fibres, the strength σ, is unequivocally connected with deformation to failure, ε.

$$P(\varepsilon, L) = 1 - \exp\left[-L\left(\frac{\varepsilon}{\varepsilon_0}\right)^m \right] \qquad (2)$$

In this case the deformation to failure of the fibres is proportional to the deformation of the material.

Then limiting deformation of a bunch of fibres will depend on fibre length L according to a formula:

$$Ln(\varepsilon) = \text{const} - \frac{Ln(L)}{m} \qquad (3)$$

According to Rosen's representation, strength of a fibrous composite material at extension it is equal to strength of a bunch of fibres of the ineffective length δ, in the same way it is possible to estimate the ultimate deformation of the composite material.

However critical length, δ, characterizing nonuniformity of the fields of tensions and deformations around the broken fibre, changes in process of static deformation of a material at its loading. It grows, from elastic deformation to plastic deformation, and then tends to infinity when delamination occurs from shear deformations.

In the elastic case, δ weakly depends on the material deformation, while during plastic deformation δ increases with deformation according to a linear relationship.

In static deformation of the unidirectional composite material, it is difficult to define whether the, elastic or plastic ineffective length δ should be used because strength of fibres also depends on length of specimen.

Use of deformation to failure instead of strength of fibres gives an opportunity to connect on one chart the deformation properties matrix (the elastic modulus, a limit of plasticity and limiting deformation) and the statistical distribution of the deformation to failure of the reinforcing fibres.

2. Estimates of inefficient length

Plastification of the polymer matrix or temperature change leads to changing of a few parameters simultaneously - i.e. the elastic modulus, E, the limit of plasticity σ and the deformation to failure, ε, of a polymeric matrix.

To consider how these changes influence δ, in advance it is very difficult for all combinations of reinforcing fibres and matrix.

For an approximate assessment of deformation to failure and the maximum strength of the unidirectional composite, it is more convenient, not to use a fixed value δ, depending on properties of polymer matrix and fibres, but a dependence of δ upon deformation of the composite material.

In our model we adopt basic provisions of Rosen's model of accumulation of damage and assume, that de-

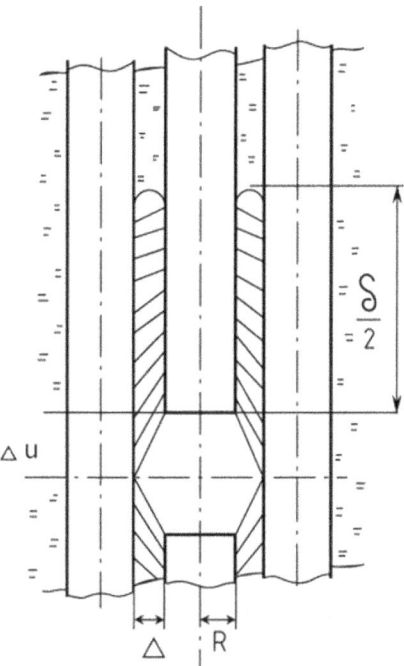

Figure 1. Scheme of "ineffective length" model.

pendence of $Ln(\varepsilon)$ from $Ln(L)$ can be defined separately and this dependence does not depend on properties of the polymer matrix.

In this work it is also supposed that the modulus of elasticity of fibres is constant, deformation of fibres is linear before destruction, and the distribution of strength is connected only with statistical distribution of deformation to failure of the fibres. The assumption of linearity of deformation before failure is valid for high-strength reinforcing fibres (glass, carbon, aramide).

For experimental determination of dependence of $Ln(\varepsilon)/Ln(L)$ or $Ln(\delta)$ turns out a settlement way on the basis of easily defined dependence of strength of fibres on length of $Ln(\sigma)/Ln(L)$ or $Ln(\delta)$, by dividing on the of values of strength into value of the modulus of elasticity of the fibres, known of literature or defined in separate experiments.

It is known (Argon, 1978) that if polymer matrix about the broken fibre it is deformed the conditions are elastic, and the ineffective length does not depend on deformation of fibres (Area I in Fig. 2). One of the possible formulas for calculation of inefficient length was suggested by Argon (Argon, 1978) and by Rosen. It is shown below

$$\frac{\delta}{d_f} = 1.15 \left[\frac{1 - k_f^{1/2}}{k_f^{1/2}} \left(\frac{E_f}{G_m} \right)^{1/2} \right] \qquad (4)$$

where d_f is diameter of fibre, E is the the modulus of elasticity of fibre, G is the shear modulus of the matrix, k_f is the parameter depending on a volume fraction of fibres in a material.

At further deformation of the matrix about an end face of the broken fibre it will be deformed as plastic material. In the a limiting case of ideally plastic behaviour of a matrix the ineffective length can be described by a formula (Kelly and Tyson, 1965) shown below in equation 5.

$$\delta = \frac{R_f}{\alpha} \frac{E_f}{\tau_{12}^M} \varepsilon_{11}^f \qquad (5)$$

where τ is a yield stress limit of matrix in shear, ε is lengthening in reinforcing fibres, α is the fibre fraction of the composite.

In this case dependence of the ineffective length δ depends upon deformation during static loading of the unidirectional composite materials.

In Fig. 2 there are areas of plastic deformation of matrix between fibres, are shown by a straight line, marked II, with a slope of less than 45°, and elastic deformation corresponding to formula (1) indicated by a vertical straight line.

As the deformation is increased the ultimate deformation of the matrix in shear will be reached and connection between the broken fibre and matrix will fail..

It will be shown below that delamination in a thin film of the matrix will begin even at a lower deformation of a polymer matrix in a composite than in neat matrix, tested separately.

There are formulae for deformation assessment in reinforcing fibres at which there is possible a delamination between fibres and polymer matrix.

For this purpose it is possible to use "the shear analysis", that is to assume that the polymer matrix was only sheared between fibres, and fibres are solely in extension.

Considering that shear stress in zones of the plasticity which has arisen about an end edge of cracked fibre, is approximately constant and equal. The balance condition near break may be written as follows:

$$\frac{\partial^2 U}{\partial x^2} = \frac{2\tau_{12}^M \alpha}{E_{11}^f R_f} \qquad (6)$$

where α is a share of perimeter of the fibre, cooperating with the next fibres.

Solving this equation under the condition of zero stretching tension on fibre in the gap plane, including a zone of plasticity with length equal to δ, it is possible to calculate shear deformation near the end fibres face.

Using an energy approach, it is possible to estimate extension of fibres ε_{11} at which the composite will be stratified for all lengths of fibre. For this purpose it is nec-

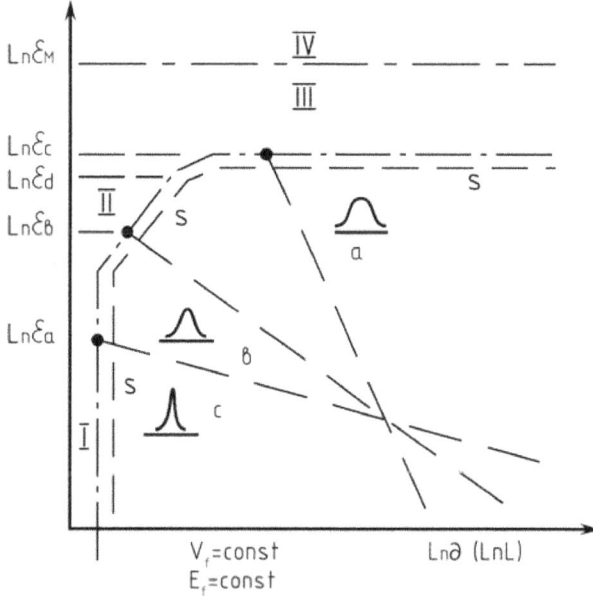

Figure 2. The incorporated dependence of ineffective length δ on deformation at static loading.

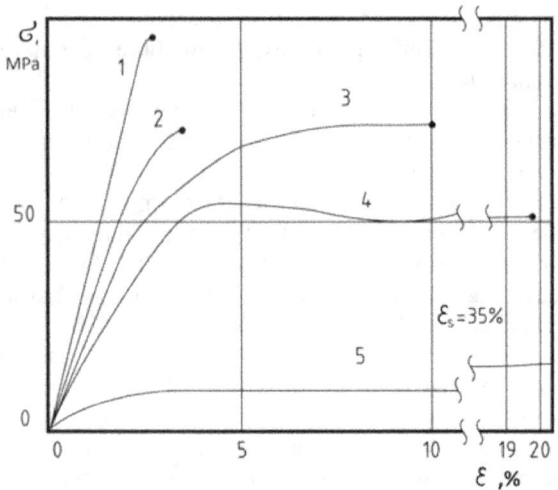

Figure 3. Curve deformations of the homogeneous epoxy resins.
1 – EKT-15E; 2 – EDT-10; 3 – EDT-10 +10% DEG;
4 – EDT-10 +20% DEG; 5 – EDT-10 +40% DEG

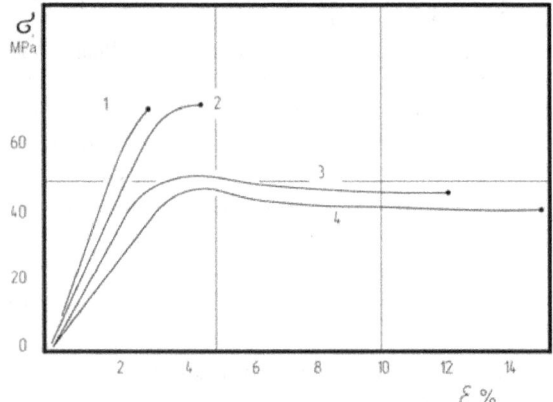

Figure 4. Curve deformations of the epoxy resins with rubber.
1 – EDT-10 +1% rubber; 2 – EDT-10 +3% rubber;
3 – EDT-10 +7% rubber

essary, that the elastic energy stored in the fibres exceeded the energy necessary for stratification.

$$\varepsilon_{11} \geq 2 \frac{\gamma^{1/2} \left(m, f, s, \sigma_0, \sigma_m \right) \alpha^{1/2}}{\left(E_{11}^f R_f \right)^{1/2}} \qquad (7)$$

where γ is work of destruction of a layer matrix.

High-strength fibres have quite high deformation to failure (up to 4.5%) and this can surpass the deformation

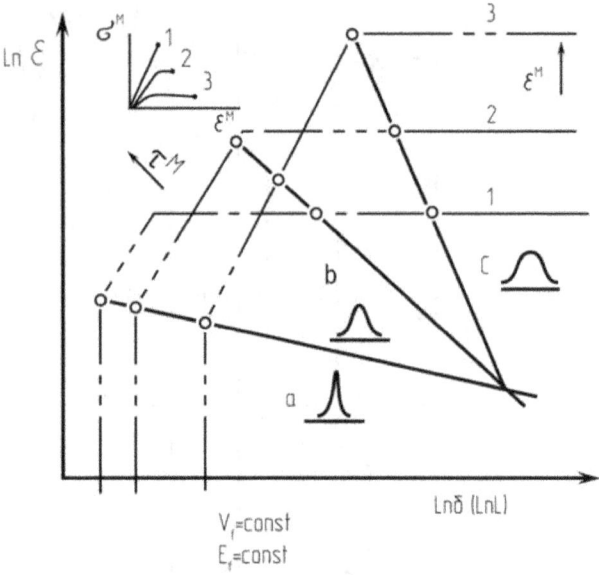

Figure 5. The scheme of influence of deficiency of fibres on limiting deformation of a composite.

to failure of the matrix, and then is possible there will be situations when the polymeric matrix will failure due to the tensile load before the fibres.

In Fig. 2 the zone of possible failure from the – tensile load is shown above the straight line, IV.

3. Forecast of influence of matrix properties

However at creation of a composite material from same matrixes, can be used fibers by different presence of defects, (that is reflected by various values of parameters σ_0 and m in Weybull's distribution).

Distinction in deficiency of fibers leads to various inclinations of schedules of dependences of a logarithm of strength or a logarithm of limiting deformation from a logarithm of fibres length.

In Fig. 2 it is schematically presented, as limiting deformation of a composite material will change at change properties of fibers.

a, b and c are examples of dependences of fibers with different deficiency but the identical module of elasticity. The disorder of strength of fibers increases from an example – c for example – a.

Table 1
Properties of the studied epoxy matrix

	Elasticity module	Plasticity limit	Proportionality limit	Proportionality limit	Limiting deformation
	GPa	MPa	MPa	%	%
EKT-15E	3.60	90	90	2.5%	2.5%
EDT-10	2.50	75	50	2.0%	3.5%
EDT -10+ 10% DEG	1.60	75	40	2.5%	10.0%
EDT -10+ 20% DEG	1.20	60	30	2.5%	19.0%
EDT -10+40% DEG	1.00	1.5	10	1.0%	35.0%
EDT -10+ 1% rubber	3.00	75	60	2.0%	3.0%
EDT -10+ 3% rubber	2.00	75	60	3.0%	4.5%
EDT -10+ 7% rubber	2.00	50	40	2.0%	12.0%
EDT -10+ 10% rubber	1.33	45	40	3.0%	16.0%

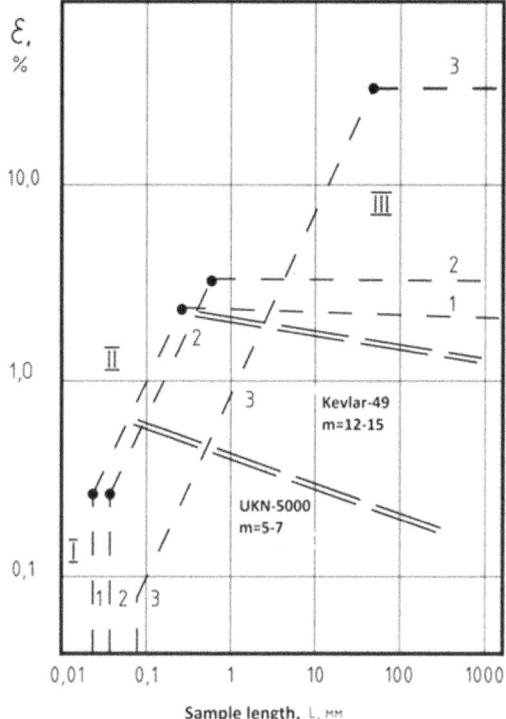

ε, %

10,0

1,0

0,1

0,01   0,1   1   10   100   1000

Sample length, L, мм

Kevlar-49
m=12-15

UKN-5000
m=5-7

Figure 6. Example of application of model for real materials.

Then dependence of strength of fibers on length turns into a straight line. In Fig. 2 to critical length δ in elastic area of deformation of the matrix there corresponds a straight line parallel to an axis of ordinates of $Ln(\varepsilon)$ (Area I), ineffective length in plastic area corresponds a straight line under an inclination 45° to axes (Area II). In the case of brittle matrix section II may be entirely absent.

To delamination of matrix (to achievement of its limiting deformation) there corresponds a straight line parallel to the $Ln(L)$ axis (Area III). Besides to matrix destruction before destruction of the weakest fiber corresponded a straight line parallel to the $Ln(L)$ axis (Area IV).

I – area of elastic deformation of the polymer matrix; II – area of shift plastic deformation of polymer matrix, III – delamination area at a fibre end face; IV – area of destruction matrix from tension.

S-curves demonstrate the dependences at the accounting of surface effects. The ineffective length δ was changed, as fibres on a surface have smaller number of the neighbouring fibres.

Table 2
Examples calculations of limiting deformation of a composite

	Kevlar-49	Terlon	UKN-5000
	Limiting deformation, %	Limiting deformation, %	Limiting deformation, %
EKT-15Э	2.8%	2.8%	1.3%
EDT-10	3.0%	2.9%	1.2%
EDT-10 + SKN (rubber)	3.1%	3.2%	-
EDT-10 + DEG	-	2.5%	-

On such chart it is possible to estimate deformation to failure ε and to predict the possible mechanism of failure of this composite material based on separately defined properties of the polymer matrix (E_m, σ_m, ε_m) and fibres.

From the value of the failure deformation it is possible to calculate strength of a composite, knowing the modulus of elasticity and volume fraction of fibres.

4. Cases of real materials

This approach is now applied to real materials. In Fig. 3 and 4 the stress-stain behavious the various epoxy matrix are presented.

In Fig. 3 the stress-strain behavior of an epoxy resin, plasticized by DEG (Diethelenglicol) which is chemically connected with epoxy is presented.

Chart 1 in Fig. 3 corresponds to a rigid epoxy resin EKT-15E, charts 2, 3, 4, 5 differ with a different amount of softener (DEG) in the epoxy matrix.

In Fig. 4 the stress-strain behavious of a heterogeneous epoxy resin in which rubber is added is presented. The proportion of rubber increases from chart 1 to chart 4.

In Table 1 values of the mechanical properties of the polymer matrix investigated in this work are presented.

Values of moduluss and strain to failure of the resin can be used to calculate ineffective lengths δ as in elastic, so in plastic cases.

Using formulae (1) (2) and (3) for dependences of ineffective length δ on static deformation of the unidirectional composite, the combined curves of ineffective length δ for different resins are presented in Fig. 5.

Curves for polymer matrixes were calculated on the basis of formulas of ineffective length δ in an elastic case (1) and formulas in a plastic case (2) and experimental values of the modulus of elasticity of E_m, limit of plasticity of the matrix σ_m and failure strain of the matrix ε_m.

The change of mechanical properties of the polymer matrix at plastification will lead to change of dependences of ineffective length δ, shown schematically on Fig. 5.

Depending on efficiency of fibres plastification of a polymer matrix it is possible to observe a decrease in lim strain to failure in case of low-defective fibres (A) or to lead to increase in strain to failure in case of defective fibres (C).

As example fibres Kevlar - 49 and UKN-5000 (Analogue of AS-4 carbon fibres) are presented.

Dependence of the strain to failure of the fibres on length was compared with strength impregnated fibres.

Dependence of the strain to failure of fibres on their length was found by calculation from dependence of fibre strength upon length divided by modulus of elasticity (120 GPa for Kevlar-49) and 220 GPa for UKN-5000 (Zelensky et al., 2001).

For manufacturing of the impregnated fibres for the experiment polymer matrices with the same proportion of softener or rubber, as in Table 1 were used (Micheev and Berlin, 1990). The stress-strain behavior of the these matrces are presented in Fig. 3 and 4 .

Samples of Kevlar contained either 300 or 1000 single fibres, samples of UKN-5000 contained 5000 single fibres. The volume fraction of fibres was 50% in both cases.

The projection of points of intersection of curves (δ from ε) were used to obtain the dependences of ε from L for given values of strain to failure ε.

In Fig. 6 the dependences of ineffective length δ on deformation of composites for Kevlar 49 and UKN 5000 are presented. Calculations are carried out using formulae 4 and 5 on the basis of data on the deformation of the matrix (Fig. 3).

Chart 1 was calculated for EKT-15E, chart 2 for EDT-10, chart 3 for EDT-10 30 % of DEG The ineffective length δ changes in a range from 0.1 mm to 0.8 mm.

In Table 2 results from interpretation of the experimental data are presented.

For carbon fibres different small values of the parameter m limit the deformation of plastics and for aramide fibres with a larger value of m the strain to failure increases, as the approximate model presented here predicts.

5.Conclusions

The proposed model of interaction of a polymeric matrix with the high-strength fibres with distribution of strength and strain to failure, gives an opportunity to determine the strain to failure of a unidirectional composite material.

On the one graph is combined the dependence of strain to failure of a bunch of fibres on length, and dependence of ineffective length δ with different static deformationsl. Dependence of the strain to failure of fibres upon their length can be easily defined on the basis of dependence of strength upon length of fibres by dividing on the elastic modulus E.

If the distribution of diameters of fibres and variation in elastic modulus can be neglected, strength of the unidirectional composite material at extension may be calculated.

The proposed model allows the effects of plastification of polymer matrix to be taken into account, and also the influence of ambient temperature on the failure properties of a material as usually properties of the polymer matrix which is usually stronger, than the fibres, depend on temperature.

There is a possibility of further specification and development of model for the purpose of receiving exact results for optimization of properties of a material on the basis of data on properties polymer matrix and reinforcing fibres tested separately.

References

Argon, A. (1978). Statistical aspects of destruction, in *Composite Materials: Destruction and Fatigue.* Moscow: Mir, pp. 201-205. (in Russian)

Fudzy, T. and Dzako, M. (1982). *Mechanics of Destruction of Composite Materials*, Moscow: Mir. (in Russian)

Kelly, A. and Tyson, W. R. (1965). Tensile properties of fibre reinforced metals: Copper/tungsten and copper/molybdenum, *Journal of the Mechanics and Physics of Solids*, Vol. 13, pp. 329-350.

Mikheev, P.V. and Berlin A.A. (1990). Empiric model of influence of properties of a polymeric matrix on strength and limiting deformation of unidirectional composites at stretching in the direction of fibers, in *Proceedings of the First Moscow International Conference on Composites*, November 14-16, Part 1, p. 232. (in Russian)

Nemez, A.C. and Strelyaev, B.C. (1970). *Strength of Plastics.* Mocsow: Mashinostroenie. (in Russian)

Zelensky, E.S., Kuperman, A.M, Gorbatkina, Yu.A., Ivanov-Mumzhiyeva V.G. and Berlin A.A. (2001). The reinforced plasticity of modern constructional materials, *Journal of Russian Chemical Society*, Vol. XLV, No 2. (in Russian)

Preparation of Cf/HfC Composites by Instantaneous Liquid Infiltration at Low Temperature Based on Alloy Design and its Ablation Property

Yicong Ye, Hong Zhang, Li'an Zhu, Yonggang Tong, Ke Chen, Shuxin Bai

College of Aerospace and Materials Engineering, National University of Defence Technology, Changsha, 410073, P.R.China

Abstract: A 50Hf10Zr40Si3Ta alloy, melting point of which is around 2440°C, is designed and used to prepare a carbon fiber reinforced HfC-based composite (C_f/HfC) by instantaneous liquid infiltration (ILI) method at 1900°C. It is proposed that the primary reason for triggering the ILI process at the temperature much lower than the alloy melting point is that the Hf component with the strongest carbide-forming ability prefers to react with the carbon in the surface layer of the C/C preform, leading to a phase composition change at the interface and generating instantaneous liquid phases. The composite exhibits excellent oxidation-resistant and ablation-resistant properties because of the behaviour of various oxides formed during the ablation process.

Key Words: Alloy design, C_f/HfC composites, Instantaneous liquid infiltration, Ablation property.

1. Introduction

Because of combination of high mechanical performance of carbon/carbon (C/C) composites at elevated temperatures and excellent anti-ablation property of ultrahigh temperature ceramics (UHTCs) (Levine et al., 2002, 2004; Scatteia et al., 2005; Wuchina et al., 2007; Schwab et al., 2004), the carbon fiber reinforced UHTCs-based composites (C_f/UHTCs) have been widely studied in recent years. For example, C/ZrC composite nozzle developed by Ultramet Company possesses excellent anti-ablation performance at high temperatures (Zou et al., 2010). Different kinds of C_f/UHTCs of the carbon fiber reinforced HfC-based composites (C_f/HfC) have attracted people's attention due to the very high melting point of HfC and its anti-ablation property (Ohlhorst and Vaughn, 2003). Reactive melt infiltration (RMI) has been a preferred method for preparing C_f/UHTCs because of the short preparation period and low cost (He et al., 2006; Krenkel, 2001; Jayaseelan et al., 2011). However, it is unfeasible to prepare C_f/HfC by RMI because the melting point of pure hafnium is as high as 2230°C and at this temperature the mechanical property of C/C composite will be weakened.

Different with the reactive melt infiltration method, by which the metal or alloy has to be melted at a very high temperature and then infiltrates into C/C preform, our research group prepared the C_f/HfC composite by instantaneous liquid infiltration (ILI) at a relative low temperature below the alloy melting point. Based on alloy design, the alloy reacts with the carbon in the surface layer of C/C preform, generating Hf-rich liquid which infiltrates into C/C preform below alloy melting point. The infiltrated Hf-rich liquid reacts with the pyrolytic carbon, forming HfC-based carbides of very high melting points, thus the C_f/HfC composite is obtained. This paper is to discuss the process of ILI method for preparing C_f/HfC composite and the ablation property of the obtained composite.

2. Experiment

2.1 Preparation of C_f/HfC composite

A 50Hf10Zr37Si3Ta alloy ingot was prepared in a vacuum arc furnace with pure hafnium (purity ≥ 99.4%), zirconium (purity ≥ 99.9%) and silicon (purity ≥ 99.8%). PAN-based carbon fiber (T300, Toray, Japan) needled felts preforms were used as reinforcements, initial density of which is 1.28g/cm³. The needled felts were prepared by the three-dimensional needling technique, starting with repeatedly overlapping the layers of 0° non-woven fiber cloth, short-cut-fiber web, and 90°non-woven fiber cloth with needle-punching step by step. Pyrolytic carbon (PyC) was then deposited on the surface of the carbon fibers by chemical vapour infiltration process.

The carbon/carbon preform (C/C preform) and the alloy ingot were put into a graphite crucible, which was then placed in a carbon tube furnace and heated up to 1900°C, holding for 30 mins, followed by furnace cooling (Fig. 1). ILI process took place in the meanwhile. The temperature changes as the curve in Fig. 2.

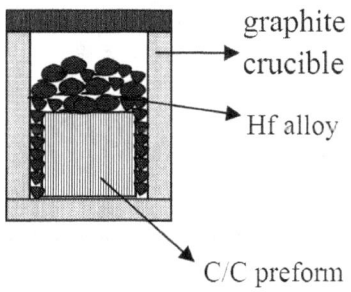

Figure1. Sketch of the instantaneous liquid infiltration experiment

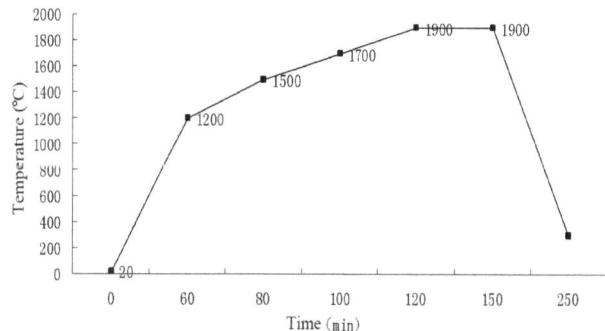

Figure 2. Temperature change of the carbon tube furnace

Figure 3. The acetylene flame test

2.2 Test and analysis of Cf/HfC composite

The instantaneous liquid infiltration (ILI) sample were obtained and examined by X-ray diffraction (XRD), field emission scanning electron microscopy (SEM), and energy dispersive spectroscopy (EDS) to determine its phase composition, constituent and microstructure.

Acetylene flame (Fig. 3) is applied to test the anti-ablation performance of the Cf/HfC composite at high temperature. The temperature of the sample as tested is around 1800°C. The test is lasted for 10 min.

3. Results and discussion

3.1 Alloy design and preparation

The melting point of hafnium is 2230°C. It is possible to design a Hf alloy, which is suitable for instantaneous liquid infiltration (ILI) process for preparing HfC-based Cf/UHTCs. The composition of Hf alloy is designed as follows.

Multiple phase ultrahigh ceramics possess excellent anti-oxidation and ablation properties. Ceramic composed of carbide, boride and silicide of hafnium, zirconium, silicon and tantalum convert to their oxide after high temperature oxidation and ablation, the behaviour of which is similar with ceramic composed of only carbides. Thus it is reasonable to design our Hf alloy basing on the relative mature ultrahigh temperature ceramic system.

Multiple phase ceramic $HfB_2(ZrB_2)$-20%SiC-20%$TaSi_2$ prepared by Levine and Opila (2003) by hot pressed sintering kept undamaged at 1627°C for a long duration. The anti-ablation property of the ceramic comes from the anti-oxidation property of oxides of different metals. Oxides of alloy elements play different roles during the ablation process. HfO_2, the melting point of which is 2810°C, is the main ultrahigh temperature phase and can bear scouring of high temperature gas flow. ZrO_2, the melting point of which is 2700°C, is also an important ultrahigh temperature phase and can form solid solution of unlimited mutual solubility with HfO_2. SiO_2, the melting point of which is 1728°C, converts to melt of certain viscosity at high temperatures. The melt of SiO_2 possesses good infiltrating performance, very low oxygen permeability and can spread out and form an intact silicate film on the surface of the matrix. This film will heal cracks formed due to thermal mismatch and pores generating as gas escape, thus make a good barrier layer of oxygen. Ta_2O_5, the melting point of which is 1872°C, can improve the solubility of HfO_2 in melt at high temperatures. Besides, tantalum element enters into the crystal lattice of HfO_2, which decrease the oxygen permeability of HfO_2.

Therefore, it is decided that the elements of the designed alloy includes Hf, Zr, Si and Ta with a mole ratio of 50:10:37:3. The Hf content is not less than 50% because the Cf/HfC composite is expected to be applied in higher temperature environment. The silicon and tantalum content should not be too high, because SiO_2 and Ta_2O_5 evaporate at very high temperature, which will do harm to the anti-ablation property of the composite. Zirconium is added into the alloy in order to decrease the melting point of the alloy and reduce the alloy density.

3.2 Structure of the ILI sample

The density of the as-received composite increased to 1.88 g/cm^3 from 1.26 g/cm^3 of the C-C perform, and the open porosity decreased to around 10% from 35% of the perform.

After the ILI process, it is seemed that part of the alloy ingot adjacent to the surface of the C-C preform underwent melt. The melting alloy flew to the side face from the top of the sample and covered the whole surface of the preform, leaving a piece of unmelted alloy on top of the preform. The cross section of the sample was observed

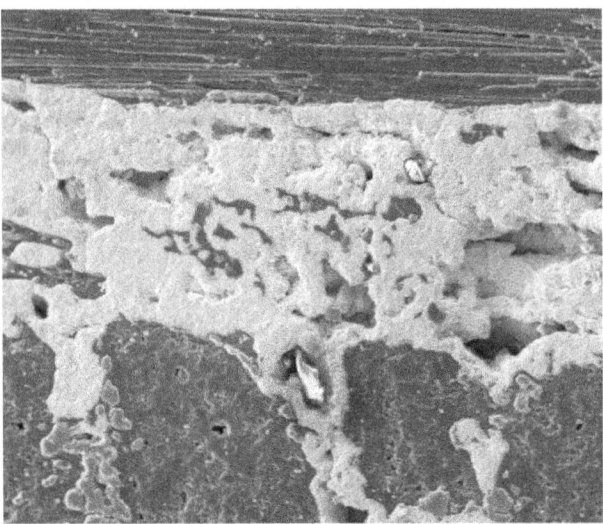

Figure 5. Cross section of the Cf/HfC composite

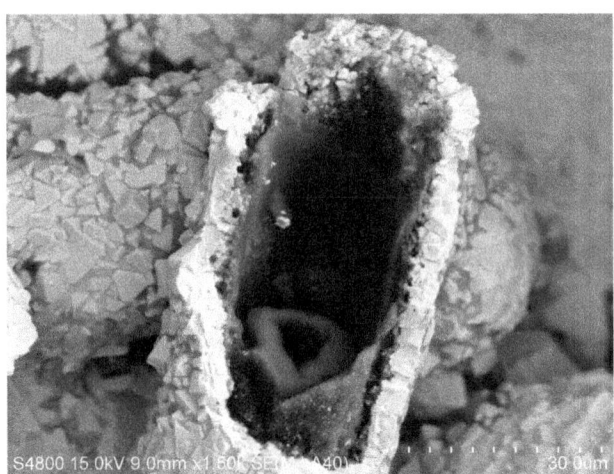

Figure 6. Fracture micrograph of the Cf/HfC composite

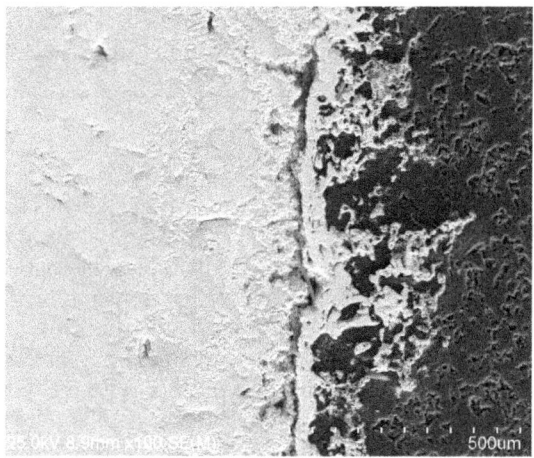

Figure 7. Interface between the alloy ingot and the graphite crucible

under SEM (Fig. 5). The fracture micrograph (Fig.6) shows an obvious characteristic of the ILI structure, i.e., there are layers surrounding the carbon fibers. It is indicated that the white fine particles are either HfC or ZrC, the medium size and large size white particles are either HfC or ZrC, and the dark substance is SiC. No dissociative component was found. A C_f/HfC composite is obtained.

3.3 Mechanism of instantaneous liquid infiltration

The 50Hf10Zr37Si4Ta alloy, whose theoretic melting point is 2440°C, does not melt at 1900°C, which is also demonstrated by the unmelted block of alloy left on top of the preform. However, it is indicated by the microstructure analysis of the cross section of the sample that the infiltration process definitely occurred. It is supposed that instantaneous liquid generated and infiltrated into the preform. A proving test was conducted as follows. An alloy ingot was placed in a graphite crucible and was heated up to 1900°C followed by furnace cooling. The alloy ingot contacted with the bottom of the crucible, giving a chance for the alloy to react with carbon. It was found that the alloy contacting with the crucible bottom appeared completely different from its initial state. The alloy had infiltrated into the graphite with a depth of around 250μm. XRD was then conducted to determine the phase composition of the infiltrated alloy. The main phase of the infiltrated alloy is HfC, the subordinate phases are $HfSi_2$, $ZrSi_2$ and HfSi, and a very small amount of Hf_5Si_3 also exists. The initial phases of the alloy consist of $(Hf,Zr)_2Si$ and $(Hf, Zr)_5Si_3$. Apparently, the HfC, $HfSi_2$, $ZrSi_2$ and

Figure 9. Micro-morphology of the film on the surface of the sample after ablation: (a) low magnification; (b) high magnification

HfSi phases were generated from the interface reactions, while the minor Hf_5Si_3 was the residual nonequilibrium phase of the initial ingot, which did not take part in the reactions.

This phenomenon may be due to carbonization reactions occurring between the alloy and the carbon in the surface layer of the preform. During the carbonization reactions, because of the very strong carbide-forming ability of Hf, perhaps none but the Hf component took part in the reactions, leading to a phase composition change of the alloy at the interface. An instantaneous liquid phase generated from the phase transformation and infiltrated into the C/C preform, thus the C_f/HfC composite was obtained. More detailed research and calculation is still needed.

3.4 Ablation property

It is seen from Fig. 8 that after ablation a relative compact film is formed on the surface of the sample, protecting the

Figure 8. The composite sample after the acetylene flame test for 10 min

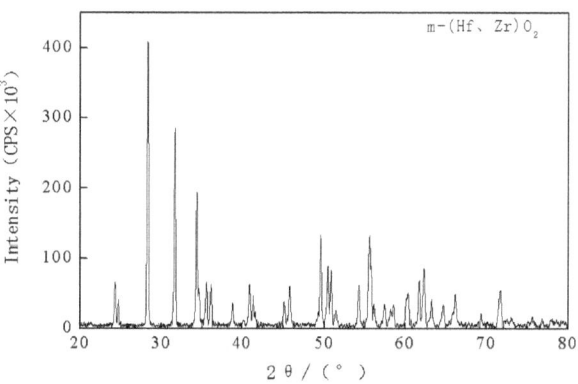

Figure 10. XRD analysis result of the film on the surface of the sample after ablation

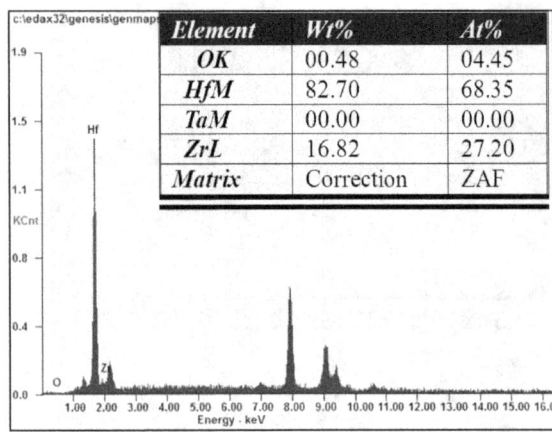

Element	Wt%	At%
OK	00.48	04.45
HfM	82.70	68.35
TaM	00.00	00.00
ZrL	16.82	27.20
Matrix	Correction	ZAF

Figure 11. EDS results of the film on the surface of the sample after ablation

sample from serious ablation. Micro-morphology of the film is shown in Fig. 9. The film is compact and uniform, few cracks can be seen and micro-crack does not extend because the film tends to convert to glass phase. X-ray analysis (Fig. 10) and EDS results (Fig. 11) indicate that the film is composed of Hf, Zr and O elements, and the phase is mainly monoclinic linear compound $(Hf, Zr)O_2$.

The composite exhibits excellent oxidation-resistant and ablation-resistant properties because of the behaviour of various oxides formed during the ablation. The obtained composite is composed of HfC, ZrC, SiC, C and minor TaC phase, whose evolution during ablation is shown as the reaction equations in Tab. 1. SiO_2, HfO_2, ZrO_2 and Ta_2O_5 generate from the oxidation of HfC, ZrC, SiC and minor TaC. No SiO_2 and Ta_2O_5 was found in the surface layer of the sample. At 1800°C, SiO_2 evaporated significantly. A large amount of heat was taken away with the melt and evaporation of SiO_2. The solubility of HfO_2 and ZrO_2 in Ta_2O_5 is very high, thus stable Ta_2O_5 mHfO_2 and Ta_2O_5 mZrO_2 phases generate as a barrier layer of oxygen. Besides, tantalum element enters into the crystal lattice of HfO_2, which decrease the oxygen permeability of HfO_2. However, due to the long duration of ablation, Ta_2O_5 phase finally exhausted. As the ablation goes by,

Table 1. Possible reactions during the ablation process of the C_f/HfC composite

No	Reaction equations
1	$HfC(s) + 3/2O_2(g) = HfO_2(s) + CO(g)$
2	$HfO_2(s) = HfO_2(l)$
3	$HfO_2(l) = HfO_2(g)$
4	$ZrC(s) + 3/2O_2(g) = ZrO_2(s) + CO(g)$
5	$ZrO_2(s) = ZrO_2(l)$
6	$ZrO_2(l) = ZrO_2(g)$
7	$C(s) + 1/2O_2(g) = CO(g)$
8	$CO(g) + 1/2O_2(g) = CO_2(g)$
9	$SiC(s) + 3/2O_2(g) = SiO_2(s) + CO(g)$
10	$SiO_2(s) = SiO_2(l)$
11	$SiO_2(l) = SiO_2(g)$
12	$TaC(s) + 5/2O_2(g) = Ta_2O_5(s) + CO(g)$
13	$Ta_2O_5(s) = Ta_2O_5(l)$
14	$Ta_2O_5(l) = Ta_2O_5(g)$

SiO_2 and Ta_2O_5 go away, leaving the stable layer of $(Hf, Zr)O_2$ on the surface of the sample and protecting it from further ablation.

4. Conclusion

A C_f/HfC composite was obtained by instantaneous liquid infiltration (ILI) using 50Hf10Zr40Si3Ta alloy and C/C preform at 1900°C. It is proposed that the primary reason for triggering the ILI process at the temperature much lower than the alloy melting point is that the Hf component with the strongest carbide-forming ability prefers to react with the carbon in the surface layer of the C-C preform, leading to a phase composition change at the interface and generating instantaneous liquid phases. Different carbide-forming abilities of the elements are vital for the ILI to proceed, because if all the components react synchronously with carbon in the surface area, only the carbide product and initial alloy will exist on the interface without any liquid phase generating, thus the ILI will not be realized and there is no chance to obtain the Cf/HfC composite. This is an important principle of alloy design for the ILI process of Cf/HfC composite. The composite exhibits excellent oxidation-resistant and ablation-resistant properties because of the behaviour of various oxides formed during the ablation. However, longer tests are necessary for unequivocal definition of heat-protective characteristics of an offered material in the conditions of approached to the natural.

References

He, H.W., Zhou, K.C., Xiong, X. and Huang, B. (2006). Investigation on decomposition mechanism of tantalum ethylate precursor during formation of TaC on C/C composite material, *Materials Letters*, Vol. 60, No 28, pp. 3409-3412.

Jayaseelan, D.D., de Sa, R. G. and Brown, P. (2011) Reactive infiltration processing (RIP) of ultra high temperature ceramics (UHTC) into porous C/C composite tubes, *Journal of the European Ceramic Society*, Vol. 31, No 3, pp. 361-368.

Krenkel, W. (2001) Cost effective processing of CMC composites by melt infiltration (LSI-process), *Ceramic Engineering and Science Proceeding*, Vol. 22, No.3, pp. 443-454.

Levine, S.R., Opila, E.J., Halbig, M.C., Kisera J.D., Singh, M. and Salem, J. A. (2002). Evaluation of ultra-high temperature ceramics for aeropropulsion use, *Journal of the European Ceramic Society*, Vol. 22, pp. 2757-2768.

Levine, S.R. and Opila E.J. (2003). *Tantalum Addition to Zirconium Diboride for Improved Oxidation Resistance*. NASA Glenn Research Centre.

Levine, S.R., Opila, E.J., Lorincz, J.A., Robinson, R.C., Singh, M., Petko, J., Ellerby, D.T. and Gasch, M.J. (2004). *UHTC Composites for Leading Edges*. NASA Glenn Research Center.

Ohlhorst, C.W., Vaughn, W.L., Lewis, R.K. and Milhoan, J.D. (2003). *Arc Jet Results on Candidate High Temperature Coatings for NASA's NGLT Refractory Composite Leading Edge Task*. NASA Langley Research Centre.

Scatteia, L., Riccio, A., Rufolo, G., Filippis, F.D., Vecchio, A.D., and Marino, G. (2005). PRORA-USV SHS: Ultra high temperature ceramic materials for sharp hot structures, *AIAA Paper*, pp. 2005-3266.

Schwab, S.T., Stewart, C.A., Dudeck, K.W., Kozmina, S.M., Katz, J.D., Bartram, B., Wuchina, E.J., Kroenke, W.J. and Courtin, G. (2004). Polymeric precursors to refractory metal borides, *Journal of Materials Science*, Vol. 39, pp. 6051-6055.

Wuchina, E., Opila, E., Fahrenholtz, W. and Talmy, I. (2007). UHTCs: Ultra-high temperature ceramic materials for extreme environment applications, *Electrochemical Society Interface*, Winter, pp. 30-36.

Zou, L.H., Wali, N., Yang, J.M. and Bansal, N.P. (2010). Microstructural development of a Cf/ZrC composite manufactured by reactive melt infiltration, *Journal of the European Ceramic Society*, Vol. 30, pp. 1527-1535.

First Ply Failure Analysis of Laminated Composite Cylindrical Shells

Prithwish Saha, Kaustav Bakshi, Dipankar Chakravorty

Civil Engineering Department, Jadavpur University, 188 Raja S.C. Mullik Road, Kolkata, 700 032, India

Abstract: Composite cylindrical shells are stiff surfaces with simple curved geometry and are extensively used to build large column free areas in shopping malls, airports and car parking lots with reasonably less material consumption. Laminated composites gained popularity in civil engineering structures as use of these materials results in reduced mass and mass induced forces like seismic forces. Failure study of these materials is necessary, which includes the load value (first ply failure load) at which failure initiates, the mode of failure and the failure propagation location. The present article aims to study first ply failure of uniformly loaded simply supported cylindrical shells using finite element method. An eight noded curved quadratic isoparametric shell element is used to develop the finite element program and validated through solution of benchmark problems. Well accepted failure criteria are used to evaluate the failure loads and failure modes. .

Key Words: Failure modes, Finite element method, First ply failure loads, Laminated composite cylindrical shell. .

1. Introduction

Cylindrical shells are used as roofs for buildings with large column free spaces such as auditoriums, airport terminals, exhibition halls and factories. The curved geometry combines bending and axial capacities of the material and greater spans may be covered with thin shells. The fuselage of an airplane, containing flight crew, passengers and payload, fuel etc. is also very often cylindrical in shape.

Laminated composites are increasingly being used as the structural units in aerospace, civil, marine and other weight-sensitive engineering applications due to their high strength/stiffness to weight ratio, long fatigue life, good corrosion resistance and dimensional stability during large temperature change in space. Composites are extensively being used in secondary structures such as rudders, elevators, landing gear doors etc in aerospace engineering. Helicopters and tiltrotors use rotor blades made of composites that not only increase the life of blades but also increase the top speeds. The stiffness parameters of the laminated composites can be altered by varying lamina stacking sequences and fiber orientations which make them more attractive option to the engineers.

The failure of laminated composites initiates with the failure of the weakest lamina and the load value is designated as the first-ply failure load. Several researchers like Singh and Kumar (1998), Akhras and Li (2007) and Ganesan and Liu (2008) reported that first ply failure load of a laminated composite is much lower than the ultimate ply failure load and hence the application of a higher safety factor leads to a highly conservative design. However if first ply failure load remain undetected, it may lead to a sudden catastrophic collapse later under service conditions. Realizing this, many researchers studied the first ply failure characteristics of laminated composites. Initial flexural failure loads of simply supported GFRP and CFRP plates subjected to lateral pressure distributions were reported by Turvey (1980). The author used symmetric cross ply lay-ups in the study. Reddy and Pandey (1987) studied first ply failure loads and locations of plates subjected to uniformly distributed loads acting along transverse and in-plane directions of the plates. Ex-

perimentally first ply failure loads for centrally loaded square plates were reported by Kam and Jan (1995). The authors proposed a layerwise linear displacement theory and validated their approach by comparing the results with the experimental values. Kam and Sher (1995) studied nonlinear first ply failure loads of centrally loaded square cross ply plates using Von Karman-Mindlin plate theory and Ritz method. Along with the initial failure loads, authors proposed a stiffness reduction model to study the plate behavior beyond the first ply failure. First ply failure loads of plates using three different finite elements were evaluated by Kam et al. (1996).The authors also compared the results with the experimental values. Prusty et al. (2001) investigated the first ply failure loads for unstiffened cylindrical shells and stiffened spherical shells under gravity loading respectively. The authors considered T300/5208 graphite- epoxy as the construction material in both of their studies. The cylindrical shell studied by the authors was supported on rigid diaphragms along the curved boundary and free along the straight edges. The authors reported the failure loads for varying radius to side ratio. Thus, it is noticed that researchers explored first ply failure of laminated plates but such study on shells lack the depth of attention. The present paper aims to study first ply failure of a simply supported shallow cylindrical shell for varying lamination which consist of symmetric and antisymmetric stacking orders of cross and angle ply laminates. Q-1115 graphite-epoxy is considered as construction material in the present study. Apart from the failure loads, failure modes are also presented.

2. Mathematical formulation

2.1 Governing equation of shell bending

The governing equation of a composite shell is derived based on the principle of minimum total potential energy where the total potential energy 'π' is expressed as sum of strain energy 'U' and work done due to external load 'V'.

$$\pi = U + V \tag{1}$$

Strain energy of the shell is expressed as,

$$U = \frac{1}{2} \int_v \{\sigma\}\{\varepsilon\} dv \qquad (2)$$

And work done by external load,

$$V = -\iint_A \{u\}^T [q] dA \qquad (3)$$

where 'v' represents shell volume and 'A' shell area. External load on shell can be expressed as,

$$\{q\} = \begin{bmatrix} 0 & 0 & q_z & 0 & 0 \end{bmatrix}^T \qquad (4)$$

where q_z represents transverse load intensity on the shell.

2.2 Finite element formulation

An eight noded curved quadratic isoparametric element with C^0 continuity is considered in the present study to formulate the bending stiffness of the cylindrical shell. Five degrees of freedom that are considered include three displacements and two transverse rotations.

The strain displacement matrix $[B]$, laminate elasticity matrix $[D]$ and the cubical shape functions used in the present study are those as were reported by Chakravorty et al. (1995).

2.3 Stiffness matrix formulation

The strain energy of the shell is expressed as

$$U_1 = \frac{1}{2} \iint_A \{d_e\}^T [B]^T [D][B] \{d_e\} dxdy \qquad (5)$$

and the work done is expressed as,

$$V = -\iint_A \{d_e\}[N]^T [q] dxdy \qquad (6)$$

To minimize the total potential energy of the shell with respect to its deformations, the shell has to satisfy the following condition,

$$\frac{\partial \pi}{\partial \{d_e\}} = 0 \qquad (7)$$

By applying Equations (5) and (6) in Equation (7) we get,

$$\left(\iint_A [B]^T [D][B] dxdy \right)\{d_e\} - \left(\iint_A [N]^T [q] dxdy \right) = 0$$

$$[K_e]\{d_e\} = \{Q_e\} \qquad (8)$$

where

$$[K_e] = \iint_A [B]^T [D][B] dxdy$$

and

$$\{Q_e\} = \iint_A [N]^T [q] dxdy$$

The element stiffness matrix $[K_e]$ and load vector $\{Q_e\}$ are transformed to isoparametric coordinates ξ and h for numerical integration by 2×2 Gauss quadrature rule. Global stiffness matrix and load vector are obtained by assembling the element matrices with proper transformations due to the curved geometry of the shell and they are expressed as,

$$[K]\{d\} = \{Q\} \qquad (9)$$

where

$$[K] = \sum_{i=1}^{ne} [K_e]$$

and

$$[Q] = \sum_{i=1}^{ne} [Q_e]$$

2.4 Lamina stress calculation

Generalized laminate midplane strains are evaluated using the strain displacement relationship. Inplane strain components for a lamina situated at a distance 'z' from the lamina midplane are evaluated in global axes as,

$$\begin{aligned} \varepsilon_x &= \varepsilon_x^0 + zk_x \\ \varepsilon_y &= \varepsilon_y^0 + zk_y \\ \gamma_{xy} &= \gamma_{xy}^0 + zk_{xy} \end{aligned} \qquad (10)$$

Lamina strains are transformed from the global axes of the shell to the local axes of the lamina using transformation matrix,

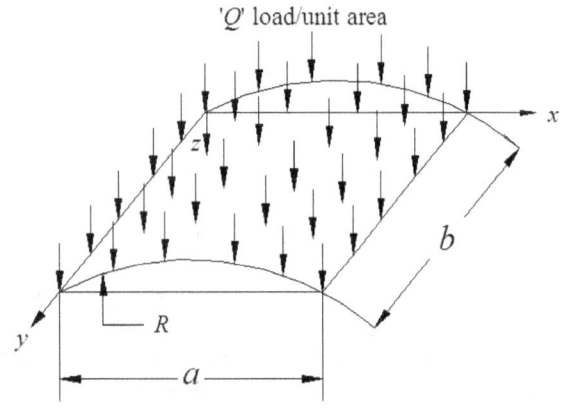

Figure 1. Cylindrical shell.

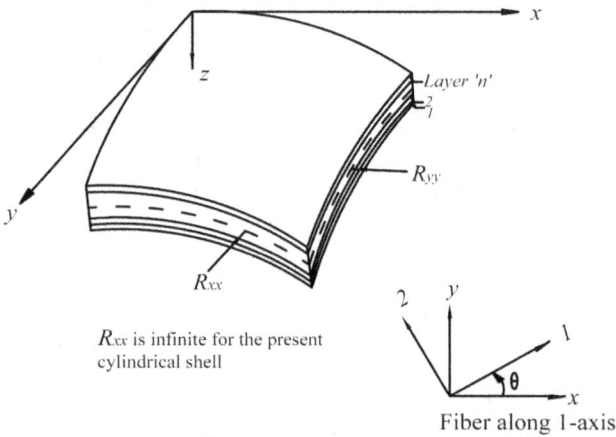

R_{xx} is infinite for the present cylindrical shell

Fiber along 1-axis

Figure 2. General doubly curved laminated composite shell element.

$$\left\{ \begin{array}{c} \varepsilon_1 \\ \varepsilon_2 \\ \dfrac{\varepsilon_6}{2} \end{array} \right\} = \left[\begin{array}{ccc} m^2 & n^2 & 2mn \\ n^2 & m^4 & -2mn \\ -mn & mn & -n^2 \end{array} \right] \left\{ \begin{array}{c} \varepsilon_x \\ \varepsilon_y \\ \dfrac{\gamma_{xy}}{2} \end{array} \right\} \qquad (11)$$

where $m = \sin\theta$ and $n = \cos\theta$.

Lamina stresses are obtained using the constitutive relation of the lamina,

$$\left\{ \begin{array}{c} \sigma_1 \\ \sigma_2 \\ \sigma_6 \end{array} \right\} = \left[\begin{array}{ccc} Q_{11} & Q_{12} & 0 \\ Q_{12} & Q_{22} & 0 \\ 0 & 0 & Q_{66} \end{array} \right] \left\{ \begin{array}{c} \varepsilon_1 \\ \varepsilon_2 \\ \varepsilon_6 \end{array} \right\} \qquad (12)$$

where

$$Q_{12} = \left(1 - v_{12}v_{21}\right)^{-1} E_{11}v_{21}$$
$$Q_{11} = \left(1 - v_{12}v_{21}\right)^{-1} E_{11}$$
$$Q_{22} = \left(1 - v_{12}v_{21}\right)^{-1} E_{22}$$
$$Q_{66} = G_{12}$$

Lamina stresses and strains are used in well accepted failure theories like maximum stress, maximum strain, Tsai-Hill, Tsai-Wu and Hoffman failure criterion to evaluate the first ply failure loads of the composite cylindrical shell under study. The expressions of the failure theories adopted here are those reported by Kam et al. (1996).

3. Failure modes

A composite lamina may fail in different ways. Under tensile stresses, a composite lamina may fail through fiber breakage, by transverse matrix cracking in the plane of the lamina or by inter-fiber shear failure of the matrix. When subjected to compressive stresses, fiber buckling dominates the fiber failure and matrix crushing leads to the failure of the composite matrix. All these failure modes can be identified through maximum stress and maximum strain failure theories (Table. 1).

In case of interactive failure theories none of the individual lamina stress component reaches the permissible value but their interaction leads to the failure of the composite lamina. In case of such failure the individual stress values developed are compared to their corresponding permissible values to investigate which stress component contributing to the interacting criteria plays the most significant role in the failure. The stress component for which the ratio of developed to permissible stress is nearest to unity is identified as the most significant contribution to the failure.

4. Numerical problems

To establish the correctness of the cylindrical shell formulation of the present finite element code, the authors compared static displacement values evaluated by current formulation with the pre-established results published by Qatu and Algothani (1994). The comparison is presented in Table 2. The material property of AS4/3501-6 Epoxy (Vinson and Sierakowski, 2002) and geometric properties of the cylindrical shell is presented as footnote with the table. First ply failure loads evaluated using the present formulation are compared with the linear failure loads reported by Kam et al (1996) for a partially clamped plate to establish the correctness of the first ply failure formulation.

Table 2
Comparison of nondimensional downward displacements of composite cylindrical shell

Lamination (Degree)	Exact solution	Static displacement reported by Qatu and Algothani (1994)	Static displacement from present formulation
0/90	0.0110	0.0109	0.0095
0/90/0	0.0086	0.0089	0.0075

$a/b = 1$; $a/h = 100$; $a/R = 0.5$; $E_{11} = 15.40E_{22}$; $G_{12} = G_{13} = 0.79E_{22}$; $G_{23} = 0.2E_{22}$; $v = 0.30$

Table 3
Material properties of graphite-epoxy (Q-1115)

Material Constants		Strength			
E_{11}	142.50 GPa	X_T	2193.5MPa	$X_{\varepsilon t}$	0.01539
E_{22}	9.79 GPa	X_C	2457.0 MPa	$X_{\varepsilon c}$	0.01724
E_{33}	9.79 GPa	$Y_T = Z_T$	41.30 MPa	$Y_{\varepsilon t} = z_{\varepsilon t}$	0.00412
$G_{12} = G_{13}$	4.72 GPa	$Y_C = Z_C$	206.80 MPa	$Y_{\varepsilon c} = z_{\varepsilon c}$	0.02112
G_{23}	1.192 GPa	R	61.28 MPa	R_ε	0.05141
$N_{12} = v_{13}$	0.27 GPa	S	78.78 MPa	S_ε	0.01669
N_{23}	0.25 GPa	T	78.78 MPa	T_ε	0.01669

Table 4
Comparison of first ply failure loads in Newton for a $(0_2°/90°)_s$ plate

Failure criteria	Side/ thickness	First ply failure loads (Kam et al., 1996)	First ply failure loads (present formulation)
Maximum stress		108.26	112.14
Maximum strain		122.86	128.56
Hoffman	105.26	106.45	98.40
Tsai-Wu		112.77	110.50
Tsai-Hill		107.06	104.40

Table 1

Maximum Stress Theory		Maximum Strain Theory	
Stress ratio	Failure mode	Strain ratio	Failure mode
$\dfrac{\sigma_1}{X_T} > 1$	Fiber breakage	$\dfrac{\varepsilon_1}{X_{\varepsilon T}} > 1$	Fiber breakage
$\dfrac{\sigma_2}{Y_T} > 1$	Transverse matrix cracking	$\dfrac{\varepsilon_2}{Y_{\varepsilon T}} > 1$	Transverse matrix cracking
$\dfrac{|\sigma_6|}{T} > 1$	Shear failure of the matrix	$\dfrac{|\varepsilon_6|}{T_\varepsilon} > 1$	Shear failure of the matrix
$\dfrac{\sigma_1}{X_C} > 1$	fiber buckling	$\dfrac{\varepsilon_1}{X_{\varepsilon C}} > 1$	fiber buckling
$\dfrac{\sigma_2}{Y_C} > 1$	Matrix crushing	$\dfrac{\varepsilon_2}{Y_{\varepsilon C}} > 1$	Matrix crushing

Table 5
Uniformly distributed failure load of simply supported cylindrical shells of radius 1000 mm for different laminates

Lamination (degree)	Failure theory	Failure load (N/mm²)	Location (x,y) (m,m)	First failed ply	Failure mode/failure tendency
0/90	Maximum stress	0.2725	(0,0)	2	Shear failure of the matrix
	Maximum strain	0.2725	(0,0)	2	Shear failure of the matrix
	Hoffman	0.2718[L]	(0,0)	2	Shear failure of the matrix
	Tsai-Hill	0.2725	(0,0)	2	Shear failure of the matrix
	Tsai-Wu	0.2718[L]	(0,0)	2	Shear failure of the matrix
0/90/0	Maximum stress	0.2961	(0,0)	3	Shear failure of the matrix
	Maximum strain	0.2961	(0,0)	3	Shear failure of the matrix
	Hoffman	0.2942[L]	(0,0)	3	Shear failure of the matrix
	Tsai-Hill	0.2961	(0,0)	3	Shear failure of the matrix
	Tsai-Wu	0.2942[L]	(0,0)	3	Shear failure of the matrix
0/90/0/90	Maximum stress	0.2897	(0,0)	4	Shear failure of the matrix
	Maximum strain	0.2897	(0,0)	4	Shear failure of the matrix
	Hoffman	0.2888[L]	(0,0)	4	Shear failure of the matrix
	Tsai-Hill	0.2897	(0,0)	4	Shear failure of the matrix
	Tsai-Wu	0.2888[L]	(0,0)	4	Shear failure of the matrix
0/90/90/0	Maximum stress	0.2999	(1,0)	4	Shear failure of the matrix
	Maximum strain	0.2999	(1,0)	4	Shear failure of the matrix
	Hoffman	0.2979[L]	(1,0)	4	Shear failure of the matrix
	Tsai-Hill	0.2999	(1,0)	4	Shear failure of the matrix
	Tsai-Wu	0.2979[L]	(1,0)	4	Shear failure of the matrix
45/-45	Maximum stress	0.2280	(1,0)	2	Transverse matrix cracking
	Maximum strain	0.1659[L]	(1,0)	2	Transverse matrix cracking
	Hoffman	0.2154	(0,1)	2	Transverse matrix cracking
	Tsai-Hill	0.2173	(0,1)	2	Transverse matrix cracking
	Tsai-Wu	0.1973	(0,0)	2	Transverse matrix cracking
45/-45/45	Maximum stress	0.2981	(0,0)	3	Transverse matrix cracking
	Maximum strain	0.2159[L]	(0,0)	3	Transverse matrix cracking
	Hoffman	0.2808	(0,0)	3	Transverse matrix cracking
	Tsai-Hill	0.2836	(0,0)	3	Transverse matrix cracking
	Tsai-Wu	0.2570	(0,0)	3	Transverse matrix cracking
45/-45/45/-45	Maximum stress	0.2857	(1,0)	4	Transverse matrix cracking
	Maximum strain	0.2070[L]	(1,0)	4	Transverse matrix cracking
	Hoffman	0.2692	(1,0)	4	Transverse matrix cracking
	Tsai-Hill	0.2718	(1,0)	4	Transverse matrix cracking
	Tsai-Wu	0.2464	(1,0)	4	Transverse matrix cracking
45/-45/-45/45	Maximum stress	0.3256	(0,0)	4	Transverse matrix cracking
	Maximum strain	0.2353[L]	(0,0)	4	Transverse matrix cracking
	Hoffman	0.3064	(0,0)	4	Transverse matrix cracking
	Tsai-Hill	0.3095	(0,0)	4	Transverse matrix cracking
	Tsai-Wu	0.2803	(0,0)	3	Transverse matrix cracking

Inplane degree of freedoms along the boundaries of the plate was released to model the partially clamped boundary condition. The material properties of the plate are presented in Tables 3 and geometric properties are presented as the footnote of the Tables 4. The radius of the principle curvature and cross curvature of the present element is assigned a high value (10^{30}) to make them effectively zero to model a plate with no curvature.

Apart from solving the bench-mark problem for verifying the finite element code proposed here, authors solve a number of cylindrical laminated composite shells under uniformly distributed pressure, with different laminations,

Table 6
Geometric dimensions of the cylindrical shell

Cylindrical shell dimensions	Values
Length (*a*)	1000 mm
Width (*b*)	1000 mm
Thickness (*h*)	10 mm
Radius(along *X* axis) Radius(along *Y* axis)	Infinite 1000 mm

stacking orders and curvatures. These practical parametric variations include both angle and cross ply lamination of both antisymmetric and symmetric stacking orders. The curvature is varied so that the shell configuration is always shallow (Rise/Span ratio less than 0.2). The results of failure pressures obtained from the numerical experimentation are presented in Table 5. Plies are started to be numbered from the top of the laminate i.e. the topmost ply is numbered one and bottommost ply has the last ply number. Material properties of the graphite-epoxy composite to fabricate the cylindrical shell are presented in Table 3 and its geometric dimensions are furnished in Table 6.

Table 1 shows good agreement of the present results with the established ones and this validate the cylindrical shell formulation. Table 3 also exhibits a very good agreement between present results and published values which validates the present first ply failure formulation.

The results furnished in Table 5 shows that in all the cases of cross ply laminate, the Hoffman and Tsai-Wu criteria yield the lowest value of failure pressure. It may be noted that the Hoffman and Tsai-Wu criteria converge to the same condition when the transverse stress matrix vanishes. For the cross-ply laminates under uniformly distributed load, the plan direction of major load transfer and the orientations of the fibers are identical and hence hardly any transverse stress develops in the matrix. This is why the two above mentioned criteria converge to give the same result. Interestingly, on the other hand, in all the cases of angle ply laminates the maximum strain criterion indicates the design failure loads. On these failure pressures coming from different criteria for cross and angle ply laminates, the factor of safety should be applied, to obtain the working pressure value.

The five failure criteria taken up here give comparable results for cross ply shells although the Hoffman and Tsai-Wu criteria consistently yield the minimum value. Contrary to this, the angle ply laminates the failure load values obtained from the criteria a part from maximum strain criterion are often quite high, when compared with the minimum pressure value. This also indicates that for cross ply laminates, all the stresses and strains contributing in the failure criteria, increase in magnitude simultaneously as the superimposed load is increased. This indicates a more efficient utilization of material strength and this is why the cross ply surfaces are stiffer.

Another point which strikes designers' attention is that, the failure loads for angle ply laminations are remarkably less than that what are observed for cross ply ones. This leads to the natural inference that for a given quantity of material consumption, cross ply cylindrical shell should always be preferred than the angle ply ones for simply supported boundary conditions under uniformly distributed load. Infact the cross ply lamination showing the least failure load (0°/90°) can support a superimposed pressure, more than what is obtained for the stiffest angle ply shells (45°/-45°/45°/-45°) by about 13.4%.

Although the failure load values for cross and angle ply cylindrical shell are markedly different, but one trend is common for both types of laminations. This is the fact that the symmetric laminations always perform better than the antisymmetric ones.

Among the cross-ply laminates, the 0°/90°/90°/0° stacking sequence yields the maximum failure pressure. For a cylindrical shell, symmetrically supported along all the four edges, the loads and moments are transferred mainly along the two plan directions. For cross-ply laminates, the on axis stiffness of the individual lamina play an important role in resisting the load because they too are aligned parallel to the direction of load transfer. This is why the cross-ply shells are significantly better in performance than the angle ply ones and for four layered cross-ply stacking sequence, (0°/90°/90°/0°) equal thickness of 0° and 90° laminae along the shell cross-sections cause a balanced load transfer mechanism along the plan direction and the failure load shows an improved value. A 90°/0°/0°/90° shell, on the other hand gives a failure load of 0.2571 N/mm^2 which is 86.3% of the failure load for a 0°/90°/90°/0° shell. The cylindrical shell configuration considered in the present study has curvature in *y*-direction and it is singly ruled in *x*-direction with no curvature. This particular shape by virtue of its geometry only, has an enhanced stiffness along the arch or *y*-direction where the bending and axial stiffness work together to resist the load. Along the *x*-axis on the other hand there is no such coupling and the load transfer is almost like that of a plate. A 0°/90°/90°/0° lamination has the 0° fibers stiffening the *x*-direction and being away from the mid plane, renders adequate bending inertia to the shell. This brings about a more coMPatible balance of stiffness along the two plan directions and the failure load reaches the peak. In contrast to this, for 90°/0°/0°/90° shell the fiber stiffening the beam direction are more towards the midplane and contribute less significantly to the bending inertia and the failure occurs for a much lower value of the superimposed load indicating a failure along the beam direction.

5. Conclusion

In all the cases of cross ply laminate, the Hoffman and Tsai-Wu criteria yield the lowest value of failure pressure. Interestingly, on the other hand, in all the cases of angle ply laminates , the maximum strain criterion indicates the design failure loads. This also indicates that for cross ply laminates, all the stresses and strains contributing in the failure criteria, increase in magnitude simultaneously as the superimposed load is increased. This indicates a more efficient utilization of material strength and this is why the cross ply surfaces are stiffer. Although the failure load values for cross and angle ply cylindrical shell are markedly different, but one trend is common for both types of laminations. This is the fact that the symmetric laminations always perform better than the antisymmetric ones.

A 0°/90°/90°/0° lamination has the 0° fibers stiffening the *x*-direction and being away from the mid plane, renders adequate bending inertia to the shell. This brings about a more compatible balance of stiffness along the two plan directions and the failure load reaches the peak. The failure modes/tendencies of the shell taken up here, it is observed in all the cases, the failure occurs at the corner of the shell where shear is critical and at the bottommost lamina.

6. Notations

A	Area of the shell.
a and b	Length and width of shell in plan respectively.
D	Flexural rigidity matrix of the laminate.
$\{d\}$	Nodal displacements of the shell.
$E_{11}, E_{22,}$	Elastic moduli along the directions 1 and 2 of a lamina respectively.
$1,2$ and 3	Local co-ordinates of a lamina respectively.
G_{12}	Shear moduli of a lamina in 1-2 plane of a lamina.
ne	Number of finite elements in the shell domain.
T	Shear strength of a lamina in its 1-2 plane.
T_ε	Allowable shear strain of a lamina in its 1-2 plane.
$\{u\}$	Generalized displacement vector.
$\overline{w}$	Nondimensional transverse displacement of shell. $=[wE^{22}h^3/(qa^4)]$
X_T and X_C	Normal strengths of a lamina along the fiber direction in tension and compression respectively.
$X_{\varepsilon T}$ and $X_{\varepsilon C}$	Allowable normal strains of a lamina along the fiber direction in tension and compression respectively.
Y_T and Y_C	Normal strengths of the matrix in tension and compression respectively.
$Y_{\varepsilon T}$ and $Y_{\varepsilon C}$	Allowable normal strains of the matrix in tension and compression respectively.
x, y and z	Global Cartesian co-ordinates of the shell.
e_x, e_y	Inplane normal strains along x and y axes respectively.
e_1, e_2	Inplane normal strains along 1 and 2 axes of a lamina respectively.
ε_6	Inplane shear strain in 1-2 plane of a lamina.
g_{xy}	Inplane shear strain in x-y plane of the shell.
n_{ij}	Poisson's ratio which characterizes compressive strain along x_j direction produced by a tensile strain applied in x_i direction.
s_1, s_2	Inplane normal stresses along 1 and 2 axes of a lamina respectively.
s_6	Inplane shear stress in 1-2 plane of a lamina.
$\kappa_x, \kappa_y, \kappa_{xy}$	Curvature changes of the shell due to loading.

7. Acknowledgement

The second author gratefully acknowledges the financial assistance of Council of Scientific and Industrial Research (India) through the Senior Research Fellowship vide Grant no. 09/096 (0686) 2k11-EMR-I.

References

Akhras, G. and Li, W.C. (2007). Progressive failure analysis of thick composite plates using the spline finite strip method. *Composite Structures*, Vol. 79, pp. 34-43.

Chakravorty, D., Sinha, P.K. and Bandyopadhyay, J.N. (1995). Finite element free vibration analysis of point supported laminated composite cylindrical shells. *Journal of Sound and Vibration*, Vol. 181, No 1, pp. 43-52.

Ganesan, R. and Liu, D.Y. (2008). Progressive failure and post buckling response of tapered composite plates under uni-axial compression. *Composite Structures*, Vol. 82, pp. 159-176.

Kam, T.Y. and Jan, T.B. (1995). First ply failure analysis of laminated composite plates based on the layerwise linear displacement theory. *Composite Structures*, Vol. 32, pp. 583-591.

Kam, T.Y. and Sher, H.F. (1995). Nonlinear and first-ply failure analysis of laminated composite cross-ply plates. *Journal of Composite Materials*, Vol. 29, pp. 463-482.

Kam, T.Y., Sher, H.F. and Chao, T.N. (1996). Predictions of deflection and first-ply failure load of thin laminated composite plates via the finite element approach. *International journal of Solids and Structures*, Vol. 33, No 3, pp. 375-398.

Prusty, B.G., Ray, C. and Satsangi, S.K. (2001). First ply failure analysis of stiffened panels-a finite element approach. *Composite Structures*, Vol. 51, pp. 73-81.

Prusty, B.G., Satsangi, S.K. and Ray, C. (2001). Firstply failure analysis of laminated panels under transverse loading. *Journal of Reinforced Plastics and Composites*, Vol. 20, No 8, pp. 671-684.

Qatu, M.S. and Algothani. (1994). A bending analysis of laminated plates and shells by different methods. *Computers and Structures*, Vol. 52, No 3, pp. 529-539.

Reddy, J.N. and Pandey, A.K. (1987). A first ply failure analysis of composite laminates. *Computers and Structures*, Vol. 25, No 3, pp. 371-393.

Singh, S.B. and Kumar, A. (1998). Postbuckling response and failure of symmetric laminates under inplane shear. *Composites Science and Technology*, Vol. 58, pp. 1949-1960.

Turvey, G.J. (1980). An initial flexural failure analysis of symmetrically laminated cross-ply rectangular plates. *International Journal of Solids Structures*, Vol. 16, pp. 451-463.

Vinson, J.R. and Sierakowski, R.L. (2002). *The Behavior of Structures Composed of Composite Materials*. 2nd ed., New York: Kluwer Academic Publishers.

C/C-Zr-Si Composites Prepared by Alloyed Reactive Melt Infiltration

Yonggang Tong, Shuxin Bai, Hong Zhang, Ke Chen, Yicong Ye, Li'an Zhu

College of Aerospace and Materials Engineering, National University of Defense Technology, Changsha, 410073, P.R.China

Abstract: High performance and low cost C/C-Zr-Si composites were prepared by Si-10Zr and Zr-9.8Si alloyed reactive melt infiltration. Carbon fiber felt was firstly densified by pyrolytic carbon using chemical vapor infiltration to obtain a porous C/C preform. The eutectic Si-10Zr and Zr-9.8Si alloy melts were then infiltrated into the porous preforms to prepare two kinds of C/C-Zr-Si composites. The composite prepared by Si-10Zr alloyed melt infiltration was composed of C, SiC and $ZrSi_2$ with a little amount of ZrC, while the composite prepared by Zr-9.8Si alloyed melt infiltration was composed of C, ZrC and Zr_2Si. Ablation properties of the two composites prepared by Si-10Zr and Zr-9.8Si alloyed melt infiltration were tested by a pulse laser. The linear ablation rates were 0.041 and 0.028 mm/s, much smaller than the linear ablation rate of the C/SiC composite, 0.107 mm/s. The good ablation resistance of the two composites was attributed to the carbides and silicides formed during reactive melt infiltration.

Key Words: C/C-Zr-Si composites, Reactive melt infiltration, Microstructure, Ablation properties.

1. Introduction

Ceramic matrix composites, reinforced by high strength continuous ceramic fibers, are the most promising materials for high temperature structural applications because of their superior high temperature strength, low density and improved damage tolerance (Van de Voorde and Nedele, 1996; Naslain, 2004). Particularly, continuous carbon fiber reinforced SiC, ZrC and SiC-ZrC binary matrix composites attract much attention owing to their low density, high hardness, excellent oxidation resistance, high strength and thermal shock resistance (Krenkel, 2003; Padmavathi et al., 2009; Zou et al., 2010). They are potential candidates for highly demanding engineering applications such as heat shields, structural components for reentry space vehicles, high performance brake discs and high temperature heat exchanger tubes (Tressler, 1999).

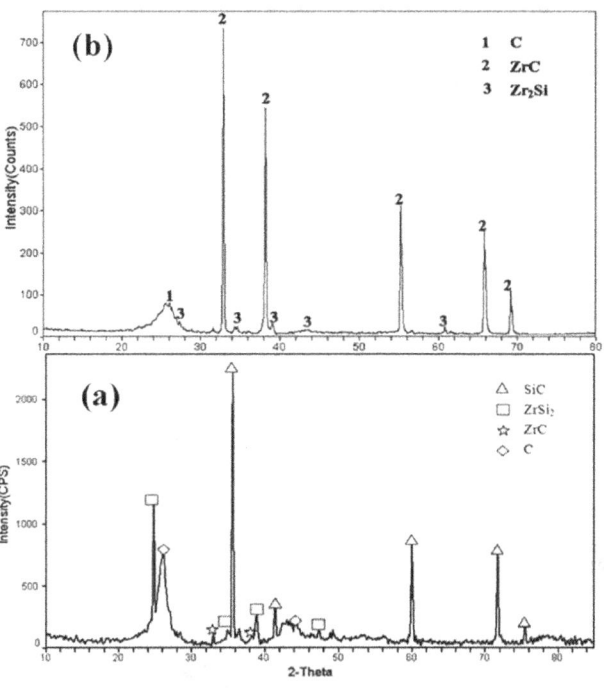

Figure 1. XRD patterns of the as-received
(a) composite A and (b) composite B.

Various techniques such as chemical vapor infiltration (CVI), polymer impregnation and pyrolysis (PIP), and reactive melt infiltration (RMI) have been developed to fabricate ceramic matrix composites. CVI and PIP are the practical processes for fabricating the composite, but they have obvious disadvantages of time-consuming and high-cost (Jiang et al. 2009). Conversely, reaction melt infiltration (RMI) does not suffer from the drawbacks of CVI and PIP and it has outstanding advantages such as short fabrication period, low cost and near net shape (Yang and Ilegbusi, 2000). It has become a commercialized method of great market competition. C/SiC and C/ZrC composite has been successfully fabricated by reactive melt infiltration (Wang et al., 2011; Kumar et al., 2009). However, few publications are available on the C/C-Zr-Si composites prepared by alloyed reactive melt infiltration. The aim of the present work is to fabricate C/C-Zr-Si composites by a low cost alloyed reactive melt infiltration. Multiple matrix composites composed of carbides and silicides were prepared by a one step alloyed reactive melt infiltration. The microstructure and composition of the C/C-Zr-Si composites were studied and the ablation performances were investigated as well.

2. Experimental

Carbon fiber needled felts were used as preforms. The carbon fibers were PAN-based. The needled felts were prepared by a three-dimensional needling technique, starting with repeatedly overlapping the layers of 0° non-woven fiber cloth, short-cut-fiber web, and 90° non-woven fiber cloth with needle-punching step by step. Pyrolytic carbon was deposited on the carbon fibers to prepare the C/C preforms by chemical vapor infiltration (CVI) process. The density of the porous C/C preforms was about 1.33 g/cm³. The porous C/C preforms were cut, polished, ultrasonically cleaned with ethanol and dried at 100°C for 4 h in an oven. Two kinds of C/C-Zr-Si composites were then prepared at 1450°C and 1800°C for 1h by infiltrating the Si-10Zr and Zr-9.8Si alloy. Pieces of Si-10Zr and Zr-9.8Si alloys were placed on top of the C/C preforms in graphite pots, and the pots were put into a high temperature furnace. The samples were heated to the RMI temperature and kept for 1h followed by furnace cooling to prepare C/C-Zr-Si composites. In the following passage, the C/C-Zr-Si composites prepared by Si-10Zr and Zr-9.8Si alloyed melt infiltration were named composite A and composite B for short.

Ablative resistance properties of the composites were tested by a pulsed laser. The laser power of 1000W/cm^2 was selected to vertically irradiate on the materials exposed in the air. The morphology of composite was observed by Hitachi-S4800 scanning electron microscope (SEM). The chemical composition was examined by energy dispersive spectroscopy (EDS). The phases in the composite were identified by X-ray diffraction (XRD, Rigaku D/Max 2550VB-) using a Ni-filtered Cu Kα radiation at a scanning rate of 5°/min and scanning from 20° to 80° of 2θ.

3. Results and discussion

XRD phase analysis of the as-received composite A and composite B are shown in Fig.1. It is indicated that the composite A is composed of SiC, ZrSi$_2$, ZrC and carbon, while the composite B is composed of carbon, ZrC and Zr$_2$Si. The broad carbon-peak refers to the carbon fibers and unreacted PyC. The phases ZrC and SiC were resulted from the reaction of Zr and Si with PyC. The phases ZrSi$_2$ and Zr$_2$Si derived from the reaction between Si and Zr in the alloyed melt.

Fig.2 shows the cross-section SEM micrographs of the composite A and B. As can be seen, both the composites show a dense morphology with few pores. The distribution of the phases in both composites can also be revealed in Fig. 2. It is indicated that the residual PyC mostly ex-

ists in the intra-fiber bundles and on the edge of the fiber bundles, while SiC, ZrSi$_2$ and ZrC mostly exist in the inter-fiber bundles in the composite A. In the inter-fiber bundles, the reaction-formed SiC and a small amount of discontinuous ZrC distribute around the PyC on the edges of the carbon fiber bundles while ZrSi$_2$ locates in the middle of the inter-fiber bundles surrounded by SiC. The residual PyC in composite B has the same distribution with composite A. Nevertheless, a continuous ZrC distributes around the PyC on the edges of the carbon fiber bundles and Zr$_2$Si locates in the middle of pores surrounded by ZrC in the composite B.

The reasons for this kind of microstructure can be explained as follows. During the CVI process, PyC firstly deposits on the surface of the carbon fibers, which then grows and fills the pores in the intra-fiber bundles. So the pores in the C/C preforms mostly exist in the inter-fiber bundles. After the RMI process, the pores are filled by the infiltrated alloyed melts, and SiC, ZrSi$_2$ ZrC and Zr$_2$Si are formed by the reactions of Si and Zr with PyC. The reactions of Si and Zr with PyC and their thermodynamic calculations are as follows:

$$C+Zr \rightarrow ZrC \quad \begin{aligned} \Delta G_{1450°C} &= -183.66 \text{ kJ/mol} \\ \Delta G_{1800°C} &= -180.09 \text{ kJ/mol} \end{aligned} \quad (1)$$

$$C+Si \rightarrow SiC \quad \begin{aligned} \Delta G_{1450°C} &= -62.520 \text{ kJ/mol} \\ \Delta G_{1800°C} &= -55.946 \text{ kJ/mol} \end{aligned} \quad (2)$$

Known from the thermodynamic calculations, the formation of ZrC according to Eq. (1) is more favorable than that of SiC according to Eq. (2) because of the much more negative Gibbs free energy. For the composite A, Si-Zr melt homogenously infiltrates into the porous C/C preform during the process of alloyed melt infiltration. Zirco-

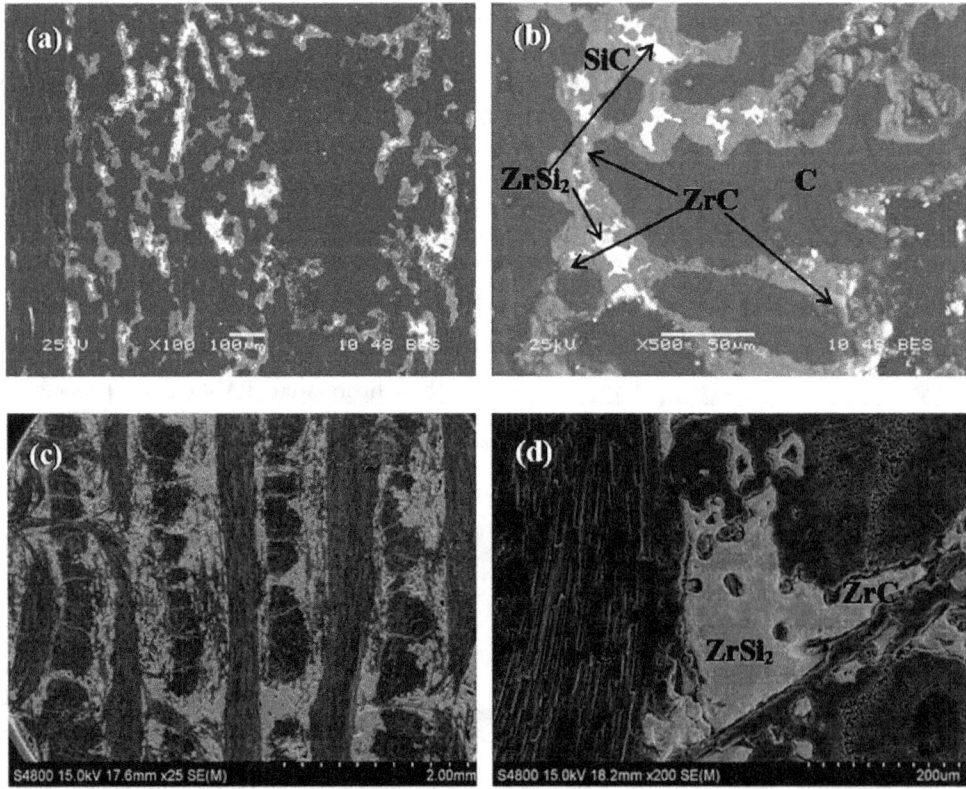

Figure 2. Cross-section SEM micrographs of (a, b) composite A and (c, d) composite B.

nium in the melt prefers to react with PyC when the melt meets PyC at the beginning. However, the zirconium (10 at.%) in the Si-Zr melt is not enough for the subsequent formation of ZrC, which leads to the small amount of discontinuous reaction-formed ZrC distributing around PyC. Here, the reaction between silicon and carbon occurs and the phase SiC is formed. There is enough silicon in the Si-Zr melt, and more and more SiC is formed. The concentration of silicon in the melt decreases due to the reaction between silicon and carbon, and $ZrSi_2$ phase begins to solidify from the Si-Zr melt at a certain time, which is indicated by the phase diagram of Si-Zr system (Fig. 3). With the proceeding of the reaction between silicon and carbon, SiC and $ZrSi_2$ phases keep forming until both silicon and zirconium in the melt are completely consumed. Finally, the continuous SiC and discontinuous ZrC and $ZrSi_2$ are developed in the composite A, in which the continuous SiC and discontinuous ZrC distribute around PyC inside the pores of the C/C preform and $ZrSi_2$ locates in the middle of pores surrounded by the SiC. However, for the preparation of composite B, 91.2% of the alloyed melt is Zr, which is sufficient for the reaction between Zr and PyC. Zr prefers to react with PyC at the beginning and a ZrC layer is formed around the PyC. With the proceeding of the reaction between zirconium and carbon, the concentration of zirconium in the melt decreases and Zr_3Si phase tends to be formed in the melt, which is indicated by the phase diagram of Zr-Si system (Fig. 3). Zr_3Si phase, with the melting point of 1650°C, is in the liquid condition at the RMI temperature. Atoms in the melt diffuse rapidly and the reaction between zirconium and carbon still progresses at a rather high speed.

Table 1
Linear ablation rates of C/SiC, composite A and composite B

Materials	Linear ablation rate, mm/s
C/SiC	0.108
Composite A	0.041
Composite B	0.028

With the further decrease of the zirconium in the melt, all the eutectic Zr-Si melt changes to Zr_3Si melt and then the Zr_3Si melt totally transfers to Zr_2Si phase duo to the significant decrease of zirconium content. Zr_2Si, with the melting point of 1925°C, is in solid state during the RMI process. Atoms in the solid Zr_2Si diffuse slowly and the formation of ZrC goes along at a rather low speed. Consequently, ZrC and Zr_2Si were formed in the composite B after RMI process. ZrC distributes around the PyC and Zr_2Si locates in the middle of pores surrounded by ZrC.

A pulsed laser was used to test the ablation properties of composite A and composite B in the air. The ablation was sustained for 20s. The linear ablation rates of composite A and composite B are listed in Table 1, which was calculated by the eroded depth at the ablation center dividing the ablation time. As comparison, C/SiC composite was also tested and the linear ablation rate is also listed in Table 1. As can be seen, the composite B presents the best ablation resistance compared with the composite A and C/SiC composite. The ablation resistance of composite B is intervenient of composite A and the C/SiC composite.

Fig. 4 shows the surface morphologies of the ablated C/SiC, composite A and composite B. As can be seen, a lot of humps composed of Zr, C and O appear on the surface

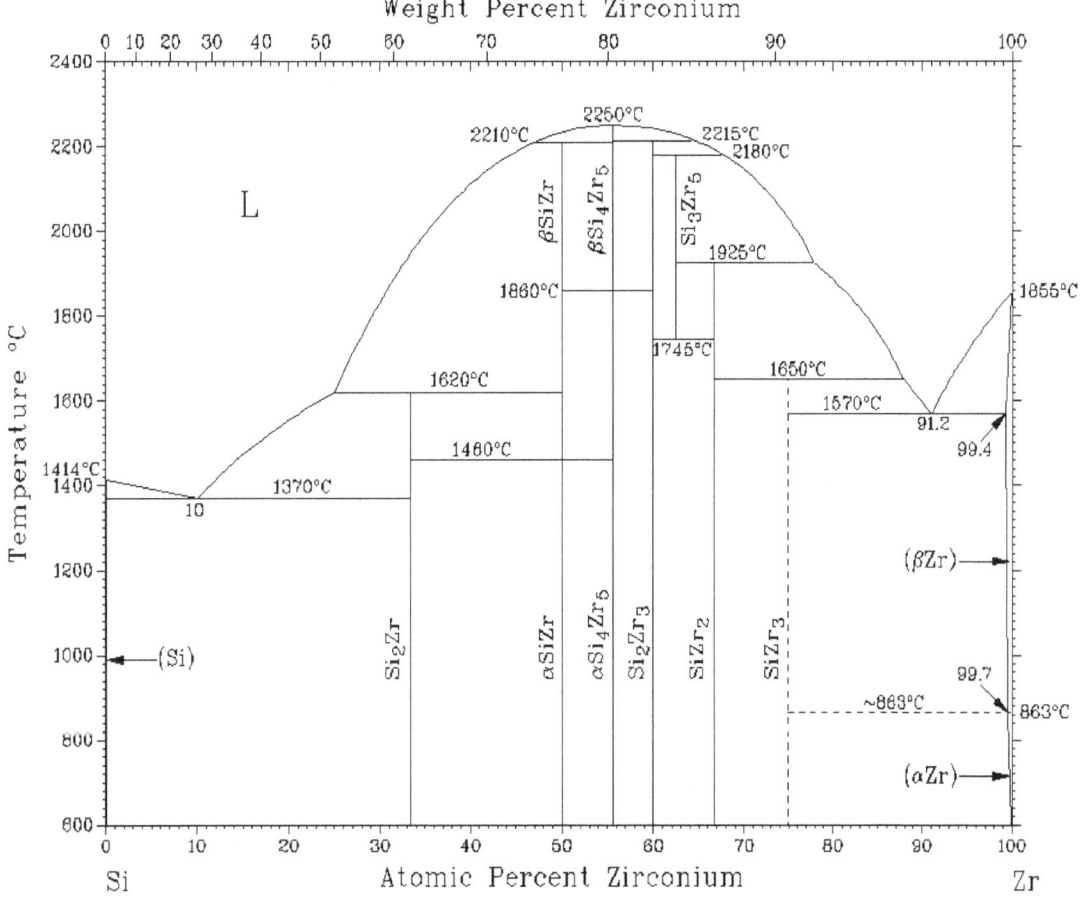

Figure 3. Phase diagram of Zr-Si system.

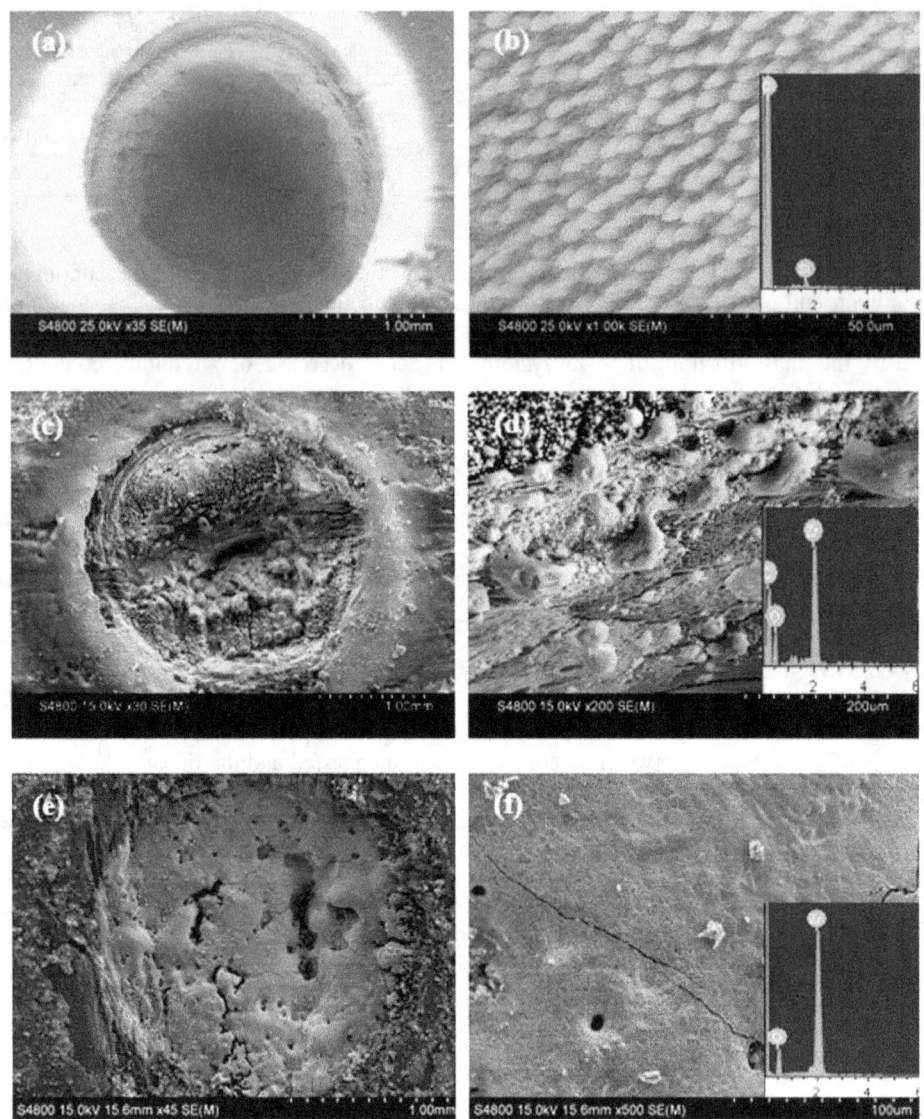

Figure 4. Surface morphologies of the ablated C/SiC, composite A and composite B, (a, b) C/SiC, (c, d) composite A, (e, f) composite B.

of the composite B after ablation (Fig. 4d). It is believed to be the phases ZrC and ZrO_2. Compared with the composite A, a continuous white layer is formed on the ablated surface of composite B (Fig. 4f). EDS analysis indicates that the layer is composed of zirconium and oxygen, which is deemed as ZrO_2 phase. ZrC and ZrO_2, known as high temperature ceramics, have excellent ablation resistance and the diffusion of the oxygen through ZrO_2 is quite low. They effectively protect the composites form being damaged. A continuous ZrO_2 layer was formed on the composite B while some discontinuous humps composed of ZrC and ZrO_2 were just found on the composite A. Thus, the linear ablation rate of composite B is much smaller than composite A. Different form the composite A and composite B, the surface of the C/SiC composite shows a needled-like structure with the composition of carbon and a very little amount of silicon (Fig. 4b). The whole surface of the C/SiC composite is exposed without any protection compared with composite A and composite B, which is the main reason for its largest linear ablation rate. No detectable SiO_2, which is always used as a good layer to block oxygen diffusion, was found on ablation surface of the C/SiC composite. The same situation occurs on the composite A and composite B, even though

both the composites contain certain amount of Si. The reasons can be attributed to the oxidation of Si-included phases and the gasification of SiO_2. The ablation testing was performed in the air. SiO_2 is formed by the oxidation of Si-included phases in the composites. With the laser power of 1000W, the ablation region was instantly heated to a very high temperature, approximately 3000°C. SiO_2 immediately gasified (boiling point of SiO_2 is 2230°C (Fang et al., 2009)) and thus, no detectable SiO_2 was found on the ablation surface.

4. Conclusion

(1) Two kinds of C/C-Zr-Si composites were prepared by Si-10Zr and Zr-9.8Si alloyed melt infiltration.

(2) The composite prepared by Si-10Zr alloyed melt infiltration was composed of C, SiC and $ZrSi_2$ with a little amount of ZrC, while the composite prepared by Zr-9.8Si alloyed melt infiltration was composed of C, ZrC and Zr_2Si.

(3) The linear ablation rates of the two C/C-Zr-Si composites were 0.041 and 0.028mm/s, much smaller than the linear ablation rate of the C/SiC composite, 0.107 mm/s.

References

Fang, D., Chen, Z.F., Song, Y.D. and Sun, Z.G. (2009). Morphology and microstructure of 2.5 dimension C/SiC composites ablated by oxya-cetylene torch, *Ceramic International*, Vol. 35, pp. 1249-1253.

Jiang, S.Z., Xiong, X., Chen, Z.K., Xiao, P. and Huang, B.Y., (2009). Influence factors of C/C–SiC dual matrix composites prepared by reactive melt infiltration, *Materials and Design*, Vol. 30, pp. 3738-3742.

Krenkel, W. (2003). Application of fiber reinforced C/C-SiC ceramic, *Ceramic Forum International*, Vol. 80, No 8, pp. 31-38.

Kumar, S., Kumar, A., Devi, R., Shukla, A. and Gupta, A.K. (2009). Capillary infiltration studies of liquids into 3D-stitched C–C performs Part B: Kinetics of silicon infiltration, *Journal of the European Ceramic Society*, Vol. 29, pp. 2651-2657.

Naslain, R. (2004). Design, preparation and properties of non-oxide CMCs for application in engines and nuclear reactors: An overview, *Composite Science and Technology*, Vol. 64, pp. 155-170.

Padmavathi, N., Kumari, S., Bhanu Prasad, V.V., Subrahmanyama, J. and Ray, K.K. (2009). Processing of carbon-fiber reinforced (SiC + ZrC) mini-composites by soft-solution approach and their characterization, *Ceramics International*, Vol. 35, pp. 3447-3454.

Tressler, R.E. (1999). Recent developments in fibers and interphases for high temperature ceramic matrix composites, *Composites A.*, Vol. 30, pp. 429-437.

Van de Voorde, M.H. and Nedele, M.R. (1996). CMC's research in Europe and the future of CMC's in industry, *Ceramic Engineering and Science Proceeding*, Vol. 17, No 4, pp. 3-7.

Wang, Y.G., Zhu. X.J., Zhang, L.T. and Cheng L.F. (2011). Reaction kinetics and ablation properties of C/C–ZrC composites fabricated by reactive melt infiltration, *Ceramics International*, Vol. 37, pp. 1277-1283.

Yang, J. and Ilegbusi, O.J. (2000). Kinetics of silicon-metal alloy infiltration into porous carbon, *Composites A*, Vol. 31, pp. 617-625.

Zou, L.H., Wali, N., Yang J.M. and Bansal N.P. (2010). Microstructural development of a Cf/ZrC composite manufactured by reactive melt infiltration. *Journal of the European Ceramic Society*, Vol. 30, pp. 1527-1535.

Interfacial Response in Glass Fiber/Polypropylene Composites

Touqeer Rasheed[1], Frank R. Jones[2], Sohaib Akbar[1], Sajid Mirza[1]

[1] Pakistan Space and Upper Atmosphere Research Commission - SUPARCO, P.O.Box 8402, Karachi, 75270, Pakistan
[2] Department of Materials Science and Engineering, The University of Sheffield, Sir Robert Hadfield Building, Mappin Street, Sheffield, S1 3JD, UK

Abstract: The use of glass fibre/polypropylene composites has grown in recent times due to their advantages over conventional material and thermoset composites. One of the major problems in the development of glass fibre/polypropylene composites is the adhesion between the components. Coupling agents such as maleic anhydride modified polypropylene (MAPP) can be used to promote adhesion. The effect of the MAPP concentration on the interfacial shear strength of glass fibre-PP composites was quantified using microbond testing. IFSS of glass fibre-PP composites increased as the concentration of MAPP. The change in the mechanical properties of the PP on addition of MAPP was also studied. At 5% MAPP the change in crystallinity was sufficient to modify the ductility of the polypropylene and influence on the interfacial shear strength of composite.

Key Words: Polypropylene, Maleic anhydride modified polypropylene, Differential scanning calorimetry, Injection molding, Mechanical properties, Interfacial shear strength, Microbond test.

1. Introduction

Thermoplastic composites have gained much attention in recent times. The use of polypropylene (PP)-glass fibre reinforced composites has gained popularity in various sectors especially in automotive because of the low cost of manufacturing of complex shapes of low weight (Astrom, 1997; Jang, 1994). However, a major issue in designing a PP-glass fibre composite is the compatibilisation of the non-polar matrix, PP with the polar glass fibre surface. Coupling agents have been employed to overcome this problem. They bridge these dissimilar materials and increase the adhesion properties. Various types of coupling agents are available for such purpose.

Maleic anhydride modified PP (MAPP) is the most effective and common type of coupling agent used for polypropylene-glass fibre composites. Maleic anhydride (MAH) is usually grafted onto the backbone of the polypropylene and formed maleic anhydride modified polypropylene (MAPP). Adhesion between fibre and matrix in composites is quantified as interfacial shear strength (IFSS). Good adhesion provides better stress transfer between matrix and fibres and improves the composite strength (Harutun, 2003). Interfacial shear strength of a fibre-matrix system can be determined by various single fibre testing techniques such as fibre pull-out, microbond and fragmentation tests. The microbond test is popular because of easy and fast sample preparation and testing so

that more results can be obtained in less time. (Karger-Kocsis, 1995; Hull and Clyne, 1996; Mukhopadhyay et al., 2003; Miller et al., 1987)

The objective of this paper is to investigate the effect of different concentration of MAPP addition onto the properties of polypropylene. The study was carried out with PP-MAPP blends with varying MAPP concentration. The study involved the thermal analysis by differential scanning calorimetry (DSC) and the tensile testing of polymer blends; microbond test was conducted to quantify the IFSS of the blends with glass fibre.

2. Experimental procedure

2.1 Material

The materials which are used as matrix resins in this study were pure polypropylene and polypropylene-maleic anhydride copolymer (MAPP)-PP blends. The Pure PP & PP-MAPP blend was supplied by SABIC plc Netherlands, with different concentrations of MAPP (wt %) as shown in Tab. 1.

Glass fibres used in this paper are Advantex which were supplied by Owens Corning. The average diameter of the Advantex used in this paper was 19±1 μm. Fibres used in this study were sized for use with polypropylene.

2.2 Mechanical testing of PP-MAPP blends

2.2.1 Sample preparation

Dumb-bell shaped samples of PP for tensile testing by injection moulding were prepared. The injection moulding machine used for moulding samples was NEGRI-BOSSI, NB-25, UK. Premixed pellets of PP-MAPP were dried in an oven at 80°C for 30 min. The barrel temperatures end set at 190°C, 210°C and 220°C.

2.2.2 Tensile testing

The dumb-bell shaped specimens were carried out according to the ASTM D638 using universal testing machine

Table 1
Matrix resin composition

S.No	Matrix Resin Type	MAPP (wt %)
01	PP	0.00
02	PP-MAPP	0.50
03	PP-MAPP	1.00
04	PP-MAPP	1.50
05	PP-MAPP	2.00
06	PP-MAPP	2.50
07	PP-MAPP	5.00

(Hounsfield Test Equipment, UK). A 10 kN load cell used in combination with a loading rate of 50 mm/min. The gauge length was 50 mm. At least 10 samples of each composition were tested at ambient temperature.

2.3 Thermal analysis

Differential scanning calorimetry (DSC) was used to investigate the effect of the addition of MAPP to PP on crystallinity. Experiments were conducted on a DSC7 (Perkin-Elmer). 10-12 mg of sample of each blend was weighed into an aluminium pan, covered with a lid and closed. An empty aluminium pan was used as a reference. They were heated from -30°C to 200°C at a heating rate of 10°C/min and cooled from 200°C to -30°C at the same rate. The analysis was carried out in argon atmosphere with the flow rate of 2×10^4 cm^3/min.

Melting and crystallization temperatures were obtained directly from the thermograms whereas degree of crystallinity (%) for each composition was calculated from Equation 1 (Kim, 2007).

$$X_C \left(100\%\right) = \frac{\Delta H_f}{\Delta H_f^{\circ}} \qquad (1)$$

where X_C is degree of crystallinity (in %); ΔH_f is heat of melting; ΔH_f° is heat of fusion (the value for 100% crystalline PP was taken as 138 J/g (Kim, 2007).

2.4 Microbond testing

2.4.1 Sample preparation

Samples for the microbond test were prepared from Advantex glass fibre and MAPP-PP blend. Single fibres were glued to a U-shaped frame using the adhesive Y6010, as shown in Fig. 1.

The adhesive was allowed to cure for 24h at room temperature. A SD 160 Digital Hotplate, (Stuart), was used to melt a PP pellet on a glass slide at 220°C. A sharp knife was used to draw a thin fibre from the molten PP which

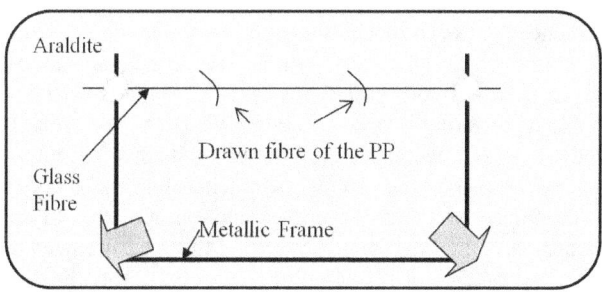

Figure 1. U-shaped frame used for microbond sample preparation

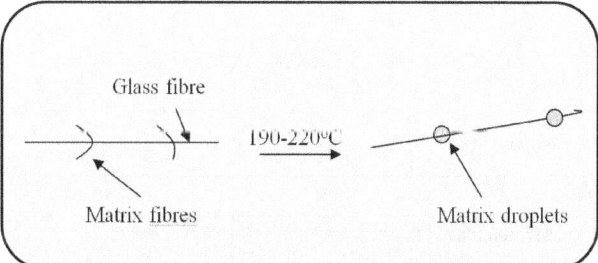

Figure 2. Formation of PP droplet on a glass fibre

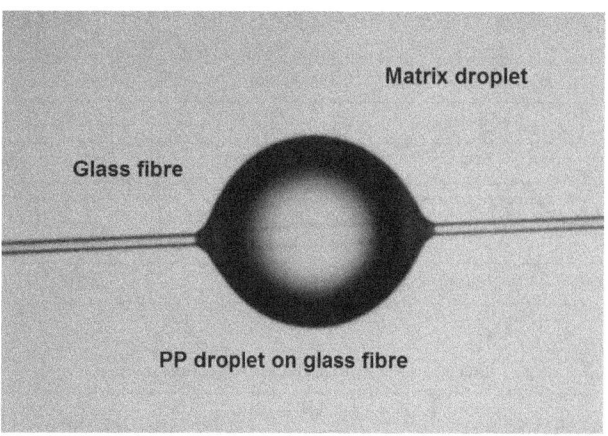

Figure 3. A typical PP-MAPP droplet on an Advantex glass fibre

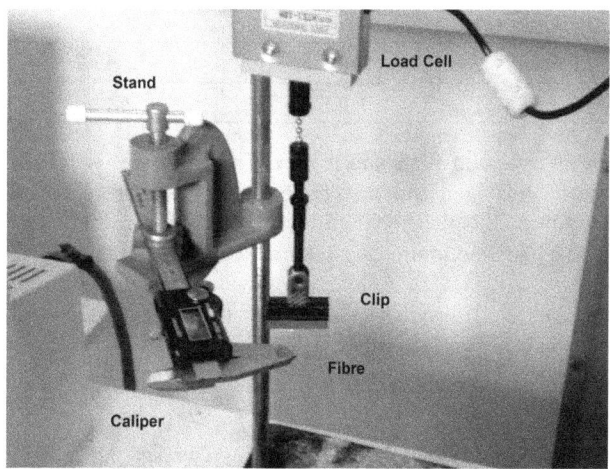

Figure 4: Microbond test configuration

was wound onto the glass fibre. PP micro droplets were formed by heating in an oven (Carbolite, ELF 11/6B) at $190-220^{\circ}$C for 5 min, as shown in Fig. 2.

The fibres with adhered micro droplets were cut and removed from the frame and a end tape at one end and stored in a petri dish.

2.4.2 Diameter measurement

The diameter of the glass fibre and embedded length of the matrix polymer of each sample were measured in an optical microscope (Polyvar).

2.4.3 Microbond testing

The maximum force required to debond the polymer from the fibre. Testing was done by using callipers as a micrometer and 10 N load cell as shown in Fig. 4.

The samples were tested at a displacement rate of 0.2 mm/min. For the test the distance between the calliper blades was kept constant at 0.04 mm or 40 μm. One end of the single fibre (Paper/masking tape attached) with microdroplet was placed into the clip attached to the load cell in the moving crosshead, as shown in Fig. 4.The free end of the fibre was placed between the calliper blades and microdroplet was positioned below the calliper. It was ensured that the fibre was aligned vertically with the test frame. When the crosshead moved up, the droplet was returned by the calliper blades introduces a shear stress at

Table 2
Melting (T_m) and Degree of Crystallinity (X_C) of
PP-MAPP Blends obtained by DSC

Composition	T_m (°C)	X_C (%)
PP	169.7	53.6
PP-MAPP (1%)	168.8	51.1
PP-MAPP (2%)	168.5	50.1
PP-MAPP (5%)	169.7	49.0

Table 3
Interfacial shear strength for each composition

Composition	Avg. Debonding force (N)	Avg. IFSS (MPa)
PP-MAPP (0%)	0.13	6.47
PP-MAPP (1%)	0.15	8.33
PP-MAPP (2%)	0.24	13.10
PP-MAPP (5%)	0.32	16.17

the interface in fibre and matrix. The force was recorded by the load cell and the HTES-Series Software recoded the force – displacement graphs. The force increased until it reached maximum and fell indicating debonding. The curve finally levelled out at a low value corresponding to the frictional force associated with the sliding of micro-droplet (Day and Rodrigez, 1998).

3. Results

3.1 The degree of crystallinity of the matrix

PP contains 0, 1, 2, and 5% MAPP. The degree of crystallinity obtained from Equation 1, are given in Tab. 2.

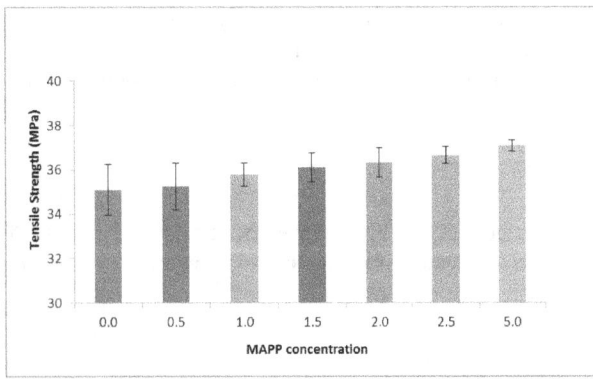

Figure 5. The tensile strength of polypropylene containing different concentrations of MAPP (wt %)

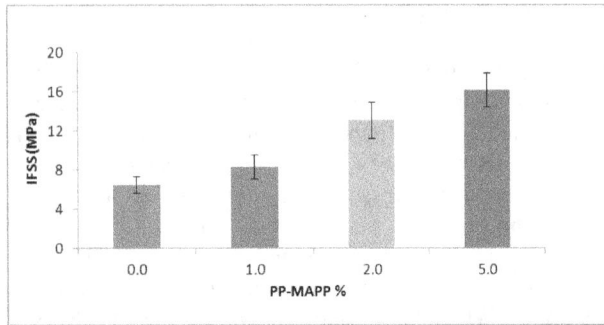

Figure 6. The microbond interfacial shear strength (IFSS) of the PP blends as a function of the concentration of MAPP

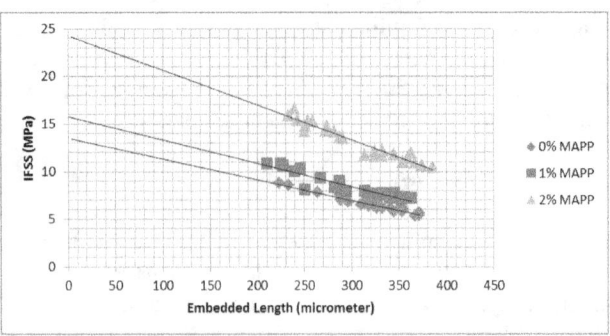

Figure 7. Extrapolated curves for each PP-MAPP (%) blends

3.2 Effect of MAPP on tensile strength of PP

The average tensile strengths of the PP blends are given in Fig. 5.

There is no significant effect of the addition of MAPP on the tensile strength of polypropylene. However there is a trend towards an increase in tensile strength. Assuming that the modulus is not significantly reduced by the small change in crystallinity it causes deduced that the ductility of PP trends to increase on addition of small quantities of MAPP.

3.3 Interfacial shear strength (IFSS)

Tab. 3 shows that the debonding force for a microdroplet of PP from the glass fibre increases with addition of MAPP from the debonding force as IFSS can be calculated

Tab. 3 clearly indicates that IFSS of PP-Advantex fibre system improves as the concentration MAPP is increased. The highest IFSS obtained was 16.17 MPa, for 5% MAPP composition. This represents an increase in IFSS from 6.47 to 16.17 MPa after addition of 5% MAPP.

4. Discussion

The calculation of interfacial shear strength from the microbond test data is complicated by the rating of interface failure. The standard analysis assumes a constant interfacial shear stress so that this assumes instantaneous failure of the interface, however when a crack propagates along the interface a more complex stress distribution exist because of differing degree of matrix yield. As the embedded length (l_e) decreases the interfacial stress distribution becomes closer to a constant shear. Therefore the analysis of the microbond data is complex and extrapolation of the data with embedded length provided the measurement of local interfacial shear strength. The data for the 5% MAPP blend indicated that the reduction of interfacial

Table 4
Comparison between average IFSS and local IFSS
(Mader's analysis)

Composition	Avg. IFSS (MPa)	Local IFSS (MPa) (Mader's Method)
PP-MAPP (0%)	6.47	13.50
PP-MAPP (1%)	8.32	15.78
PP-MAPP (2%)	13.10	24.29
PP-MAPP (5%)	16.17	17.37

failure changed because the lower crystallinity as the 5% MAPP PP increased the ductility of the polymer. The local interfacial shear strength obtained by extrapolating the IFSS data to zero embedded length as described by Maeder et al. (2001) provides a maximum value which is constant to give a better quantification of interfacial shear strength. Fig. 7 shows the plot of IFSS versus embedded length. The extrapolated values are given in Tab. 4 of particular interest is that the extrapolated local IFSS using 5% MAPP is similar to the average value. This is can be understood by the 5% MAPP/PP blend exhibiting same elastic-plastic behavior. The reduction in crystallinity from 54% to 49% supports this mechanism. Therefore the trend to higher matrix strength also adventure of a change in failure mechanism of the interface.

5. Conclusion

The performance of composites depends upon the properties of its individual components and interfacial adhesion. MAPP is the most common coupling agent for PP-glass fibre composites. In current study, PP was blended with MAPP and the interfacial strength measured.

Interfacial adhesion of MAPP-PP matrix to Advantex fibre was investigated by microbond testing. The interfacial shear strength was significantly improved with increase in concentration of MAPP. The highest average IFSS was reached for the 5% MAPP addition at 16.17 MPa compared to 6.47 MPa for 0% MAPP composition. These results clearly indicate the effectiveness of MAPP in improving the adhesion between PP and glass fibre.

The results indicate that up to 5% MAPP can be added to PP to increase the interfacial bond with significantly influencing the thermomechanical properties of the matrix.

References

Astrom, B. T. (1997). *Manufacturing of Polymer Composites*. London: Chapman & Hall.

Day, R.J. and Rodrigez, J.V.C. (1998). Investigation of the micromechanics of the microbond test, *Composites Science and Technology*, Vol. 58, pp. 907-914.

Harutun, G. K. (2003). *Handbook of Polypropylene and Polypropylene Composites*. 2nd edn., New York: Marcel Dekker.

Hull, D. and Clyne, T.W. (1996). *Introduction to Composites Materials*. 2nd edn., Cambridge: Cambridge University Press.

Jang, B. Z. (1994). *Advanced Polymer Composites: Principles and Applications*. Materials Park: ASM International.

Karger-Kocsis, J. (1995). Polypropylene: *Structure, Blends and Composites*. London: Chapman & Hall.

Khalil, R. (2007). Effect of coupling agents on crystallinity and viscoelastic properties of composites of rice hull ash filled polypropylene, *Journal of Material Science*, Vol. 42, pp. 10219-10227.

Kim, H.S. (2007). The effect of types of Maleic anhydride-grafted polypropylene (MAPP) on the interfacial adhesion properties of bio-flour filled polypropylene composites, *Composites: Part A*, Vol. 38, pp. 1473-1482.

Mader, E. and Zhandarov, S. (2001). How can adhesion be determined from micromechanical tests?, *Composites: Part A*, Vol. 32, pp. 425-434.

Miller, B., Muri, P. and Rebenfeld, L. (1987). A microbond method for determination of the shear strength of a fibre/resin interface, *Composites Science and Technology*, Vol. 28, pp. 17-32.

Mukhopadhyay, S., Deopura, B. L. and Alagiruswamy, R. (2003). Interface behavior in polypropylene composites, *Journal of Thermoplastic Composite Materials*, Vol. 16, pp. 479-495.

Pitkethly, M.J. (1993). A Round-Robin programme of interfacial test method, *Composites Science and Technology*, Vol. 48, pp. 05-214.

Seo, Y. and Kim, J. (2000). Study of crystallization behavior of polypropylene and Maleic anhydride grafted polypropylene, *Polymers*, Vol. 41, No 7, pp. 2639-2646.

Tripathi, D., Chen, F. and Jones, F.R. (1996). A comprehensive model to predict stress field in single fibre composite, *Journal of Composite Materials*, Vol. 30, pp. 1514-1538.

Tripathi, D., Lopattananon, N. and Jones, F.R. (1998). A technological solution to the testing and data reduction of single fibre fragmentation tests, *Composites: Part A*, Vol. 29, pp. 1099-1109.

Zhandarov, S., and Maeder, E. (2005). Characterization of fibre/matrix interface strength: applicability of different tests, approaches and parameters, *Composites Science and Technology*, Vol. 65, pp. 149-160.